FARM AND RANCH SAFETY MANAGEMENT

FARM BUSINESS MANAGEMENT
FBM18105NC

A vital tool for creating a safe environment within your agricultural enterprises.

JOHN DEERE

FREE CATALOG
Call 1–800–522–7448

**Check Out All Of Our Titles In The
FARM BUSINESS MANAGEMENT Series!**

Here are the titles in this series:

Farm and Ranch Safety Management
A vital management tool for creating a safe working environment within your agricultural enterprise.

Machinery Management
How to select machinery to fit the needs of today's farm managers.

Farm and Ranch Business Management
The definitive text on managing a farm or ranch on sound business principles.

Our FARM BUSINESS MANAGEMENT Series
examines "real-world" problems and offers practical solutions in the areas of marketing, financing, equipment selection, and crop/livestock/wildlife management.

Call 1–800–522–7448 to order; to inquire into prices; or to receive our free catalog.

FARM AND RANCH SAFETY MANAGEMENT

You can't afford one accident! Careful planning can help avoid tragic — and costly — accidents...

Why do you need this book? The answer is simple: agriculture is a hazardous industry. Farming and ranching can be uniquely rewarding occupations, but if you don't follow safe work practices they can be dangerous. According to the Bureau of Labor Statistics (BLS), there were an estimated 750 agricultural work-related deaths and 150,000 disabling work-related injuries in the United States in 2002. But you already knew farming and ranching are filled with danger. What you maybe do not know so well is what you can do about it. This book will tell you.

We will tell and show you 21 ways to avoid the No. 1 fatal farm accident...and that's just for starters. We show you very practical ways to avoid pinch points; wrapping, shearing, crushing, and pull-in hazards; thrown objects and free-wheeling parts; electrical and stored energy; and much, much more.

Farm and Ranch Safety Management is **the** definitive reference work in rural safety. This book is filled with EPA, OSHA, and other regulatory agency standards and suggested ways to meet or exceed those standards. Topics covered in its simple, straight forward language include safe machine operation, handling toxic chemicals, hazardous waste disposal, and developing an overall safety management plan.

JOHN DEERE

Deere & Company

ISBN 0-86691-352-1

PUBLISHER
DEERE & COMPANY
JOHN DEERE PUBLISHING
One John Deere Place
Moline, IL 61265
http://www.johndeere.com/publications
1–800–522–7448

Farm Business Management is a series of manuals created by Deere & Company. Each book in the series is conceived, researched, outlined, edited, and published by Deere & Company, John Deere Publishing. Authors are selected to provide a basic technical manuscript that could be edited and rewritten by staff editors.

HOW TO USE THE MANUAL: This manual can be used by anyone — experienced mechanics, shop trainees, vocational students, and lay readers.

Persons not familiar with the topics discussed in this book should begin with Chapter 1 and then study the chapters in sequence. The experienced person can find what is needed on the "Contents" page.

This current EDITION was produced by the technical writers, illustrators, and editors of Almon, Inc. — a full-service technical publications company headquartered in Waukesha, Wisconsin (www.almoninc.com).

FOR MORE INFORMATION: This book is one of many books published on agricultural and related subjects. For more information or to request a FREE CATALOG, call 1–800–522–7448 or send your request to address above or:

Visit Us on the Internet
http://www.johndeere.com/
publications

Acknowledgements:

AUTHOR: Page L. Bellinger, Product Safety And Standards Consultant, retired from Deere & Company in 1992. His 25 years of service with the Company included 16 years as Product Safety Engineer and as Manager, Product Safety And Engineering Standards. Previous positions included Associate Engineering Editor for Successful Farming Magazine, and Technical Coordinator for The American Society Of Agricultural Engineers. He has B.S. and M.S. degrees in agricultural engineering from Michigan State University.

COPY EDITOR: Robert H. Gunter is a John Deere retiree with 27 years of experience in service publications including stints as factory writer, factory supervisor, and staff editor. One of his last active primary responsibilities was the training of factory writers.

CONSULTING EDITOR/CONTRIBUTING WRITER: David E. Baker, MS, CSP, is Associate Professor and Extension Safety & Occupational Health Specialist in the Food Science & Engineering Unit at the University of Missouri. He has authored numerous manuals, articles. guide sheets, technical papers and audio visual materials on various safety and health topics during his 28 years in Extension.

SPECIAL ACKNOWLEDGEMENT: Manufacturing organizations: L. Dale Baker, J.I. Case Co.; Dwight Benninga, Hutchinson Wil-Rich Mfg. Co.; Charles Brundage, AGCO Corp.; Dow Elanco; Equipment Manufacturers Institute; Tom Ihringer, Melroe Co.; Thomas Ogle, Ag Equipment Group LP (Farmhand, Inc.); Kawasaki Motors Corp.; Ronald McAllister, Ford New Holland, Inc.; Outdoor Power Equipment Institute, Margaret Speich, National Agricultural Chemical Association.

Safety, agricultural, and educational organizations: Marilyn Adams. Farm Safety For "Just Kids"; Thomas Bean, The Ohio State University; Curry Environmental Services, Inc.; Wm. E. Field, Purdue University; C. Lance Fluegal, University Of Arizona; Alan Hoskins, National Safety Council; University Of Illinois Cooperative Extension Service; Carol Lehtola and Cheryl Hawk, University Of Iowa; Dennis Murphy, Pennsylvania State University; Charles Schwab, Iowa State University; John Shutske, University Of Minnesota.

PUBLISHER: Farm Business Management (FBM) texts and visuals are published by John Deere Publishing, ALMON-TIAC Building, Suite 140, 1300 19th Street, East Moline, IL 61244. This text and its supporting materials are intended only as a educational media and should not be considered a substitute for operator's manuals for specific operating procedures and safety precautions.

This text is part of a complete series of texts and visuals on agricultural management entitled Farm Business Management (FBM), An instructor's kit is also available for each subject. For information, request a free catalog of educational materials. Send your request to: John Deere Publishing, ALMON-TIAC Building, Suite 140, 1300 19th Street, East Moline, IL 61244 or call 1-800-522-7448.

 We have a
long-range interest in
Your Farming Success

1 SAFE FARM MACHINERY OPERATION

2 HUMAN FACTORS AND ERGONOMICS

3 RECOGNIZING COMMON MACHINE HAZARDS

4 MANAGING FARM/RANCH SAFETY

5 EQUIPMENT SERVICE AND MAINTENANCE

6 WASTE RECYCLING AND DISPOSAL

7 TRACTORS AND IMPLEMENTS

8 TILLAGE AND PLANTING EQUIPMENT

9 CHEMICAL SAFETY

10 HAY AND FORAGE EQUIPMENT

11 GRAIN AND COTTON HARVESTING EQUIPMENT

12 MATERIAL HANDLING EQUIPMENT

13 FARM MAINTENANCE EQUIPMENT

14 APPENDIX

Safe Farm Machinery Operation

1

Fig. 1 — If You Follow Safe Work Practices, Farming and Ranching Can Be Rewarding

Introduction

Farming and ranching can be uniquely rewarding occupations, but if you don't follow safe work practices, they can be dangerous (Fig. 1). According to the Bureau of Labor Statistics (BLS), there were an estimated 730 agricultural work-related deaths and another 150,000 disabling work-related injuries in the United States in 2002. The deaths involved both farm residents and non-farm residents working on farms and in other agricultural industries.

Agriculture Is a Hazardous Industry

Historically, the National Safety Council data have identified agriculture as one of the three most hazardous industries in the United States, along with mining and construction, as indicated by accidental deaths per 100,000 workers. The agriculture industry is second only to mining in worker death rate, reporting 23 accidental deaths per 100,000 agricultural workers (Fig. 2).

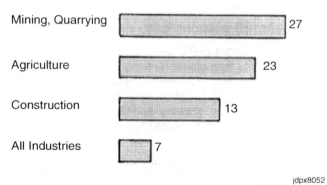

Work Accidents in USA, 2002
Accidental Deaths per 100,000 Workers

Mining, Quarrying	27
Agriculture	23
Construction	13
All Industries	7

Fig. 2 — Agriculture Is One of the Three Most Hazardous Industries (NSC, 2002)

The National Safety Council's agricultural injury estimates include not only crop and livestock production, but also forestry, fishing, and agricultural services. The actual number of farming and ranching accidents would be different from those reported totals. However, regardless of the accuracy of information, agricultural safety specialists all over the United States agree that there are too many injuries and deaths.

The safety of children and youth in agriculture is also a concern. Accurate statistics are not available, but agricultural safety specialists have estimated that an average of 160 or more children age 14 and younger are killed each year in work or worksite-related accidents. Some reports suggest that 300 children under age 16 are killed each year. However, such data often include non-work and non-worksite accidents.

The National Safety Council's 35-State Survey Summary, 1988, revealed more specific information about agricultural work-related accidents. For example, persons age 15 to 64 were involved in most of the reported work accidents (Fig. 3). That isn't surprising since they also account for most of the total hours worked. However, the report explains that children age 5 to 14, and adults age 65 and over, actually had more work injuries per million hours worked than persons age 14 to 64.

AGE OF PERSONS INJURED IN AGRICULTURAL WORK ACCIDENTS		
	Percent Reported	
Age	**Overall**	**Tractors**
5–14	6	6
15–24	22	21
25–44	35	31
45–64	32	33
65+	5	9
	100%	100%

Fig. 3 — Persons Age 14 to 64 Are Involved in Most of the Reported Agricultural Accidents (NSC 35-State Study, 1988)

Similarly, a 1972 study in Michigan and Ohio reported tractor operators under age 14 and over 64 had several times more accidents per million hours of work than operators age 25 to 64 (Fig. 4). Such reports highlight concerns for the safety of younger and older workers.

TRACTOR ACCIDENT RATE	
(Michigan and Ohio, 1972 Study)	
Operator Age	**Frequency** **(Accidents per Million Hours of Use)**
10–14	43.0
15–24	9.6
25–44	4.5
45–64	5.6
65	29.7

Fig. 4 — Tractor Operators Under Age 15 and Over 64 Had the Highest Accident Frequency Rate (Michigan State University, 1972)

The highest percentage of work-related injuries in the 1988 study occurred when persons were doing routine chores and field work (Fig. 5). They involved cuts, burns, fractures, amputations, and other painful and serious injuries.

WORK INJURY ACTIVITIES	
Activity	**Percent**
Routine Chores	25
Field Work	20
Machinery Maintenance	13
Treating Livestock	11
Building Maintenance	6
Other	25
	100%

Fig. 5 — Many Farm Work Accidents Occur During Chores and Field Work (NSC 35-State Study, 1988)

Machinery Accidents

Farm machinery, tractors, and animals are involved in many reported accidents (Fig. 6). Hand and power tools, as well as trucks and other vehicles also contribute to the totals. Excluding tractors, the farm machines most frequently involved in reported accidents were combines, balers, auger-elevators, and hay harvesting equipment, according to the National Safety Council (Table 1). However, you can be seriously injured with essentially every machine or tool on a farm or ranch if you don't follow safe work practices.

MAJOR FACTORS IN AGRICULTURAL ACCIDENTS	
Factors	Percent
Machinery	18
Tractors	8
Animals	17
Hand Tools	8
Power Tools	5
Trucks, Other Vehicles	14
All Other	30

Fig. 6 — Machinery, Tractors, and Animals are Involved in Many Work-Related Accidents (NSC 35-State Study, 1988)

MACHINERY FREQUENTLY INVOLVED IN REPORTED ACCIDENTS (Not Including Tractors)

1. Combine with Grain Head
2. Baler
3. Auger-Elevator
4. Hay Harvesting Machinery
5. Combine with Corn Head
6. Harrow
7. Cultivator
8. Seed Planter
9. Flatbed Wagon
10. Grain Grinder-Mixer
11. Corn Picker
12. Irrigation Equipment
13. Seed Drill
14. Self-Unloading Wagon
15. Manure Loader

Table 1 — Machinery Frequently Involved in Reported Accidents (Not Including Tractors)
   Note: Machines are listed according to number of accidents reported. (No. 1 is highest; No. 15 is lowest.) (NSC 35-State Study, 1988)

Tractor Accidents

Tractors are involved in about two-thirds of the reported machinery-related deaths. The No. 1 fatal farm machinery accident involves overturn of tractors not equipped with rollover protective structures (ROPS) and seat belts. However, chances of surviving a tractor upset are excellent if the tractor is equipped with a ROPS and the operator wears a seat belt.

The Most Dangerous Times of the Year

There are studies that show more accidents occur in some months than others. The monthly numbers vary by type of farming and ranching operation. Generally, accident frequency is highest when people are busiest during the planting, growing, and harvesting seasons. That is when they are exposed to more opportunities for injury.

The busiest and most stressful months will be different for fruit farms in California and Michigan, beef ranches in New Mexico, corn and hog operations in Iowa, and dairy farms in Wisconsin.

An important point for you to recognize is that when you are busy, in a hurry, and under stress, you are more likely to take shortcuts that can lead to injury or death. Don't do it! Always take time to follow safe practices. One mistake could lead to an accident that will drastically change your life . . . or end it.

Why So Hazardous?

Farm and ranch families and workers probably face a greater variety of potential hazards than persons in any other occupation. They encounter a wide range of situations involving machinery operation, crop handling and storage, as well as livestock, electricity, and chemicals.

Farm and ranch people work long hours in all kinds of weather to plant and harvest crops on time. They usually learn their work by trial and error instead of through training. Also, children are often part of the work force. Farmers and ranchers also face the pressures of financial management and decision-making. It is especially important that farmers and ranchers recognize the many potential hazards and follow safe practices to avoid injury.

Zeroing In on Safety

- Accidents hurt and kill!
- Accidents cost!
- Accidents can be avoided!

Accidents Hurt and Kill

Below are listed a few of the tractor and machinery accidents that have been reported from various sources. The names are fictitious, but the accident victims were real people, just like you and your family.

- Jeff, age 20, was standing near brush and weeds being cleared by a rotary cutter operated by his father. A solid object, thrown by the cutter, struck John's left eye. Loss of eyesight.

- David, age 60, was hand-feeding hay into a round baler while it was running. His right arm was pulled into the machine. Arm amputated.

- Marvin, age 31, was driving his car too fast on a narrow gravel road. As he went over a hill, he collided head-on with a self-propelled combine. Broken arm and hip, head injuries.

- Gary, age 18, entered a partially filled silo to repair an unloading auger. He was asphyxiated by silo gas. Fatal.

- Mary Ann, age 4, was riding on a wagon being pulled by a tractor driven by her father. She fell from the wagon and was run over by the wagon wheels. Fatal.

- Gerald, age 56, was cutting weeds near a drainage ditch with a rotary cutter and tractor without a rollover protective structure (ROPS) because the ROPS had been removed. The ditch bank collapsed. The tractor overturned, crushing Gerald. Fatal.

- Michael, age 14, was driving a tractor that slid off an embankment and overturned. The tractor had a ROPS but Michael did not use a seat belt. He was crushed between the ROPS and the ground. Fatal.

- Harold, age 48, attempted to start his tractor while standing on the ground. It was in gear and lurched forward, crushing him beneath the rear wheel. Fatal.

- Jason, age 17, was unloading shelled corn from a wagon using a tractor-powered auger-elevator. His jacket became entangled in the unshielded PTO shaft. Both arms torn off.

Accidents Cost

Safety is too expensive and important to learn by accident. Consider the impact of personal injury and loss of health on you, on your family, and on your farming or ranching operation.

Medical Care = Money

Hospital and doctor bills, medical supplies, and rehabilitation costs can be staggering. They could result in loss of the farm or ranch. In 1993, a major insurance company reported the average cost for hospital care in the U.S. in 1992 was $1,405 per day, not including doctors' fees (Fig. 7).

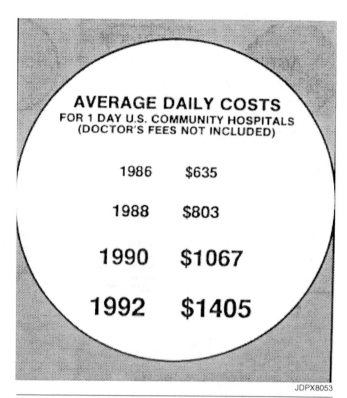

Fig. 7 — Can You Afford an Injury?

Time = Money

An accident costs lost time. What happens to your crops or livestock when you are unable to work?

Hired Help = Money

Will you have to hire someone to do your work? How difficult will it be to locate help? How much will it cost?

Property Damage = Money

Nobody knows the cost of farm machinery, equipment, and buildings and their repair better than you.

Your Health and Future = MORE than Money

A permanent disability or loss of your health can't be measured just in dollars. What effect would a disability have on you? Could you still farm? Could you go to college and pursue your chosen career? Could you join in activities with friends and family?

Accidents Can Be Avoided

Agricultural work-related accidents can be avoided. That can be done by incorporating safe work practices into your daily farming and ranching activities and management. Safety must be a normal part of the management process, just as important as managing finances, buying seed and fertilizer, planting, harvesting, and selling crops and livestock.

Almost every accident situation involves someone failing to follow safe work practices, not keeping safety features in place and in good working condition, or not using appropriate personal protective equipment. You must recognize the potential hazards and know how to avoid injury. This book is intended to help you do that.

Zeroing In on Communication

Clear communication is a key to safe work, especially when using tractors, machinery, chemicals, and tools. For example, you need information that clearly tells you how to operate a machine and how to service it for efficiency and safety. You need to be able to quickly recognize controls and what to expect when you move them. You also need to communicate accurately with other people working with you to work efficiently and safely together.

People who operate farm equipment in different parts of the world speak and read different languages. Some cannot read well enough to understand safety messages. Engineers and safety specialists have developed systems of symbols, colors, and shapes that can be understood throughout the U.S. and the rest of the world. Tractors and machinery have safety signs and operator's manuals that use those symbols, colors, and shapes along with carefully worded safety warnings and instructions. Let's look at some of the safety communications.

JDPX7494, JDPX7493

Fig. 8 — How Many Symbols Do You Recognize?

Symbols That Communicate

Almost everywhere there are symbols, colors, and shapes that have special messages. Take a look at Fig. 8. Do you recognize the highway safety symbols?

Note that they are combinations of symbols, colors, and shapes. Try to identify and understand each one. Compare your answers to those below:

Answers: (1) Stop. (2) Yield right-of-way. (3) Railroad crossing ahead. (4) Five-sided yellow sign shaped like an old schoolhouse means school zone or school crossing. (5) Bicycles not allowed. (6) No U-turn. (7) Children in cross-walk near school. Slow down. (8) Slippery pavement. (9) Downgrade or dangerous hill. (10) Two-way highway.

SMV Emblem

Another symbol for safety on roads and highways is the slow-moving vehicle (SMV) emblem Fig. 9. It is attached to the rear of agricultural machinery and other vehicles that normally travel slower than 25 mph (40 km/h). It tells motorists the machine ahead of them is moving slowly, so they can slow down to avoid hitting it from the rear.

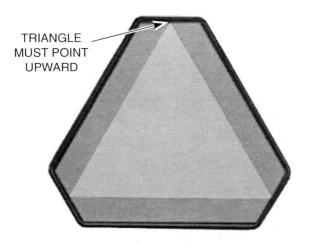

TRIANGLE MUST POINT UPWARD

JDPX7693

Fig. 9 — Slow-Moving Vehicle Emblem Identifies Machinery that Travels Slower than 25 mph (40 km/h)

Keep your SMV emblems clean, bright, and properly positioned. Replace them when they become faded. The center and border are made of special reflective and fluorescent materials for day and night visibility. When they are faded, they are not as easy to see or recognize. Check your state regulations about SMV emblems, lights, reflectors, and other requirements for safe movement of farm equipment on public roads.

The safety-alert symbol (Fig. 10) is an international symbol. It says: Attention! Be alert! Your safety is involved! You'll find this safety triangle with the exclamation point on safety signs of agricultural machines and in operator's manuals all over the world. Regardless of what language people speak or read, this symbol can alert them to safety information or hazardous situations.

JDPX7400

Fig. 10 — Safety-Alert Symbol Means: Attention! Be Alert! Your Safety Is Involved!

Signal Words and Colors

The safety-alert symbol is often used with the signal words Danger, Warning, and Caution to draw attention in safety signs and operator's manuals to potentially unsafe situations. Learn these signal words and colors and let them become your "think triggers."

DANGER means one of the most serious potential hazards is present. Exposure to these hazards would result in a high probability of death or severe injury if proper precautions are not taken. RED is used in safety signs with the signal word DANGER.

WARNING means the hazard presents a lesser degree of risk of injury or death than that associated with danger. Warning signs are usually ORANGE or YELLOW, depending on the age of the machine.

CAUTION is used to remind people of safety instructions that must be followed and to identify some of the less serious hazards. Caution signs are YELLOW.

Safety Signs and Pictorials

Fig. 11 and Fig 12 show how symbols, signal words, and colors are used together to provide safety signs that can be recognized and understood worldwide. Note that safety signs also use pictorial illustrations to help people quickly recognize specific hazards, even if they can't read. Whenever you see a safety sign or instruction, consider the hazard and follow the message. It could save your life.

JDPX7558

Fig. 11 — Safety Signs Use Signal Words and Colors to Alert

Control Symbols

Most agricultural equipment uses universal symbols to help identify operator controls. Fig. 13 shows some of the symbols designed for recognition all over the world.

Color is often used with symbols on machine control displays to indicate operating status or condition (Fig. 14). SAR standard J1362 and ISO standard 3767-1 guide manufacturers on what colors to use.

RED: Failure or serious malfunction.

YELLOW or AMBER: Condition outside normal.

GREEN: Normal operating condition.

BLUE: Headlight high/upper beam.

GREEN: Turn signal display.

RED: Hazard warning display.

RED: Hot temperature.

BLUE: Cold temperature.

Learn about the symbols in the operator's manuals of your machines. They will help you operate machines more efficiently and safely.

JDPX7402, JDPX7401

Fig. 12 — Safety Signs Also Use Pictorials, Along with Signal Words and Colors, to Warn

Several organizations worked together to develop standards that guide the use of symbols, colors, shapes, and pictorials so safety communications can be understood worldwide. They include the following: American Society of Agricultural and Biological Engineers (ASABE), Society of Automotive Engineers (SAE), American National Standards Institute (ANSI), and International Organization for Standards (ISO).

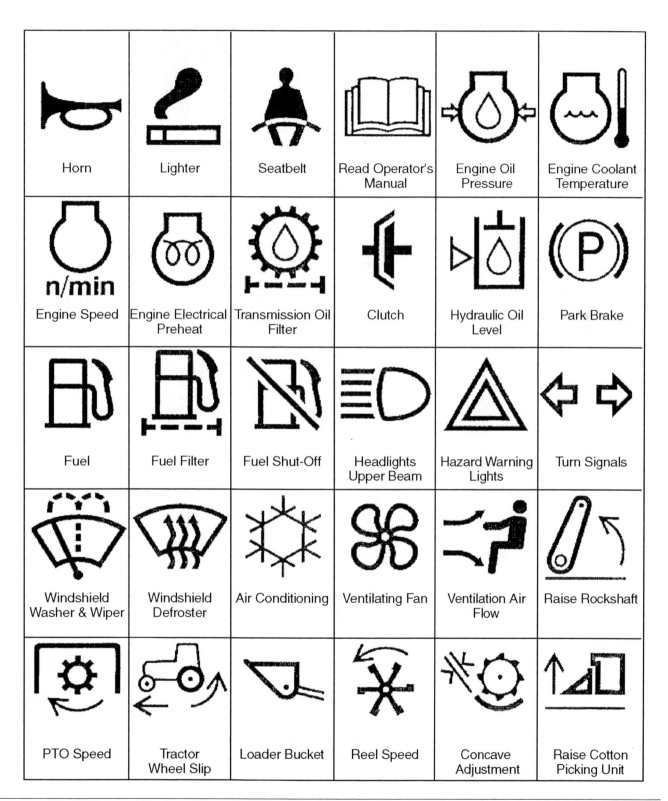

Fig. 13 — Typical International Symbols for Machinery Controls

| Park Brake (Red) | Engine Oil Pressure (Red) | Headlight Upper Beam (Blue) | Turn Signals (Green) |

JDPX7515

Fig. 14 — Color Is Often Used with Control Symbols

Shape and Color-Coded Controls

Many modern agricultural machines have color-coded controls to help you recognize what control knob, lever, or button to move (Fig. 15). ASABE published Engineering Practice "Color Coding Hand Controls" to help all equipment manufacturers provide uniform color coding. This is what the colors mean:

RED: Controls that stop the engine. Red may also mark the stop or off position of throttles or key switches.

ORANGE: Machine ground motion controls, such as engine speed controls, transmission controls, or park locks.

YELLOW: Controls that engage power mechanisms, such as PTO cutter heads, or unloading augers.

BLACK (or other dark colors): All other controls that adjust a machine function such as remote hydraulic control, header height, blade lift, or headlights.

Some manufacturers also provide controls with special shapes for buttons, toggle switches, knobs, and levers (Fig. 15). The unique shape or "feel" of a control can help you avoid confusing one control with another when you reach for a control without looking at it. You should gat acquainted with the controls of any machine before you operate it. The colors and shapes can help you operate the controls safely.

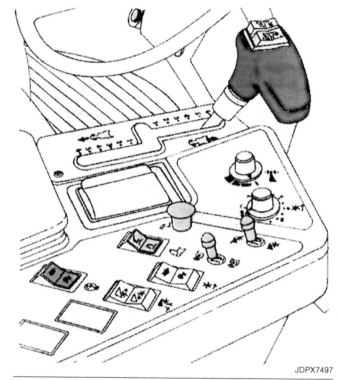

JDPX7497

Fig. 15 — Shape and Color-Coded Controls Can Help People and Machines Communicate

Hand Signals

Most farmers use various hand signals to communicate with other machinery operators or helpers. Those signals work fine if they are understood. However, if they are not understood, they could lead to problems. For instance, when two persons are hitching a tractor to an implement, a misunderstood signal could cause an accident.

To help encourage the use of hand signals that will be understood, ASABE (American Society of Agricultural and Biological Engineers) published standard S351, "Hand Signals for Agriculture" (Fig. 16 through Fig. 26). Use the ASABE signals and teach others to use then. They can help you save time and avoid misunderstandings and accidents when noise or distance prevents talking.

The ASABE hand signals are in general agreement with the U.S. Army's signals, and some of them are used in industry. You may already be using some of them.

Hand Signals

JDPX7701

Fig. 17 — Move Toward Me — Follow Me

Look toward the person or vehicle you want to move. Hold one hand in front of you, palm facing you, and move your forearm back and forth.

JDPX7698

Fig. 18 — Start the Engine

Move arm in a circle at waist level, as though you were cranking an engine.

JDPX7699

Fig. 16 — Stop the Engine

Move your right arm across your neck from left to right in a "throat-cutting" motion.

JDPX7698

Fig. 19 — Come to Me (May Mean "Come Help Me" in an Emergency)

Raise arm straight up, palm to the front, move arm around in a large circle.

Fig. 20 — Move Out — Take Off

Face desired direction of movement. Extend arm straight out behind you, then swing it overhead and forward until it is straight out in front of you with palm down.

Fig. 21 — Speed It Up

Clenching your fist, bend your arm so your hand is at shoulder level. Thrust arm rapidly straight up and down several times.

Fig. 22 — Raise Equipment

Point upward with forefinger while making a circle at head level with your hand.

Fig. 23 — This Far to Go

Put hands in front of face, palms facing each other. Move hands together or farther apart to indicate how far to go.

Fig. 24 — Slow It Down

Extend arm straight out to the side, palm down. Keeping arm straight, move it up and down several times.

Fig. 25 — Lower Equipment

Point toward the ground with the forefinger of one hand while moving the hand in a circle.

JDPX7708

Fig. 26 — Stop

Raise arm straight up, palm to the front.

Labels

Labels communicate safety, too. For example, by reading information and following instructions on pesticide labels, you can avoid serious injury to yourself and others. The chemical user who does not read and follow the instructions on the label is headed for trouble. If you do not follow safety instructions, you and others could suffer permanent damage. You can also harm crops, livestock, and the environment. Or you may waste a chemical if you don't apply it properly. Take time to read labels.

Read the Operator's Manual

The equipment operator's manual is perhaps the most important item of safety communication available (Fig. 27). It tells you what you need to know to operate and service your machines efficiently and safely. It discusses safety hazards in detail and describes safe practices to follow to avoid serious injury (Fig. 28).

JDPX7517

Fig. 27 — Read the Safety Instructions in the Operators Manual

JDPX7422

Fig. 28 — Typical Safety Instructions from a Tractor Operator's Manual

Read the operator's manual, especially the safety instructions, and especially before you operate or service a machine for the first time or at the start of each season. You may think you don't have time. However, if you take time read it, you will undoubtedly save time in the long run, and you may avoid serious injury to yourself or someone else.

Be Your Own Safety Director

Many industrial organizations have safety managers or directors who work full time to keep the workshop safe and healthy. There is no safety director on "the back forty." You must be your own safety director on a farm or ranch. Start by recognizing hazards and safe work practices. Finish by following safe work practices.

Safety Is Everyone's Job

Usually, a number of people are involved in farming and ranching operations, including family, hired help, and neighbors. In order to avoid accidents and injuries, everyone involved needs to recognize the hazards and how to avoid injury. Safety must be everyone's business (Fig. 29).

JDPX7516

Fig. 29 — Safety Must Be Everyone's Job

In addition to looking out for your own safety, you can also help make safety everyone's business. First, you can learn to recognize the hazards and follow safe work practices. This book will help you do that. Second, you can also help others recognize that they have a part in preventing accidents.

It is often awkward or difficult to tell someone else how to be safe, but it could save someone's life. Here are some of the things you can do to help prevent injury to others around machinery:

1. Set a good example. Always follow safe work practices. Others learn more from what you do than from what you say.

2. If you remove a guard or other safety device for service, be sure to replace it so someone else will not get hurt. Keep safety features in good working condition.)

3. Keep children away from tractors and machines. Don't let them get the idea that farm equipment is something to play with or ride on. (Remember, many children are killed each year in agricultural work-related accidents.

4. Stay clear of machinery being operated by someone else. Don't bother operators or interfere with their work. If you need to talk, signal the operator to stop (hand straight up) and approach the machine after it stops.

5. Always disengage tractor or machine power, shut off the engine, and remove the key before unplugging or working on farm equipment.

6. If you think something is dangerous, correct the problem or tell your parents or your supervisors so they can correct it.

Test Yourself

Questions

1. Agriculture ranks _____ in work-related fatality rates among U.S. industries.

   a. First

   b. Second

   c. Third

   d. Fourth

2. Which of the following costs could result from farm injury?

   a. Medical costs

   b. Lost time

   c. Property damage

   d. All of the above

3. Which signal word indicates the most serious hazard?

   a. Danger

   b. Warning

   c. Careful

   d. Caution

4. _____ are the most frequent cause of farm work-related fatalities.

   a. Flowing grain entrapments

   b. PTO entanglements

   c. Tractor rollovers

   d. Electrocutions

5. (T/F) SMV emblems are long-lasting and very rarely need replacing.

6. (T/F) Tractor operators under age 15 and over age 64 have the highest accident frequency rates.

7. (T/F) The highest percentage of farm work-related injuries occur during routine, everyday chores and tasks.

8. (T/F) The pictorial symbols and signs for warnings and controls vary greatly from one farm equipment manufacturer to another.

9. Which signal word/color combination is correct?

   a. Caution/blue

   b. Warning/red

   c. Danger/red

   d. Danger/yellow

10. The equipment operator's manual:

    a. Contains important safety information.

    b. Should be read carefully before using a piece of equipment.

    c. Provides important maintenance and service information.

    d. All of the above

11. (T/F) Hand signals are an appropriate and approved form of communication when working in noisy environments.

12. (T/F) If children are not actually working on the farm, they are in no danger of being injured or killed in a farm-related accident.

References

1. Accident Facts, 1993. National Safety Council, Itasca, Illinois 60143-3201.

2. 35-State Survey Summaries. 1988. Occupational Injuries in Agriculture; Agricultural Machinery-Related Injuries in Agriculture; and Agricultural Tractor-Related Injuries. National Safety Council, Itasca, Illinois 60143-3201.

3. Nature and Extent of Use of Farm Machinery in Relation to Frequency of Accidents In Michigan and Ohio. 1972. Doss, Howard J. and Pfister, Richard G., Michigan State University, East Lansing, Michigan 48824.

4. Injury in the Agricultural Work Place. 1987. Field, Wm. E., and Parschwitz, M.A., Purdue Univ., West Lafayette, Indiana 47907.

5. Slow-Moving Vehicle Identification Emblem (ANSI/ASABE S276/SAE J943). American National Standards Institute, New York, New York 10036; American Society of Agricultural and Biological Engineers, St. Joseph, Michigan, 49085; Society of Automotive Engineers, Warrendale, Pennsylvania 15096.

6. Safety Alert Symbol for Agricultural, Construction, and Industrial Equipment (ANSI/ASABE S350/SAE J284). American National Standards Institute, New York, New York 10036; American Society of Agricultural Engineers, St. Joseph, Michigan 49085; or Society of Automotive Engineers, Warrendale, Pennsylvania 15096.

7. Common Symbols (ISO 3767-1), and Symbols for Agricultural Tractors and Machinery (ISO 3767-2/SAE J1362). American National Standards Institute, New York, New York 10036; Society of Automotive Engineers, Warrendale, Pennsylvania 15096.

8. Hand Signals for Agriculture (ASABE S351). American Society of Agricultural and Biological Engineers, St. Joseph, Michigan 49085.

9. Safety Signs. (ASABE S441/SAE J115). American Society of Agricultural and Biological Engineers, St. Joseph, Michigan, 49085; or Society of Automotive Engineers, Warrendale, Pennsylvania 15096.

10. Color Coding Hand Controls (ASABE EP443). American Society of Agricultural and Biological Engineers, St. Joseph, Michigan 49085.

Human Factors and Ergonomics

2

Introduction

The human body is a marvelous creation. It has amazing capabilities; however, it also has limitations. Sometimes people make errors. Many factors contribute to human errors. Examples: Fatigue, distractions, thinking about several things at once, or incompatibility between people and machines.

Often human error results in costly damage to equipment and lost time for repairs. If an error causes injury or death, the costs and future impact can be tremendous (Fig. 1).

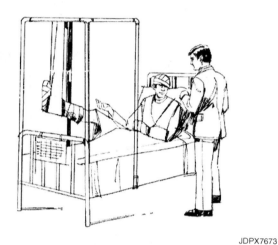

JDPX7673

Fig. 1 — Human Error Can Cause Accidents and Have Lasting Effects

Human Limitations and Capabilities

If an electrical circuit is overloaded, a fuse will blow or a circuit breaker will trip before the system is damaged. Machines have safety devices like slip clutches and shear-pins to protect against overload.

The human body also protects itself in many ways. Fatigue, pain, increased heart rate, perspiration, or fear may be signals of danger. But humans are also driven by goals, such as harvesting a crop before it rains. Such goals can drive us to continue working after the body gives us a warning signal. When we continue to push toward a goal after being warned of possible danger, we increase the risk of making mistakes, and accidents are more likely to occur.

Different people have different capabilities and limitations (Fig. 2). For your safety, it is important to know your own limitations.

Human factors, our capabilities and limitations, can be listed in three groups of characteristics:

- Physical

- Physiological

- Psychological

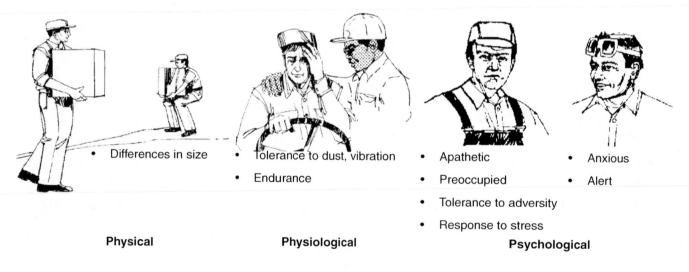

- Differences in size
- Tolerance to dust, vibration
- Endurance
- Apathetic
- Preoccupied
- Tolerance to adversity
- Response to stress
- Anxious
- Alert

Physical   **Physiological**   **Psychological**

JDP7536, JDPX7405, JDPX7630X

Fig. 2 — Physical, Physiological, and Psychological Limitations and Capabilities

Physical

Physical characteristics include size, weight, and strength. If you recognize your physical limitations and work within them, you can reduce your chances of accidents and injuries. You'll have better control over your work.

Strength

Some people are stronger than others. One person might be able to move a heavy object that would be difficult for another person. Know your strength limits and get help if you need it (Fig. 3).

WAIT! I'LL HELP YOU!

JDPX7697

Fig. 3 — Moving a Heavy Ladder Takes a Lot of Muscle. You May Need Help

Muscles make up about 45% of the body weight. When working, they convert the chemical energy of food to mechanical energy. The amount of energy used depends on the size and condition of the muscles, blood supply, heat, respiration, and type of work (Table 1).

ENERGY CONSUMPTION IN WORK CALORIES			
(Total metabolism during work minus the base metabolism)			
Activity	**Condition of Work**	**Travel Speed**	**Calories per Minute**
Walking without load	Level, smooth surface	1.86 mph (3.46 km/h)	1.7
		2.48 mph (3.96 km/h)	2.1
	Heavy, deeply plowed soil	1.86 mph (3.46 km/h)	5.2
	Country road, heavy shoes	2.48 mph (3.96 km/h)	3.1
Walking with load carried on the back	Level, firm surface 22 lb load (10 kg)	2.48 mph (3.96 km/h)	3.6
	66 lb load (30 kg)	2.48 mph (3.96 km/h)	5.3
	110 lb load (50 kg)	2.48 mph (3.96 km/h)	8.1
Climbing	16.5° gradient, climbing speed 32.72 ft/min (9.97 m/min)		
	Without load		8.3
	44 lb load (20 kg)		10.5
	110 lb load (50 kg)		16.0
Climbing stairs	30.5° gradient, climbing speed 56.42 ft/min. (18 m/min.)		
	Without load		13.7
	44 lb load (20 kg)		18.4
	110 lb load (50 kg)		26.3
Pulling cart	Level, firm surface, pulling energy 25.52 ft/min.	2.23 mph (3.56 km/h)	8.5
Working with axe	Two-handed strokes, 35 strokes/min.		
	Horizontal stroke		9.5–11.0
	Vertical stroke		10.0–11.5
Working with hammer	9.56 lb (4.38 kg) weight of hammer, 15 strokes/min.		
	Lifting stroke		7.3
	Circular stroke		6.7
Working with shovel	Throwing distance 6.56 ft (2 m)		
	Throwing height 3.28 ft (1 m)	10 liftings/min.	7.8
	Throwing height 4.92 ft (1.5 m)	10 liftings/min.	9.0
	Throwing height 6.56 ft (2 m)	10 liftings/min.	10.0
Sawing wood	Two-man cross cut saw, 60 double pulls		9.0
Bricklaying	Normal output, 1435 cu. ft/min. (67.72 cm³/min.)		3.0
Digging	Garden spade in clay soil		7.5–8.7
Mowing	Clover		8.3
Milking	Normal milking cows, 180 pulls/min.		2.2

Table 1 — How Much Energy Is Used in Your Work?

Experience tells you that muscles become tired or fatigued quickly at tasks that consume lots of energy (Fig. 4). A sprinter uses a lot of energy in a short time. That athlete works at the rate of 4 to 5 horsepower (3 to 3.8 kW), using a lot of energy for about a second at the start of a race. A person's capacity to work continuously is only about 0.1 to 0.2 horsepower (0.07 to 0.15 kW), but that worker would not tire as quickly as a sprinter.

To work safely, avoid muscle fatigue:

1. Work in a comfortable position. An awkward position, like working overhead, is far more tiring than one in which the work is done at a more comfortable height.

2. Work within your limitations. Don't demand too much of your muscles. A muscle strained to its maximum capacity tires quickly. It may not respond readily when needed to avoid an accident.

3. Keep moving. The body movement in dynamic work aids blood circulation, uses a wider variety of muscles, and is less tiring than static work. Dynamic work, such as shoveling, may not fatigue muscles as much as a static job like holding a board while someone nails it to the ceiling.

4. Take frequent, short rest breaks (10 to 15 minutes every 2 hours). They are more effective in recovering working ability than longer, infrequent breaks.

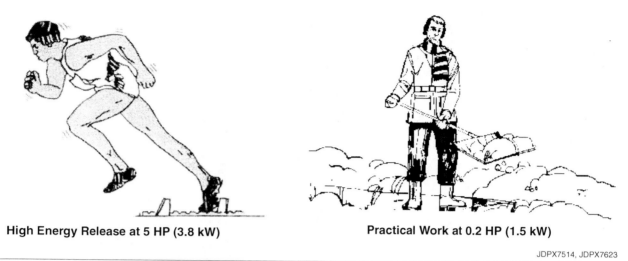

High Energy Release at 5 HP (3.8 kW)   Practical Work at 0.2 HP (1.5 kW)

JDPX7514, JDPX7623

Fig. 4 — High Energy Release vs. Continuous Work

Reaction Time

Reaction time is made up of a series of events beginning with a message to the brain and ending with a body response. The message comes from a sensor, such as eyes, ears, skin, or nose. For example: When a tractor operator sees a drainage ditch (the message), the operator's brain analyzes the message, then it tells muscles to turn the steering wheel to avoid the ditch. That all takes time (Fig. 5).

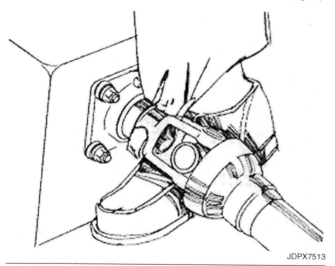

JDPX7513

Fig. 5 — Machines are Faster and Stronger Than You Are — Keep Away From Moving Parts

The best human reaction time is slow compared to powered machinery, and it depends on the type of signal and the response required. Example: When a person driving a car is surprised by a hazard in the road ahead, it will typically take 2.8 seconds to move a foot from the accelerator pedal to the brake pedal. Apply that information to these situations:

- If the snapping rolls of a corn picker are moving material at 12 feet (3.7 meters) per second, the stalks would travel more than 33 feet (10 meters) before a typical person could let go of them. A person trying to unplug a picker could be pulled into the picker rolls before he or she could let go of the stalks. Disengage the PTO, turn off the engine, and take the key with you before working on a machine.

- By the time a tractor operator realizes the tractor and front-end loader are turning over and begins to lower the bucket to avoid an upset, it will probably be too late. A tractor overturns so quickly that human reaction time is usually too slow to stop the overturn. Prevent the problem by keeping the bucket low.

Your reaction time is slower than normal when you are affected by fatigue, medicine, drugs, and alcohol. Also, total time to act may be lengthened even more by shock in a panic or emergency situation. That could cost someone's life, perhaps yours.

An airplane pilot always plans ahead for a place to land in case of emergency. Likewise, you should think about how you will react to emergencies BEFORE you're exposed to them. Also, take advantage of education and training in machinery operation. Then you can react more quickly and effectively in an emergency. Recognize what can cause problems and take steps to avoid the causes.

Body Size

A person's body size often determines the kinds of jobs that can be performed safely (Fig. 6).

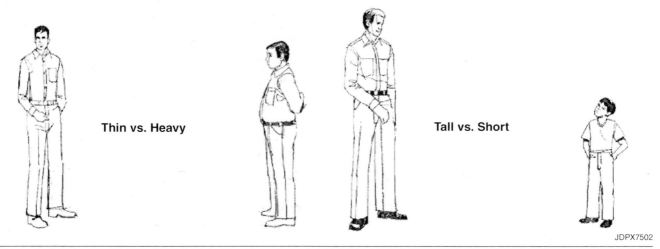

Thin vs. Heavy   Tall vs. Short

JDPX7502

Fig. 6 — What Are Your Dimensions?

What one person can do easily because of body size could be very difficult or even dangerous for someone else. A tall person can place a heavy box on a high shelf easier than a shorter person can.

A lightweight person can walk across thin ice or stand on a weak stool that would collapse under a heavier person. A tractor seat suspension adjusted for a heavy person will be too stiff for a light person.

Operator control location affects operating ability and safety. Controls adjusted for an average-sized person might be beyond the reach of a small person and could cause an accident (Fig. 7). On the other hand, a very large person might be uncomfortably cramped and restricted in the same operator station.

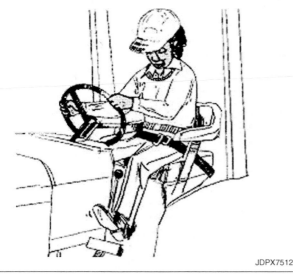

JDPX7512

Fig. 7 — A Small Person May Not Be Able To Reach Controls

What are your size limitations? Can you properly reach the clutch and brakes on your tractor? How heavy are you? Will a seat, ladder, or other structure support you properly? Your size and ability to reach machine controls and perform other tasks can affect your safety and the safety of others.

Age

Physical capabilities reach their optimum at about 25 to 30 years of age and then slowly decline. Judgment and skill, based on experience, usually continue to increase beyond age 30.

Improvement in judgment and skills can offset the decreases in reaction time and muscle strength that occur later in life. However, somewhere between 55 and 70 the typical person's eyesight, hearing, and strength deteriorate to the point that performance is poorer than it was at younger ages. Respiration ability decreases, and cardiovascular disorders are common. People who reach this stage often fail to recognize or admit their decreased abilities and continue to try to work as they did years before. These people are at greater risk of going beyond their capabilities and having or causing accidents.

Vision

The eye can focus on objects and track them. It is protected by an eyelid and is self-lubricated and self-cleaned. It automatically adjusts to light conditions and is a sensitive detector of color shades, alignment, and movement.

More than 90% of our work involves using our eyes, and so do many of our leisure-time activities.

As wonderful as the eye is, it too has physical limitations and needs protection and care. Eyes can adapt to many conditions, but eyes can be strained and even permanently damaged when conditions aren't good. Good vision depends on:

• Adequate lighting

• Size of objects viewed

• Color and contrast between an object and its background

• Steadiness of the object viewed

• Clarity and distinctness of the object

If eyes are strained because one or more of these factors are inadequate, headache, fatigue, and weariness result.

The eye is equipped with three types of color receptors (green, red, and blue) and can detect very subtle changes in color hue. However, defective color vision is fairly common. It involves some difficulty seeing a specific color, often red or green. To some degree about 8% of the male population has defective color vision. It is generally hereditary; however, females seldom have color deficiency.

Failure to recognize color deficiency could cause you to miss a traffic signal or to not recognize a safety sign. Have a doctor check your vision so you'll know if you have vision limitations.

The normal horizontal field of vision is about 188° (Fig. 8). You can detect objects and movement at the edge of your field of vision, but can't focus detail. The angle of focused vision is quite narrow (about 5°) and the eyes must be moved constantly to read or to see detail. This means that a tractor operator must turn his head 120° to 180° many times a day to see how a towed implement is doing. If he uses a rear-view mirror, he can watch these implements without the excessive head movement and should be less tired.

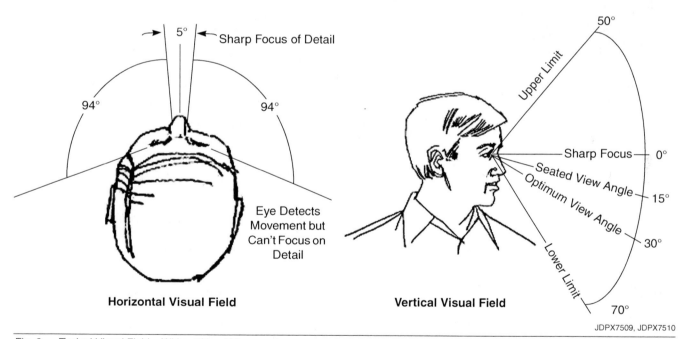

Fig. 8 — Typical Visual Field – Without Head Movement

People's ability to judge distance varies. A person with sight in only one eye can judge distance only by relative size and experience. Objects tend to look like flat photographs. Imagine trying to judge the distance between the car you're driving and one stopped on the road ahead of you if you could not judge distance.

It has been estimated that about 80% of our knowledge comes to us through the eye. Eyes are amazing organs, but they can be easily damaged, sometimes permanently. Since your sight is so important, take extra precautions to keep your eyes healthy and protect them from injury (Fig. 9).

JDPX7508

Fig. 9 — Keep Eye Protection Handy and Wear It Whenever Fumes or Particles Are Apt to Get Into Your Eyes

Wear eye protection to keep chemicals, fumes, or particles out of your eyes. Protect your eyes from bright sunlight by wearing tinted glasses. When exposed to ultraviolet rays from arc welding, wear protective goggles with darkened lenses or turn your head away. Closing your eyes doesn't completely prevent invisible ultraviolet rays from passing through your eyelids and damaging the retina. Never look at an electric welding arc without proper eye protection — not even for an instant! (See personal protective equipment recommendations later in this chapter.)

Hearing

How's your hearing? Is it as good as it was a year ago? How do you know? Hearing loss is less obvious than the losses of other senses. It occurs slowly over the years. You may not even realize your hearing is gradually decreasing because there is no pain. Studies show that one-fourth of young midwestern farmers suffer noise-induced hearing loss.

The normal human ear is adapted to wide differences in sound. It can hear the faint peep of a newly hatched chick or the loud blast of a shotgun.

The middle and inner ears are made up of sensitive organs or mechanisms. Loud sounds damage fragile hair cells in the cochlea of the inner ear. Their inability to recover from repeated high-intensity sounds results in hearing loss. You become "hard of hearing."

How loud is too loud? If your ears "ring" or seem "full," the sound is probably causing damage. Sound is usually measured in decibels by a sound level meter. Values for common sounds range from zero to 140 (Table 2). Decibels are logarithmic units of pressure and can't be added or subtracted. Doubling the amount of sound pressure energy results in a three-decibel increase. For example, if two tractors are running side by side and each produces 90 decibels of sound, the combined sound level would be about 93 decibels — not 180.

DECIBEL LEVELS OF COMMON SOUNDS, dB(A)	
0	Acute threshold of hearing
15	Average threshold of hearing
20	Whisper
30	Leaves rustling, very soft music
40	Average residence
60	Normal speech, background music
70	Automobile, 50 mph (80 km/h) at 50 ft (15 m)
80	Busy office, crowded restaurant
90	OSHA limit — hearing damage on excess exposure to noise above 90 dB, 8 hours
100	Noisy tractor, subway car, heavy city traffic
120	Thunder clap, jack-hammer, basketball crowd, amplified rock music
140	Threshold of pain — shot gun, near jet taking off, 50 hp (37 kW) at 100 ft (30 m)

Table 2 — Decibel Levels of common Sounds

Because the ear also hears logarithmically, the two tractors would not sound twice as loud as one. Many people mistakenly assume that a few decibels increase is not significant when, in fact, every three decibels increase in the sound represents a doubling of the sound pressure energy that the ear receives. It is energy received that can damage hearing.

The ear can detect changes in sound frequency or pitch. The limit of human hearing is approximately 20 to 20,000 hertz (or cycles per second). The low notes of a tuba or organ would be approximately 20 hertz.

The 20,000 hertz sound might be reached by a violin. As people age, the higher hearing range usually declines to 12,000 or 15,000 hertz.

Low frequency sounds (below 1,000 hertz) are common in agriculture. Low frequency sounds are measured on the "A" scale of a sound level meter. The "A" scale responds to sound more or less the same as the human ear. The values are given in units of dB(A), i.e., decibels on the "A" scale. The "A" scale is also used by OSHA to monitor and enforce regulatory compliance. The "B" and "C" scales on a sound meter correspond to higher frequencies.

The hearing threshold is the lowest sound in decibels that a person can hear. Abusing ears with loud sound shifts the hearing threshold upward so a person can hear only louder sounds. Maybe you've experienced this after listening to loud music or operating noisy equipment for several hours. When you stop, your ears ring and your hearing doesn't seem normal. Hearing will usually return to normal again overnight. However, repeated exposure will eventually result in a permanent threshold shift or hearing loss.

Hearing losses occur first, with increasing age, above the frequency of 1,000 hertz (cps), with maximum loss occurring at the 4,000 hertz level. High-frequency losses are more common in people between 30 and 60.

Farmers have greater hearing losses than people in other occupations. This may be partly due to the frequent and continuous operation of noisy equipment (Fig. 10).

JDPX7505

Fig. 10 — Use a Hearing Protective Device. Continuous Exposure to Excessive Noise Results in Hearing Loss.

Use of hearing protection in noisy situations helps prevent hearing loss. Protection can help you hear spoken words in a noisy environment, since the protection reduces other sound coming to the ears.

Hearing protection is necessary when the sound level exceeds 90 dB(A) (decibels on the "A" scale) and you're exposed 8 hours or more. Some experts recommend hearing protection when sound levels exceed 85 dB(A) for 8 hours or more.

Here are some general recommendations to help protect your hearing:

1. Consider quiet operation when buying farm equipment. A machine that makes a lot of noise doesn't necessarily have more power or do more work than a quieter unit.

2. Use hearing protection for all noisy jobs. Hearing protection is required for workers covered by OSHA regulations if noise exceeds 90 dB(A) for 8 hours or more. You should use hearing protection the minute noisy activity starts, even if your work site is not regulated by OSHA. (See personal protective equipment recommendations later in this chapter.)

3. Keep equipment well maintained and lubricated for quiet operation. Replace worn or damaged engine exhaust system parts promptly. Don't use a straight pipe. You'll get damaging sound but no significant power increase.

4. Build acoustic sound barriers to block out loud noise from stationary equipment, such as compressors.

5. Limit the time of exposure to loud noises (Table 3).

PERMISSIBLE NOISE EXPOSURE Hours per Day That You Can Safely Be Exposed to These Sound Levels (as specified by the Occupational Safety and Health Act for industrial situations)	
Duration per Day (Hours)	Sound Level (Decibels)
8	90
6	92
4	95
3	97
2	100
1 1/2	102
1	105
1/2	110
1/4 or less	115

Table 3 — Permissible Noise Exposure

6. Stay as far away from loud noises as possible. Doubling the distance from a noise source reduces the sound pressure level to one-fourth.

Physiological

Your body has certain physiological capabilities and limitations. Some of these are muscle tone and strength, your metabolism efficiency, how much food it takes to keep you going, your resistance to illness, and the amount of sleep and rest your body requires.

Physiological capabilities and limitations vary widely between different people and can change for the same person from day to day. They are affected by:

• Fatigue or exertion

• Drugs, alcohol, and tobacco

• Medication (prescription and over-the-counter)

• Chemicals, pesticides

• Illness

• Environmental conditions, such as temperature, humidity and dust.

Fatigue

Everyone knows about fatigue. You've probably had the experience of working or exercising to the point that your body could no longer continue and you had to rest. Fatigue may also affect you psychologically, but you may not recognize it the way you do physical fatigue. Many things can cause fatigue (Fig. 11).

Causes of Fatigue

• Monotony

• Intensity of Work

• Psychological Worries

• Illness—Malnutrition

**The Rate Of Recovery Must
Balance The Rate Of Fatigue**

Recovery:

• Sleep   • Comfortable Environment

• Short Rest   • Happy (Not Worried)

• Good Health   • Change of Task

• Excitement   • Food

JDPX7504

Fig. 11 — Fatigue Is Caused By Many Factors

When fatigue is due to physical work, typical results are muscle tightness and cramping. Work requires that chemical energy be converted to mechanical energy in the muscles. Blood flow and respiration must increase to supply the muscles with the needed energy and oxygen and to carry off carbon dioxide and chemical waste.

When muscles work at a rate exceeding the ability of the heart and lungs to supply the necessary oxygen and chemical food, aching in the muscles, cramping, tremor, and finally, loss of control may result. You have to stop and rest to recover. Muscle fatigue is like a safety device that doesn't let a muscle work beyond its capacity. Individual motivation or desire plays a very significant role in determining this limit. Strong motivation will push the limit far beyond the normal one, but there is a limit.

If you reach this limit and continue to work, you are more likely to make mistakes. Because of loss of muscle power and control, lessening of attention, slow-down of reactions, and loss of sensitivity, you become more susceptible to injury. Exhausted legs may tremble and fail to operate the brakes or clutch promptly. A load that can normally be lifted easily may be far too heavy for you to budge.

To avoid general fatigue and muscle fatigue, rest regularly. Frequent short pauses are more effective than longer rests at wider intervals (Fig. 12). Take rests in the late morning and late afternoon. Also, eating nutritional foods (such as fruit, fruit juice, and crackers) throughout the day will help maintain strength and prevent fatigue and dehydration.

JDPX7712

Fig. 12 — Frequent Short Rests Are More Effective Than Longer Rests

Drugs, Alcohol, and Tobacco

The use of mood-enhancing drugs, alcohol, and tobacco is usually defended on emotion rather than logic. Some people may feel they are exceptions and that drugs, stimulants, and depressants don't seriously affect their ability to perform. However, all those substances definitely have adverse effects on health, and also on safety.

Tobacco

The use of tobacco probably does not have a close correlation to farm work accidents. However, as discussed earlier in this chapter, when you are fatigued and reach the limits of physical endurance, you are more likely to lose some muscle power and control. Consequently, you are more susceptible to accidents or failure to perform up to par.

It has been well documented that use of tobacco over the long term reduces the ability and capacity of your body to perform. Hence, tobacco is one more negative influence on your overall safety as well as your health and longevity.

Smoking reduces work capacity as much as 10% because of carbon monoxide in the smoker's blood (Fig. 13). Tobacco smoke contains up to 4% carbon monoxide. Carbon monoxide is absorbed by the blood. Carbon monoxide has a greater attraction to the hemoglobin of blood than oxygen does, so it's not expelled during respiration. Blood carrying carbon monoxide can't be used for respiration. And because of the decreased oxygen-carrying capacity, a smoker becomes short-winded and has reduced work capacity, especially right after smoking.

Smoke? No, Thanks!
JDPX7503

Fig. 13 — Smoking Reduces Work Capacity as Much as Five to 10%

Shortness of breath can be observed immediately. Long-term effects such as the loss of taste sensitivity, sense of smell and the development of lung cancer, require a little more time to develop. Medical research has now proved beyond a question that lung cancer and smoking are directly related. Also, lung ailments from exposure to dust and toxic gases are more prevalent among smokers, because their lungs are already irritated and weakened.

Alcohol

Any amount of alcohol in the blood affects human coordination and reflexes (Fig. 14). As the amount of alcohol in the blood goes up, performance goes down (Fig. 15) and reactions are slower. Even low alcohol levels affect judgment. Alcohol is a contributing cause in thousands of fatal accidents each year, including agricultural accidents.

JDPX

Fig. 14 — Alcohol in the Blood Affects Judgment and Coordination. The Drinking Driver of This Tractor Was Killed When His Tractor Went Over a Bridge Abutment

Compare Times and Distances Needed to React

At 20 MPH (32 Km/h) Your Machine Will Travel 7 Feet
(2.13 m) for Every 1/4 Second Delay in Reaction

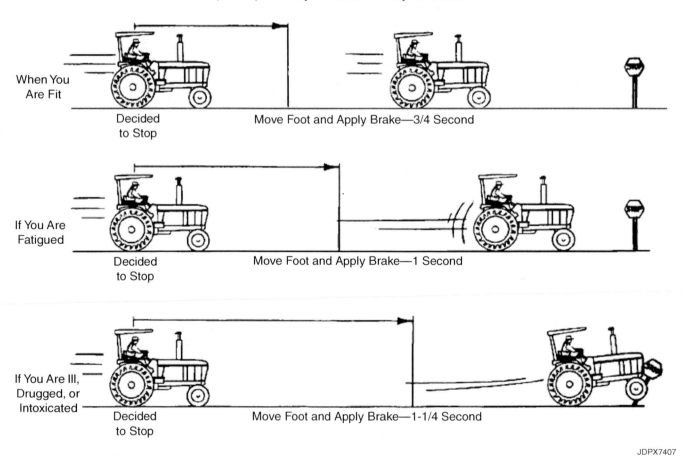

When You Are Fit

Decided to Stop — Move Foot and Apply Brake—3/4 Second

If You Are Fatigued

Decided to Stop — Move Foot and Apply Brake—1 Second

If You Are Ill, Drugged, or Intoxicated

Decided to Stop — Move Foot and Apply Brake—1-1/4 Second

JDPX7407

Fig. 15 — Reaction Time is Affected by Fatigue, Illness, Alcohol, Drugs, and Medicines

Very high levels of alcohol result in complete loss of coordination and judgment. Loss of consciousness is nature's survival switch. The level of alcohol in the blood that causes loss of consciousness is so high that a slight increase in intake can result in death.

Experiments in England relating driving skill and alcohol levels showed that bus drivers with high levels of alcohol were often able to steer a bus through a narrow space successfully. But their ability to judge whether a space was wide enough was very poor and they often tried to drive buses through spaces that were too narrow for them. Alcohol affected their judgment ability before their coordination dropped off.

Studies have shown that alcohol physically destroys some of the brain and slows the thinking process. The evidence that a small amount of alcohol impairs the ability to drive is so overwhelming that a legal intoxication point of 0.08% alcohol in the blood has been established in all 50 states of the U.S., the District of Columbia, and Puerto Rico. Some European countries have gone even lower. In Spain, Austria, and Germany, 0.05% alcohol in the blood constitutes under-the-influence. In Norway and Sweden, a person with 0.02% is legally drunk.

The effects of alcohol on body functions are temporary at first. However, toxins can build up in the liver, causing hepatitis and eventual permanent damage. Also, dependency can produce terrible problems for the drinker and even more difficult ones for those around the drinker. The danger lies in the loss of judgment and coordination that the drinker either doesn't recognize or refuses to admit.

Drugs

Drugs range from aspirin to heroin. They all have some chemical effect on the body. Hard drugs and hallucinatory drugs have direct and devastating effects on the person using them. But there is also a secondary danger that a drugged person will not recognize a dangerous situation.

The narcotic analgesics, heroin, morphine, etc., depress brain functions. Although skeletal muscle coordination may not deteriorate significantly, the user may be lulled into a false sense of euphoria and ignore signs of impending danger.

Hallucinogens such as LSD, cocaine, and other mood-enhancing substances can distort sensory perception so the user cannot properly interact with surroundings. For example, the drug user may walk off a high building, thinking it's only a step down. In addition, such substances may disrupt emotional balance and thought processes. Marijuana has similar effects. It causes a decrease in body functioning ability and muscle coordination, making the operation of machinery potentially hazardous.

Medication

Sedatives, such as sleeping pills (barbiturates) and tranquilizers (Librium™), cause a general depression of brain function. A person may become sleepy and lose the ability to concentrate, and coordination becomes poor. For example, coordination of the foot on the clutch and hand on the gearshift lever may be disrupted.

A more subtle danger lies in common antihistamine medicines like cold pills, pain killers, allergy remedies, and pep pills. These drugs can cause sleepiness, drowsiness, lowered reflexes, or increased respiration and heart rate. Those that have such an effect carry a warning on the label for the user to avoid driving or operating dangerous machinery. Any drowsiness, dizziness, or loss of coordination or muscle response can be potentially dangerous, depending on your job demands. If you don't know the side effects of a certain drug or medicine you are taking, check with a physician before you operate equipment. Always follow the advice of the physician and heed the warnings on the labels. If they warn against operating machinery, don't do it. Never mix drugs and alcohol.

Chemicals

The earth and atmosphere are made of chemicals. The nature of the chemical environment must be considered. The question is not only whether a material is harmful, but also how much of it can accumulate in the body before harm results. Small doses of some chemicals that show no immediate symptoms may build up over the years to a fatal level.

Chemicals, including pesticides, can enter the body by:

- Ocular exposure (through the eyes)

- Oral ingestion (through the mouth)

- Absorption through the pores of the skin, cuts, or scratches

- Injection under pressure into the skin

- Inhalation (breathing dust or vapors such as aerosol propellants, glues, and spray paints)

Certain changes can indicate dangerous exposure to chemicals long before clinical signs of poisoning are present. These include changes in:

- Blood pressure

- Urine concentration

- Nervous tension

- Equilibrium

- Work performance

- Reaction time

One indication of harmful exposure to chemicals is that your heartbeat does not return to normal as quickly as it should after a routine job is finished.

People vary widely in their reaction. Chemicals may have very little or no effect, or they may cause a violent reaction. A modest exposure may irritate a person or impair performance, while a higher dose could cause permanent damage.

Eyes

Eyes are very sensitive to dust and chemicals, including particle drift from crop sprays. Even small amounts of anhydrous ammonia can cause blindness. Wear unvented or chemical splash goggles when transferring anhydrous ammonia from one tank to another or when unplugging knives. If you're hit directly in the eyes with it, flood your eyes with water immediately. A few seconds' delay could mean loss of sight. Continue flushing for at least 15 to 20 minutes and then go to a doctor without delay.

Skin

Skin, particularly on arms and hands, is frequently exposed to chemicals on farms and ranches. Exposure to irritating chemicals can cause burning, itching, hives, rash, and blisters. Some chemicals may not cause such discomfort, but are absorbed by the skin. Some absorbed chemicals can cause very serious illness and sometimes permanent damage.

When working with chemicals, you should avoid contact with them by wearing protective clothing, including rubber gloves and a face shield. If you are exposed, quickly wash with soap and water.

Respiratory System

The respiratory system is sensitive to chemicals and also to dust and mold from grain and forage material (farmers' lung disease). Complications can arise from even the most innocent-appearing substances:

- Propellant from aerosol cans
- Silo gas (NO_2). Bleach-like odor and yellowish or reddish-brown vapor on top of silage. Sometimes difficult to smell or see.
- Engine exhaust — carbon monoxide (CO)
- Crop and fruit sprays
- Chemical dusts
- Household insect sprays
- Sprays for animals

These are only a few substances that can cause respiratory problems. Safety measures include:

- Avoid or reduce exposure when possible
- Work outside or in a well-ventilated room
- Use an approved respirator if you can't avoid exposure to the chemical
- Avoid drift — stay upwind

Nervous System and Vital Organs

Organic phosphates can be taken in through the mouth or nose, or absorbed through the skin. They can cause nausea, headache, and even death. Other chemicals affect the heart and vital organs. They are usually taken into the body through the mouth or a cut in the skin.

Misuse of chemicals can be disastrous. Keep chemicals away from children and pets. Read the label before you use chemicals so you know how to safely use the chemicals and the antidotes required. Be prepared to call your doctor or the nearest poison center for help in case of accidental poisoning. See chapter 9 "Chemical Safety" for more information on the safe use of chemicals.

Illness

Illness impairs human performance.

Even minor ailments like headaches and colds can reduce performance (Fig. 16).

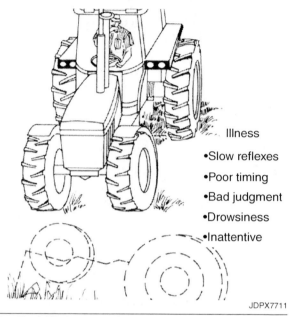

Illness
•Slow reflexes
•Poor timing
•Bad judgment
•Drowsiness
•Inattentive

JDPX7711

Fig. 16 — Don't Work Around Machinery When You're Sick

A serious illness that requires bed recovery leaves people weak for several days after they are back on their feet. Don't demand too much of yourself if you're in this situation.

Environmental Conditions

The most common environmental factors that affect farmers are:

• Temperature, humidity, and sun

• Vibration

• Noise

• Dust and mold

Temperature and Humidity

Extreme temperatures reduce work efficiency (Fig. 17). Touch sensitivity decreases at low temperatures.

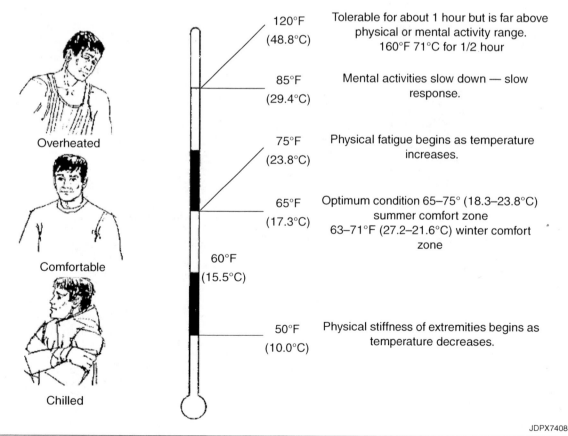

Overheated

Comfortable

Chilled

120°F
(48.8°C) Tolerable for about 1 hour but is far above physical or mental activity range.
160°F 71°C for 1/2 hour

85°F
(29.4°C) Mental activities slow down — slow response.

75°F
(23.8°C) Physical fatigue begins as temperature increases.

65°F
(17.3°C) Optimum condition 65–75° (18.3–23.8°C) summer comfort zone
63–71°F (27.2–21.6°C) winter comfort zone

60°F
(15.5°C)

50°F
(10.0°C) Physical stiffness of extremities begins as temperature decreases.

JDPX7408

Fig. 17 — Temperature and Humidity Affect Your Comfort and Work Performance

Heat

Water is very important for the body to adjust to high temperatures. The rate of water intake must equal the rate of water loss by perspiring and body functions to keep body temperature normal. When it's hot, drink plenty of water!

The older and less physically fit you are, the less you'll be able to work in the heat.

Heat leaves the body in several ways (Fig. 18):

- Transfer of heat from skin to air (convection) (Table 4)

- Evaporation of perspiration

- Exhaling hot air while breathing

- Touching a cooler object (conduction)

- Radiating from skin to air (radiation)

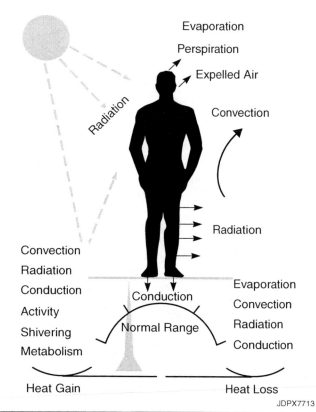

Fig. 18 — How Heat Leaves the Body

		AIR TEMPERATURES REQUIRED		
	Body Heat Produced	Air Temperature Necessary °F (°C) (assuming appropriate clothing) for Body Heat Balance at Air Movement Rates of:		
Type of Activity	BTU (kcal) per Hour	20 fpm (6 m/min) (indoors)	100 fpm (31 m/min)	20 mph (32 km/h) (outdoors)
At rest	400 (100)	70° (21°)	75° (24°)	78° (26°)
Moderate activity	1000 (252)	58° (14.4°)	60° (21°)	63° (17°)
Vigorous activity	4000 (1008)	28° (−2°)	30° (−1°)	35° (2°)

Table 4 — Air Temperatures and Velocities Required to Balance Your Body Heat Production

If both humidity and temperature are high, perspiration can't evaporate. The physiological effect of humidity is insignificant for people doing light work at temperatures below 80°F (26°C). However, at higher temperatures, body temperature rises and heart rate and blood circulation increase although work level and oxygen consumption may stay constant. You will normally stop work voluntarily when your body temperature exceeds 102°F (39°C).

Some individuals can reach a high level of oxygen intake, maintain a satisfactory heat balance, and work efficiently in the heat. But some individuals cannot. If you can't, don't push yourself beyond your limits. It can be dangerous for you around machinery and harmful to your health.

Some ways to reduce heat exposure are:

- Shading (sunshield or umbrella)

- Increased air speed (fan)

- Air conditioning (cooling)

- Frequent rests

Here are some heat wave guidelines:

1. Slow down in hot weather. Your body has a greater physiological work load when temperature and humidity are high.

2. Heed early warnings of heat stress, such as headache, heavy perspiration, high pulse rate, and shallow breathing. Take a break immediately and get to a cooler place.

3. Dress for hot weather. Lightweight, light-colored clothing reflects heat and sunlight and helps you maintain normal body temperature. Cotton is cooler than polyester clothing.

4. Eat carbohydrates in hot weather. Avoid foods high in fat. Fat needs more oxygen to metabolize and produces higher body heat than carbohydrates. Proteins increase water loss.

5. Drink plenty of water. Heat wave weather can wring you out before you know it. Don't dry out!

6. Get plenty of foods and liquids that are good sources of potassium and sodium unless a restricted diet prevents it. Most fruits and fruit juices are rich in potassium. Saltine crackers and pretzels are good sources of sodium.

7. Avoid thermal shock. Get used to warmer weather gradually. Take it easy those first two or three hot days. Your body will probably adjust if you take it slow.

8. Get out of the heat occasionally. Physical stress increases with time in hot weather. Try to get out of the heat for at least a few hours each day.

9. Don't get too much sun. Sunburn makes the job of keeping cool much more difficult. Ultraviolet rays of the sun also cause skin cancer.

Sun and Skin Cancer

Skin cancer is a special concern for farmers and ranchers, because they spend many hours in the outdoors exposed to ultraviolet rays of the sun. The most common cause of skin cancer is overexposure to the sun. The National Safety Council reported in 1992 that there are about 450,000 new cases of skin cancer each year in America.

People most susceptible have fair skin, blue eyes, and red or blond hair. However, Mayo Clinic Health Letter in June 1990 stated that virtually everyone is susceptible to some degree to ultraviolet ray damage, even those with the darkest skin.

Ninety percent of skin cancers occur on parts of the body not usually covered by clothing, including face, ears, hands, and forearms. But it is reported that there are increasing cases of cancer on shoulders, backs, and chests of men, and the lower legs of women from deliberate exposure to ultraviolet radiation for suntanning.

American Cancer Society reports that skin cancer is one of the most curable forms of cancer if it is discovered and treated early. If not treated early, it can spread throughout the body. Fortunately, most skin cancer can be prevented by following these common-sense rules:

1. Avoid overexposure to the sun, especially between 11 a.m. and 2 p.m. when the sun's rays are most intense.

2. Wear protective clothing: long-sleeved shirts, long trousers, wide-brimmed hats, and a bandanna to cover your neck (Fig. 19).

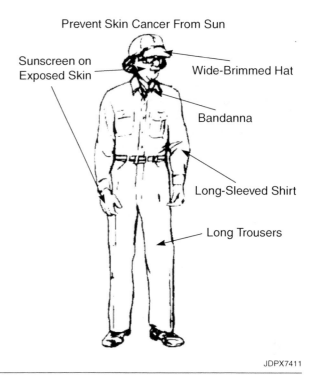

Prevent Skin Cancer From Sun

Sunscreen on Exposed Skin
Wide-Brimmed Hat
Bandanna
Long-Sleeved Shirt
Long Trousers

JDPX7411

Fig. 19 — Protective Clothing and Sunscreen Help Prevent Skin Cancer

3. Apply sunscreens or sun blocks to exposed skin. They either absorb ultraviolet radiation or physically block it from reaching your skin. Specialists recommend sunscreens with a sun protection factor (SPF) of 15 or more.

4. Use a sunblock such as zinc oxide ointment for extremely sensitive skin, or for skin already sunburned, as total protection or to prevent further damage.

5. Use a sunscreen even on cloudy or hazy days, since the harmful rays penetrate cloud cover.

6. Apply sunscreen an hour before going into the sun, and re-apply for extended exposure.

7. Examine your skin each month. Check moles, birthmarks, or other blemishes. If they change in size, shape, or color, see your physician promptly.

Cold

People can deal with low temperatures much better than high temperatures. Just add clothing. Suitable clothing can protect you against cold. But select clothing carefully for the job you'll be doing.

Heavy, bulky clothes, such as snowmobile suits, are great for snowmobiling, ice fishing, and other inactive sports. However, for active sports or work, they are much too warm. Heavy, bulky clothes can't be ventilated well or removed in layers. They may actually be hazardous if they restrict movement necessary to safely accomplish the task. Also, the body perspires in heavy clothing. Moisture is trapped, and the wearer can easily chill. Clothing for the active person should be layered so pieces can be removed or added, depending on body temperature (Fig. 20).

Your body supplies the head and vital organs with warm blood to keep these areas warm. Reducing heat loss from the head makes the heat available for other parts of the body. That's why wearing a hat helps keep your hands and feet warm.

A high calorie, high protein diet is required to give a person enough energy to work and keep warm in cold weather.

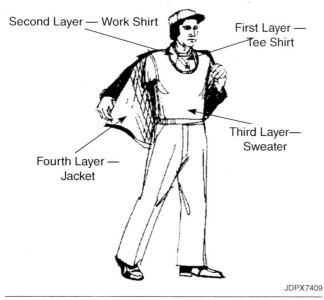

Fig. 20 — Clothing for the Active Person Should Be Layered to Allow Adding or Removing for Comfort

Vibration

The vibration of a massage chair can be soothing. The vibration of a chain saw, however, can be irritating, even damaging. Equipment that shakes or vibrates severely may damage kidneys, spine, and stomach.

Your body deals with vibration by a constant contraction and relaxation of the muscular system. Over a number of hours, this causes a change in the response of the self-regulating nervous system. This in turn affects the muscle systems of the intestinal tract and interferes with normal digestion.

Tractor vibration depends on the type of machine, seat suspension, ground, speed, and power output of the vehicle.

Vibration can affect the ability to steer or follow the movement of an object, apply foot pedal pressure, and vision. Respiratory control, coordination, and concentration are also affected (Table 5). If you are particularly sensitive to vibration, don't continue to expose yourself to it.

EFFECTS OF VIBRATION ON HUMANS			
Your Physical Activity	Effect of the Vibration	Vibration Frequency Cycles per Second	Vibration Displacement in Inches* (centimeters)
Respiration Control	Degrades Degrades	3.5–6.0 4.0–8.0	0.75 (1.9) 0.14–.61 (0.35–1.54)
Body tremor	Increases Increases	40.0 70.0	0.065 (0.165) 0.03 (0.07)
Hand tremor	Increases Increases Increases Increases	20.0 25.0 30–300 1000	0.015–0.035 (0.03–0.08) 0.035–0.055 (0.08–0.13) 0.02–0.20 (0.05–0.508) 0.008 (0.02)
Hand coordination	Degrades	2.5–3.5	0.50 (1.27)
Foot pressure consisten[cy	Degrades	2.5–3.5	0.50 (1.27)
Hand reaction	Increases	2.5–3.5	0.50 (1.27)
Visual acuity	Degrades Degrades Degrades Degrades Degrades	1.0–24 35.0 40.0 70.0 2.5–3.5	0.024–0.588 (0.60–1.49) 0.03–.05 (0.07–0.27) 0.65 (0.165) 0.03 (0.076) 0.5 (12.7)
Tracking (Steering)	Degrades Degrades	1.0–50 2.5–3.5	0.05–0.18 (0.07–0.45) 0.5 (12.7)
Attention	Degrades Degrades	2.5–3.5 30–300	0.5 (12.7) 0.02–.20 (0.05–0.508)

*Distance of movement by vibration

Table 5 — Effects of Vibration on Humans

Noise

Noise is defined as unwanted sound. A certain sound might be noise to some ears and music to others. For example, a loud, snarling sports car might sound great to its owner, but very unpleasant to neighbors.

Different people may react differently to the same sound. For example, the sound of rain on a roof may soothe some people to sleep but keep others awake.

The relative loudness of sound also affects people. A dripping faucet, unnoticed in the daytime, becomes unbearable at night.

The degree to which noise bothers people depends on several factors:

- The greater the noise intensity and the higher its pitch, the more people it will annoy.

- Unusual, intermittent noises are more upsetting than familiar, constant noise.

- Past experience associated with a given sound may determine the emotional reaction it evokes. Sounds that disturb sleep and are associated with fear are most unpleasant.

- Personal attitudes toward the noise source are important. The sound of a motorcycle engine affects people in different ways, depending on their attitude toward motorcycles.

- The schedule of a person can determine how much annoyance a sound causes. A housewife may be much less disturbed by daytime traffic noise than a person who works at night and sleeps during the day.

When a particular noise level exceeds background noise by more than three decibels at night or by five decibels during the day, it is classified as intolerable.

Loud noises affect a person. Heart rate increases with increased noise. Energy use increases. This can contribute considerably to fatigue, discomfort, and mental ease or anxiety.

When noise exposure goes down, people often:

- Have less hearing loss

- Show fewer signs of stress

- Have more energy

- Are happier

- Seem to have fewer neurotic problems

- Have fewer accidents

Psychological

Personal safety and performance depend on psychological factors. People have emotions and moods.

Psychological problems result from:

- Personal conflict — confusion and uncertainty in a person's mind

- Personal tragedy — loss of a friend or relative

- Problems in the home, disagreement, and friction between people

- Vocational problems — difficulties on the job

- Financial difficulties

- Insecurity

The effects of these psychological problems show up in a number of ways.

Emotional problems may also contribute to accidents. Learn to recognize when you are upset. Take a break when you feel you are upset.

Temper

An angry person tends to overreact and may take frustration out on people, animals, or objects that happen to be handy (Fig. 21). A person who is upset is often a risk because of poor judgment and a tendency to take chances.

JDPX7689

Fig. 21 — An Angry Person Is Dangerous

Everyone gets angry. It is a normal release. The way to handle anger safely is to stay off farm equipment, and any other potentially hazardous equipment, when you're angry. Wait until you've calmed down.

Anxiety

Anxiety makes you do things you wouldn't normally do. For example, a farmer concerned about getting a crop in before a threatening storm breaks may try to save time by not securing shields on machinery and thereby setting up an accident situation.

Apathy

A person with a "who cares" attitude toward a job will let details slip by. Sloppiness promotes accidents.

Not many of us would like to be operated on by an apathetic surgeon. There is no place for apathy around farm machinery, either. Attention to detail and concern for the safety of self and others is important.

Preoccupation and Distraction

- Suppose a person driving a tractor to the field is deep in thought about a business deal, or is distracted by some nearby activity. Because of that preoccupation or distraction, the tractor operator drives into a stump (Fig. 22).

JDPX7694

Fig. 22 — Preoccupation and Distraction Cause Accidents

- A woman is worried about difficulties her son is having in school. While driving home, she fails to recognize the slow-moving-vehicle emblem on a tractor ahead of her and runs into it.

- A young man, anxious to finish work in time for a date, neglects to shut off the power to a plugged corn picker and is caught and mangled while attempting to clear it.

Investigating It Yourself

What makes a person touch a newly painted surface when a sign says "Wet Paint"? Why do people have a tendency to check to see if something is hot or to reach for a moving part? Why does a farmer reach into a dangerous place for a dropped ear of corn, lean against a machine, pat it approvingly, or touch a moving belt? The answer is not simple and involves a deep investigation into the study of psychology. Realize people have these tendencies and recognize the potential hazards they create.

Psychological factors, attitudes, emotions, and moods are not usually changed easily or quickly. However, recognize these characteristics and understand the safety hazard and accident potential they possess. If you recognize dangerous traits in your own actions, try to take corrective measures. If you are angry, take a break and calm down before you are involved in an accident. If anxiety or preoccupation continue to bother you, some counseling or otherwise dealing directly with the underlying problem may remedy the situation completely.

If you are aware of psychological problems in other people, try to keep them away from danger. For example, the person with an apathetic attitude should not be allowed to work where errors and carelessness could cause a serious accident. Do not permit "show-offs" to operate equipment where they might injure themselves and others, or damage the machinery.

The "safe worker" is generally one who is happy, alert, content, and well adjusted. He or she works with skill and has concern for the job, for others, and for personal safety.

Personal Protective Equipment

Many farm injuries can be prevented by personal protective equipment. Devices are available to protect the:

- Head

- Eyes

- Ears

- Hands

- Feet

- Body

- Respiratory system

But they must be used in order to be effective.

Head Protection

The most effective head protection is a hard hat (Fig. 23). It protects against bumps and falling objects. Some hard hats have warm liners for cold weather. Nonconductive hard hats are good for electrical work.

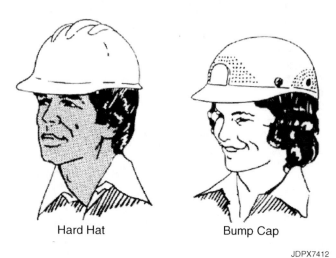

Hard Hat   Bump Cap

JDPX7412

Fig. 23 — Hard Hats and Bump Caps Protect Against Head Injuries

Use hard hats for building work, trimming trees, or other jobs where falling objects could cause head injuries. Lightweight bump caps are intended only to protect against bumping the head against objects, not as a substitute for hard hats. Approved hard hats are identified by the ANSI 289.1 code.

Eye Protection

Eyes are very sensitive. Protect them from impact, chemicals, dust, and chaff. Wear eye protection when spray painting, grinding, drilling, welding, sawing, and working in dust or chemicals. Approved eye protection is marked with the ANSI 287.1 coding. There are several types of protective devices for the eyes (Fig. 24).

Common Types Of Safety Eyeware

Spectacle Full Side Shield   Spectacles   Cover Goggles   Face Shield

JDPX7413

Fig. 24 — Common Types of Safety Eyeware

Safety Glasses

If you wear eyeglasses, including sunglasses, be sure they have impact-resistant lenses. The lenses of approved (ANSI 287.1) safety glasses are made of special impact-resistant materials in order to withstand more shock than ordinary lenses. Note that partial or full-side shields protect against particles from the sides (Fig. 25).

Impact-Resistant Lenses Meet ANSI Z87.1

JDPX7714

Fig. 25 — Safety Glasses Have Impact-Resistant Lenses

Goggles

Goggles protect eyes from impact from the front and sides. Unvented or chemical splash goggles also offer some protection against chemical vapors and liquids (Fig. 26).

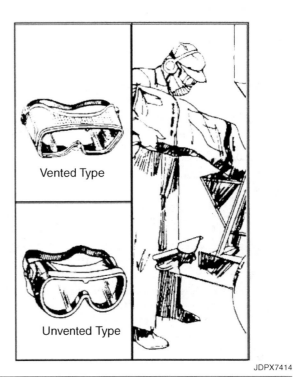

Vented Type

Unvented Type

JDPX7414

Fig. 26 — Safety Goggles Shield the Eyes From Front and Sides

Face Shields

Face shields protect the face from splashing, dust, and chaff (Fig. 35). But they offer very little protection against impact.

Face Shield

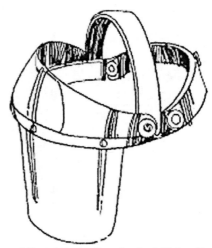

For Eye and Facial Protection Against Flying Particles Heat, Chemical Splash, and Glare

JDPX7500

Fig. 27 — Face Shields Keep Splashes Away From Your Eyes and Face

If you need impact protection, wear safety glasses or goggles under the shield or get a special impact-resistant shield that's fitted to a hard hat.

Hearing Protection

Farmers are generally exposed to considerable noise and have higher than average hearing loss.

Why? Hearing damage can begin at sound levels as low as 85 to 90 decibels, and many farm machines are louder. Wear earmuffs or at least earplugs whenever you're exposed to a noise level of 90 decibels dB(A) or higher for more than eight hours. (See previous discussion of HEARING in this chapter.) Some experts recommend hearing protection for 85 decibels or more.

In selecting hearing protection (earplugs or earmuffs), look for the Noise Reduction Rating (NRR) label on the device. Be sure the rating is high enough to reduce sound pressure to acceptable levels. Example: The NRR number is an estimate of the protection provided. A device marked "NRR 26" is intended to reduce a 100 dB(A) noise to 74 db(A) (100 − 26) under ideal conditions. In actual use, it may provide only about 50% of that protection, or 13 dB(A), resulting in a 87db(A) (100 − 13) sound level. Contact state or federal Occupational Safety and Health Administration for regulatory requirements.

Earplugs

Rubber or plastic earplugs fit into the ear canal and are effective noise suppressors. A snug fit is important, so have them custom-fit for comfort and protection (Fig. 28).

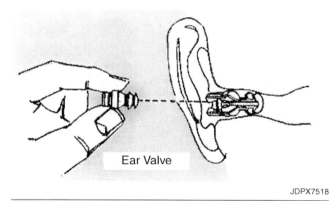

Ear Valve

JDPX7518

Fig. 28 — Use Plastic Foam or Rubber Earplugs That Fit Snugly

Don't use cotton plugs. They don't block any high frequency sounds and only a few low frequency sounds.

Earmuffs

Earmuffs (Fig. 29) are the most effective; protection against noise. They don't contribute to infection and discomfort as can earplugs, which fit tightly and carry dirt to the ear canal. Earmuffs block more noise than earplugs because they cover the sound-conducting bones around ears as well as the ears themselves. They are also comfortable and they keep your ears warm.

Ear Plug

Acoustical Earmuffs

JDPX7415

Fig. 29 — Acoustical Earmuffs Are More Effective Than earplugs

Hand Protection

Gloves can't always prevent a finger amputation, but they can guard against cuts, abrasions, chemicals, and skin irritation (Fig. 30).

JDPX7499

Fig. 30 — Wear Gloves

Leather gloves protect hands against rough or sharp objects, and they give good gripping power.

Use rubber gloves when working with fertilizers.

Wear unlined neoprene or nitrile gloves when using pesticides. Lining tends to capture and trap chemicals. Use protection recommended by the chemical label.

Canvas or cotton gloves offer some protection, but never wear these when working around pesticides. They absorb or trap chemicals.

Wear gloves that fit. If they're too big, they could easily get caught in moving parts that could seriously injure a hand.

Foot Protection

Wear safety shoes. Their steel toes and puncture-proof, skid-resistant soles protect your feet (Fig. 31). Metatarsal shoe guards will also protect the insteps of your feet against serious injury. All safety shoes should meet ANSI 741 standard.

JDPX7498

Fig. 31 — Wear Safety Boots With Safety Steel Toes and Metatarsal Guards

Body Protection

Aprons and padding protect your body, But the most important protective item is your everyday clothing. It should fit comfortably but not loosely (Fig. 32). Loose clothing near moving parts is hazardous:

1. Zip up or button your jacket. Remove all drawstrings.

2. Wear short-sleeved shirts or button the cuffs on long-sleeved shirts. Avoid rolled-up sleeves.

3. Wear pants or overalls with straight legs that don't drag on the ground.

JDPX7688

Fig. 32 — Clothing Should Be Snug but Comfortable

Respiratory Protection

There are several types of respiratory protective devices (Fig. 33). Each does a certain job. Use the right one.

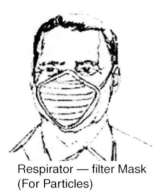 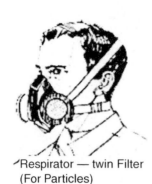  

Respirator — filter Mask
(For Particles)

Respirator — twin Filter
(For Particles)

Respirator — chemical
Cartridge

Gas Mask

JDPX7416

Fig. 33 — Good Respiratory Protectors—Except Where Oxygen Might Be Deficient

Dust Filter Respirators

These are soft fiber facepieces that filter air. The filter traps dust, chaff, and other particles. If you work where dust contains mold, a dust respirator with a tight seal around nose and mouth is required. There can be severe allergic reactions to mold in hay and forage material.

Asbestos fibers in airborne dust may cause lung cancer. So, it is important that you wear a dust respirator approved for asbestos when working on parts that contain asbestos, including brake pads, brake band and lining assemblies, clutch plate linings, and gaskets.

Read and follow labels on protective equipment to make sure you are protected for your specific situation. Dust respirators will not help if you are working with chemicals or toxic gases, or if there is an oxygen deficiency,

Chemical Cartridge Respirators

Cartridge respirators have a partial face mask fitted with one or two replaceable cartridges. The cartridges contain an absorbent material (often activated charcoal) that purifies inhaled air and dust filters to trap particles.

Make sure you have the correct cartridge for the material you want to filter out! Read the label on the MSDS (Material Safety Data Sheet).

Also check to make sure the cartridge "life" has not expired. Follow the manufacturer's directions. Make sure respirators fit snugly at all face-contact points to prevent chemical vapors from entering. Give special attention to proper fit for a bearded face, because facial hair makes a snug fit more difficult.

Cartridge respirators are effective against all but the most toxic vapors. Use cartridge respirators in open areas where there is plenty of oxygen. Don't use them in silos or manure pits where there might be a lack of oxygen.

Gas Masks

Gas masks are heavy-duty cartridge respirators. They use a chemical filter to remove toxic vapors and particles from the air. They have a greater capacity to protect against high concentrations of toxic gases than chemical respirators. But they, too, aren't effective when oxygen is lacking.

Supplied-Air Respirators

Supplied-air respirators bring compressed air or outside air in through an air hose (Fig. 34). They are used either with a blower or with compressed air.

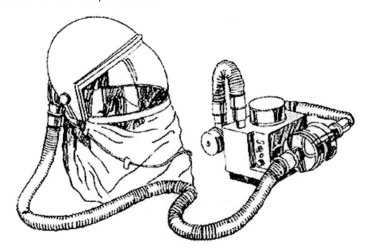 

Supplied-Air Respirator With Blower-Filter and Hood

Self-Contained Breathing Apparatus

JDPX7519

Fig. 34 — Use These Respirators When You Must Go Into a Manure Tank or Silo When There's a Lack of Oxygen

Supplied-air respirators can be used when there is an oxygen deficiency or in some contaminated atmospheres. They must have an emergency escape air supply.

Self-Contained Breathing Apparatus

A self-contained device has its own air supply in a cylinder (Fig. 34). The apparatus works in silos, grain bins, manure pits, and other places where other respirators are inadequate because of an oxygen limitation.

Lifting

The finest safety equipment available is of little value if you ignore it, use it incorrectly, or do not use correct work methods. A back injury resulting from any improper lifting technique is an excellent example.

About 75% to 80% of us are affected by back disorders sometime during our lives. Back disorders are common, even in the 20-to-30 age group. People engaged in heavy, active work, like dock hands, farmers, and ranchers, are affected more than those who do office work. Back disorders often last for life.

Back problems occur when spinal disks are damaged or diseased. The fibrous outer edges become weak and brittle. A sudden heavy load can rupture them, resulting in pain and possible paralysis or muscular problems. The injury is often referred to as a slipped disk, even though the disk does not actually slip.

Back problems often begin with incorrect lifting practices. A lot of pressure is put on the spine when a person bends the back without bending the knees when lifting (Fig. 35).

JDPX7523

Fig. 35 — Lift Heavy Objects Using Your Leg Muscles, or Use Shop Hoists or Tractor Lifts

The leverage of a poor mechanical position in lifting can create a force of as much as 1500 pounds (700 kilograms) on the spinal disks when lifting only 110 pounds (50 kilograms). This often causes a ruptured disk. Sometimes the injury is permanent.

Use the proper procedures and body position when lifting to reduce the risk of injury (Fig. 36 and 37).

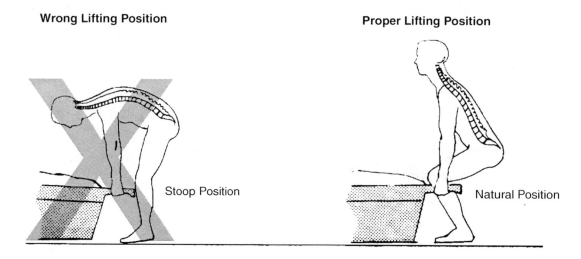

JDPX7417

Fig. 36 — Improper Lifting Can Injure the Back

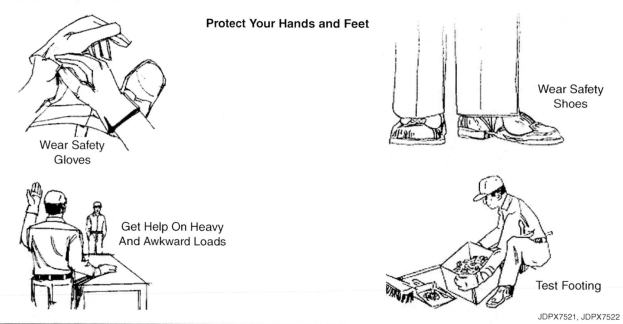

JDPX7521, JDPX7522

Fig. 37 — Practice Personal Protection When Lifting

1. Protect your hands and feet. Gloves can protect hands from snags from barbed wire, wood splinters, and sharp edges. Steel-toed safety shoes prevent many painful experiences. Dropping a heavy object on your foot is just one example.

2. Get a good footing. Remove obstacles and debris that could cause a fall. Position your feet slightly apart for balance.

3. Bend your knees. Get close to the object. Keep your back in a natural position. Do not force your back to be abnormally straight or curved.

4. Get a good grip. Grasp the object with a full hand grip instead of your fingertips. It's less likely to slip and fall. Tilt boxes and get one hand under them. Place the other hand diagonally opposite. This principle of gripping diagonally also applies to sacks of fertilizer, grain, etc.

5. Avoid twisting as you lift. Twisting or bending sideways increases the risk of injury.

6. Lift with your leg muscles. Lift smoothly, not with a jerk. Keep the load close to your body (Fig. 38).

7. Carry loads at the center of your body to balance the weight.

8. Reverse the procedure to set an object down. Keep your back in a natural position and bend your knees to lower an object.

9. Use a hoist or hooks, straps, and pulleys to lift a heavy load from the floor.

10. A lever and fulcrum can make many heavy lifting jobs easy. Hydraulic jacks may be required. Get the proper help (person or equipment). Don't be laid up for a week or a lifetime with a bad back because you tried to save a few minutes.

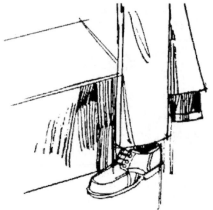

Stand Close to Object to Be Lifted

Bend the Knees

Tilt the Box Placing One Hand Under

Place Hand Diagonally Opposite

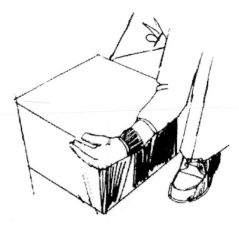

Get a Full Hand Grip — not Just Fingertips

Lift With Legs and Keep Back Natural

JDPX

Fig. 38 — Practice Proper Lifting Procedures To Avoid Serious Back Injury

Sometimes a barrier, such as a partition or part of a machine, such as the vertical wall of a pickup truck, may prevent you from using the lifting procedures already discussed. You may be able to use the assisted one-hand lifting method (Fig. 39). Push downward on the top of the barrier to help raise your upper body to a vertical position as you lift. This helps distribute the stress over the large shoulder and arm muscles while reducing stress on the lower back.

JDPX7419

Fig. 39 — Assisted One-hand Lift Helps Reduce Stress on The Lower Back

Man-Machine Systems

Compatibility

People use tools and machines as extensions of their bodies in order to do more work. The people, tools, and machines function together as a "system" often called a man-machine system. A good man-machine system is natural and easy for the people involved.

Comfort, productivity, and safety are improved when the tools and machines are compatible with the capabilities and limitations of the people who use them. They work well together. That is called "man-machine compatibility." When man-machine compatibility is not good, people must be extra careful to avoid personal injury accidents.

Performance Breakdown

You perform best when you're challenged by a task, but you shouldn't be overburdened. When a job is too difficult or involves too many controls, gauges, and observations to coordinate, it can overload an operator. You can operate only for short periods at such a task. Breakdown often occurs in the form of mistakes, frustration, and anger.

Bales coming up an elevator faster than a person can stack them is an example of overload. They may pile up and result in a messed-up bale stack, an angry worker, and possibly an accident.

Any system that forces a person to work much faster than is comfortable will result in an overburdened operator, low-quality performance, and possibly personal injury.

Machines have definite upper limits of capacity. If these limits are exceeded, the machine stops or fails. In contrast, man's upper limit is variable. It depends on environmental conditions, motivation, and physical condition. People can work in an overloaded situation for a while, but eventually will reach a failure point. On the other hand, if a job is too easy or routine, monotony, boredom, errors, and eventually accidents can result.

Cultivating row crops is a common example (Fig. 40). The tractor operator in a long field has little to do and can become bored. Some operators even fall asleep and have serious accidents.

JDPX7710

Fig. 40 — Tedious Work Can Lead to Accidents

Mental fatigue as well as physical fatigue contributes to a decline in a person's performance. Long periods of careful thinking, such as repeated calculations or reading, are mentally tiring. The mind begins to slow down and make errors. Short periods of activity are more effective. Take a safety break when you're sleepy, fatigued, or mentally tired.

Fig. 41 — Comfort and Convenience Help Reduce the Workload and Improve Safety in Machinery Operation

Factors that increase a person's physical and mental fatigue also increase workload (Fig. 41). When buying new or used equipment, try to reduce the workload by selecting equipment with features, such as:

- Good visibility

- Convenient and easy-to-operate controls

- Comfortable seating and position

- Minimum vibration

- Shielding from heat of the machine (or cold from exposure)

- Quietness

- Protection from dust and fumes

- Easy-to-read instruments

- Controls that are compatible with the operator and machine responses

- Adequate ventilation

- Adequate illumination and protection from glare

Communications

In most man-machine systems, the machine communicates its operating conditions and need for control changes to the operator in several ways. Unless the machine is computer controlled, the operator makes the changes to keep the machine operating correctly.

Machine-to-man communication includes such things as:

- General observation — Is the machine on the row? Is the material being fed into the machine at the correct rate?

- Gauges and visual indicators — Engine temperature, engine rpm, vehicle velocity, hydraulic or air pressure, malfunction alarm light on oil pressure and alternator.

- Sound — Overload slows the engine and strains machine parts. Plugged augers, hay pickups, etc., cause slip clutch drives to chatter. Underload may allow excessive engine speed. The experienced operator understands these sound communications from the machine. Some machines are equipped with sound warning alarms such as the malfunction alarm on planters.

- Feeling — A machine's vibrations often change with a change in load. You can feel a plow digging into wet ground. You can feel a plow sliding or tipping.

Adaptability and Controls

Simple mechanical machines do not think, learn, or adapt. You must do the adapting when new circumstances develop. The more unnatural the circumstances, the longer you will take to adapt and react. In an emergency you may make an error if your instinctive reaction is different from the required reaction. If dials and gauges are difficult to read and controls are awkward, you will take even longer to adapt and react (Fig. 42).

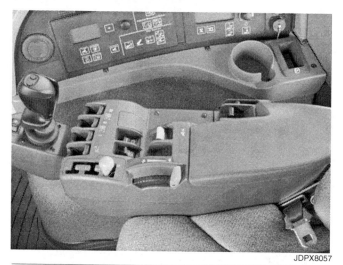

Fig. 42 — Controls Designed, Located, and Identified for Man-Machine Compatibility Help Improve Safety

Ask yourself the following questions when evaluating a machine:

- Where are the controls located? Can they be reached easily? Are they spaced for convenience? Are they adequately labeled and coded?

- What types of controls are used? Are they compatible with the operator and with the machine response?

- Do the controls themselves present a hazard? Is each one clearly identified by sight and touch? Can one be mistaken for another? There should be no confusion about which lever controls a hydraulic cylinder or which knob on a combine controls the header, or the speed, etc.

- Which arm(s) and leg(s) are involved? Are all controls operated by one arm with the other limbs doing nothing?

A better, less fatiguing arrangement is to use both arms and legs.

- Are the controls standard? Are controls common, or are they unnecessarily "unique" and possibly incompatible with other systems?

Obsolete Equipment

Machines grow old and become obsolete (Fig. 43).

JDPX7709

Fig. 43 — Machine Designs Change

Equipment may not always be worn out before it is obsolete. Hay rakes, for example, have changed from cylinder to parallel bar because parallel bar rakes handle hay more gently.

New farming technology often causes obsolescence, not age alone. Combines have replaced grain binders and threshing machines; herbicides and modern tillage practices have made checkrow corn planters a thing of the past.

Machines have been improved over the years and have modern safety features. Some of the improvements that have been made are:

- Reverser mechanisms to discharge clogs

- Quiet, dust-free operator stations

- Belt and chain drive shields

- PTO shaft shields

- Neutral start interlock switches to keep engines from starting when a machine is in gear

- Seat interlock switches to stop the engine or components if the operator leaves the seat

- Controls that are easier to reach and operate

- Rollover protective structures (ROPS) and falling object protective structure (FOPS). (Note: Not all ROPS are strong enough to handle falling objects, and not all FOPS are strong enough to protect the operator in a tractor roll-over accident).

- Safety signs with pictorial hazard illustrations that have been added to machines

If you use older equipment, be especially safety conscious. Reasonable safety can still be achieved if you maintain the equipment properly, follow good safety practices, stay extra alert, and follow the safety signs and operator manuals that have been developed for your machinery.

Test Yourself

Questions

1. Effective rest breaks from strenuous work should be:

   a. Rather short (1 to 2 minutes) and frequent

   b. Longer (10 to 15 minutes) about every 2 hours

   c. Not usually necessary for a person in good physical condition

   d. none of the above

2. With two tractors operating side-by-side, each producing a sound pressure level of 80 decibels, what is the combined sound pressure level?

   a. 80 decibels

   b. 83 decibels

   c. 86 decibels

   d. 160 decibels

3. In general, hearing protection is recommended when a person is exposed to which of the following?

   a. 70 decibels for more than 9 hours

   b. 80 decibels for 8 hours

   c. 90 decibels for 8 hours

   d. 95 decibels for 1 hour

4. A typical reaction time required for a person to receive an unexpected signal, decide what to do, and do it is:

   a. 0.5 seconds

   b. 1 second

   c. 2.8 seconds

   d. 6 seconds

5. (T/F) One principle of man-machine compatibility is this: "The machine controls must move in ways that correspond to the machine movement and are natural for the operator to use."

6. Heat stress can be relieved by:

   a. Drinking plenty of water

   b. Taking frequent rest breaks

   c. Working in the shade if possible

   d. All of the above

7. Exposure to ultraviolet sun rays is most intense:

   a. In the early morning hours

   b. Around noon

   c. In the late afternoon

   d. There is no real difference in exposure during daylight hours

8. (T/F) For cold weather, farmers should dress in layers to allow for changing temperatures and work tasks.

9. (T/F) Prescription or over-the-counter drugs will not affect a person's reaction time or coordination.

10. (T/F) Alcohol is never a factor in farm-related injuries.

11. Which of these provides the best protection against head injury from falling objects?

    a. Hard hat

    b. Bump cap

    c. Baseball cap

    d. Welding helmet

12. To prevent allergic reactions while working around moldy grain, what type of respirator should you wear?

    a. Supplied air respirator

    b. Self-contained breathing apparatus

    c. Chemical cartridge respirator

    d. Dust filter respirator

13. (T/F) Back injuries caused by improper lifting rarely result in a permanent impairment.

14. The most common problem with the use of personal protective equipment is:

    a. Proper fit

    b. Finding the proper equipment

    c. Knowing what equipment to use

    d. Getting workers to wear protective equipment

15. (T/F) A comfortable seat and easy-to-use controls can make a tractor safer and more efficient to operate.

References

1. Mayo Clinic Health Letter. May 1984. June 1990. P.O. Box 53889, Boulder, Colorado, 80322-3889.

2. Farm Safety Fact Sheet. June 1992. U.S. Department of Agriculture, Extension Service, Washington, D.C., 20250.

3. Accident Prevention Manual. National Safety Council, Chicago, Illinois, 60611.

4. Practice for Occupational and Educational Eye Protection (ANSI Z87. 1). American National Standards Institute, New York, New York 10036.

5. Personal Protection — Protective Footwear (ANSI Z41). American National Standards Institute, New York, New York 10036.

6. Protective Headwear for Industrial Workers (ANSI Z89.1). American National Standards Institute, New York, New York 10036.

7. Work Practice Guide for Manual Lifting, Publication DHHS No 81-122. National Institute for Occupational Safety and Health, 4676 Columbia Parkway, Cincinnati, Ohio, 45226.

Recognizing Common Machine Hazards

3

Using Tools Can Be Hazardous

Being Injured Is An Accident

JDPX7715

Fig. 1 — What Is a Hazard? What Is an Accident?

Introduction

What is a hazard? A hazard is simply an object or situation that has the potential to cause injury. When a person comes into contact with a hazard, an accident may occur (Fig. 1). An accident, for our purposes, is an event that occurs when a person is injured unintentionally as a result of either machine failure or human error. Let's look more closely at human and machine factors in accidents.

People and Hazards

Man's greatest ability is to think and take action according to knowledge, experience, and judgment. In contrast, man's greatest limitation is the tendency to make errors. When an accident occurs due to human error, it is usually because the individual made one of the following mistakes:

- Forgot something — such as not setting the brake or shifting the transmission into the park position before dismounting.

- Took a shortcut — such as trying to start or operate a tractor from the ground.

- Took a calculated risk — such as stepping over a rotating PTO shaft.

- Ignored a warning — such as "Disengage power, shut off engine, and take the key before adjusting or lubricating a machine."

- Used unsafe practices — such as refueling while smoking.

- Was preoccupied — such as worrying about losing time while making repairs.

- Failed to recognize the hazard — this failure results in no corrective or preventive action.

The environment may add even more risks which can affect situations. For example, the weather can contribute rain, snow, cold, and heat. In addition, the terrain might be hazardous if it is rough or hilly.

One of the most dangerous mistakes humans make is failing to recognize a hazard. Recognizing hazards and understanding them is the only way you can avoid some accidents.

Remember: Many hazards can be avoided if you always disengage the power, shut off the engine, take the key, and wait for all parts to stop moving before working on a machine.

Machines and Hazards

Machines are designed to work. Because a machine uses power, motion, and energy to do work, it presents a number of potential hazards that cannot be eliminated completely. For example, the blades on a rotary mower must rotate at high speeds to cut grass and cornstalks. The blades have the potential to injure you if you come in contact with them or if they strike an object and throw it.

Manufacturers build many safety features into their machines, but you must remember that not all potential hazards can be eliminated (Fig. 2). A mower may have a shield over the powershaft. But, if the blades were shielded completely they would not be able to cut the material.

JDPX8058

Fig. 2 — Manufacturers Eliminate Hazardous Parts or Build Safety Shields, but All Hazards Cannot Be Eliminated

When it is impossible to eliminate or fully shield a hazard, safety signs may be used to warn you. You must recognize the potential hazards and take necessary action to avoid injury.

Design Safety

Manufacturers eliminate hazards through good design. Some designs eliminate the need for shielding and complex safety training. Here are some of the safety considerations manufacturers have when designing a machine.

1, Make it easier for the operator to do the same thing. Minimize fatigue and maximize convenience, comfort, and safety.

2. Prevent the operator from coming in contact with a hazard by shielding hazardous machine parts which cannot be eliminated.

3. Protect the operator.

4. Provide safety instructions in operator's manuals and in signs on machines to warn of hazards and explain how to avoid them.

Common Machine Hazards

This chapter deals with machine hazards. It will help you recognize hazards and understand why they are hazardous. You will know what to avoid. The information in this chapter will help you develop a safe attitude.

Here are some of the most common machine hazards

- Pinch points
- Wrap points
- Shear points
- Crush points
- Pull-in points
- Freewheeling parts
- Thrown objects
- Stored energy
- Slips and falls
- Slow-moving vehicle
- Second party

Pinch Points

Pinch points are where two parts move together and at least one of them moves in a circle. Pinch points are also referred to as mesh points, run-on points, and entry points. There are pinch points in power transmission devices such as belt drives, chain drives, gear drives, and feed rolls (Fig. 3).

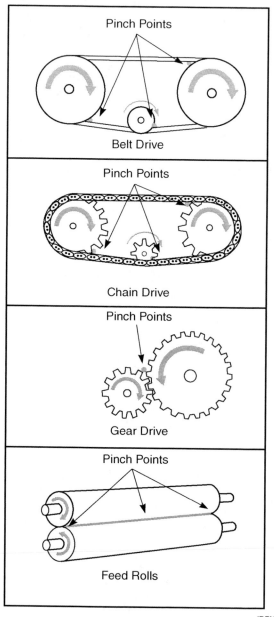

JDPX7421

Fig. 3 — Pinch Points on Rotating Parts Can Catch Clothing, Hands, Arms, and Feet

You can be caught and drawn into the pinch points by loose clothing. Contact with pinch points may be made if you brush against unshielded rotating parts, or slip and fall against them. Farmers can get tangled in pinch points if they deliberately take chances and reach near rotating parts. Machines operate too fast for persons to withdraw from a pinch point once they are caught.

Avoiding Pinch Points

Manufacturers build shields for pinch points. But some pinch points, such as feeder rollers on corn heads and silage choppers, must be open so crops can enter. Always replace shields if you must remove them to repair or adjust a machine. Remember that a portion of the money you paid for the machine went for safety research and design and for the actual hardware involved in shielding. Get your money's worth and protect yourself.

For pinch points that cannot be shielded, the best protection available is operator awareness. Know the location of the pinch points on your machinery. Avoid them when the machine is operating. And above all, never attempt to service or unclog a machine until you have disengaged all power, shut off the engine, removed the key, and all parts have stopped moving.

Wrap Points

Any exposed component that rotates is a potential wrap point. Rotating shafts are usually involved in wrap-point accidents. Often, the wrapping begins with just a thread or frayed piece of cloth catching on the rotating part (Fig. 4). More fibers wrap around the shaft.

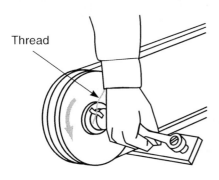

Thread

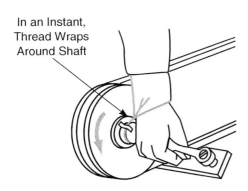

In an Instant, Thread Wraps Around Shaft

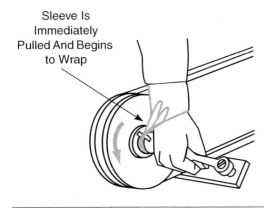

Sleeve Is Immediately Pulled And Begins to Wrap

JDPX7423

Fig. 4 — Wrapping May Begin with Just a Thread. In an Instant the Victim Is Entangled with Little Chance to Escape Injury.

The shaft continues to rotate, pulling you into the machine in a split second. The more you pull the tighter the wrap. If the clothing would tear away, a person might escape serious injury, but work clothes are usually too rugged to tear away safely. Long hair can also catch and wrap, causing serious, permanent injury.

Smooth shafts often appear harmless, but they too can wrap and wind clothing. Rust, nicks, and dried mud or manure make them rough enough to catch clothing (Fig. 5). Even shafts that rotate slowly must be regarded as potential wrapping points.

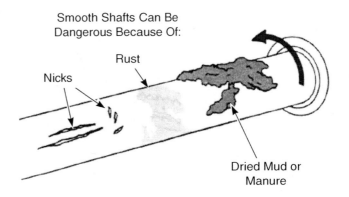

Smooth Shafts Can Be Dangerous Because Of:

Nicks

Rust

Dried Mud or Manure

JDPX7538

Fig. 5 — Even Seemingly Smooth Shafts Can Catch And Wrap Clothing

Ends of shafts which protrude beyond bearings can wrap up clothing (Fig. 6). Splined, square, and hexagon-shaped shafts are more likely to wrap than smooth shafts (Fig. 7).

Leaning against Protruding Shafts Is Dangerous

JDPX7617

Fig. 6 — Shafts that Extend Much Beyond Bearings or Sprockets Can Be Dangerous

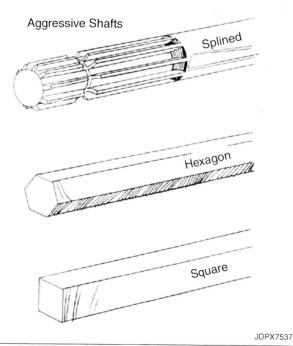

Aggressive Shafts

Splined

Hexagon

Square

JDPX7537

Fig. 7 — Splined, Square, and Hexagon-Shaped Shafts Are More Likely to Entangle than Round Shafts

Couplers, universal joints, keys, keyways, pins, and other fastening devices on rotating components will wrap clothing (Fig. 8).

Aggressive Components

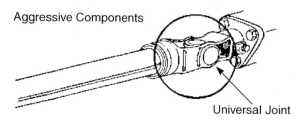

Universal Joint

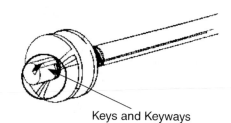

Keys and Keyways

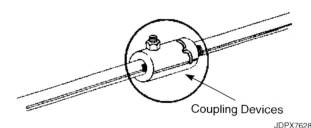

Coupling Devices

JDPX7628

Fig. 8 — Some Rotating Components Are Even More Aggressive than Their Rotating Shafts.

Exposed beater mechanisms will entangle and wrap clothing (Fig. 9).

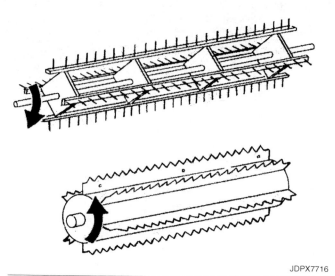

JDPX7716

Fig. 9 — Beater or Feeder Mechanisms Can Tangle or Wrap Clothing

Avoiding Wrap Points

The best protection against wrap-point hazards is to be aware of them, understand how you can become entangled, and follow common-sense, safe practices. Be sure to keep all shields in place and in good working condition; wear snug-fitting clothing without frayed edges or drawstrings; wear fairly close-cut hair or keep long hair covered and away from moving parts; shut off machines before working on them to clean, unplug, grease or adjust them; stay away from unshielded rotating parts; and use only the proper replacement bolts, pins, or keys on rotating parts. Replacements that are too long or large can increase the potential for wrapping accidents.

Shear Points

Shear points are where the edges of two moving parts move across one another. For example a set of hedge-trimming shears (Fig. 10) or a cutter head (Fig. 11).

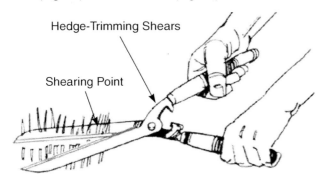

Hedge-Trimming Shears

Shearing Point

Two Objects Moving Closely
Toward or Across One Another

JDPX7539

Fig. 10 — Shear Points Are Created When Two Parts Move Across One Another

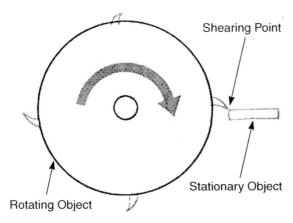

Shearing Point

Rotating Object

Stationary Object

One Object Moving Near a
Stationary Object

JDPX7717

Fig. 11 — Shear Points Are Extremely Dangerous on a Cutter Head

Cutting Points

Cutting points are created by a single object moving rapidly or forcefully enough to cut a relatively soft object. A table knife, a hand-held grass sickle, and a rotary lawn mower blade are examples of single objects that cut because of their sharpness and force (Fig. 12 and Fig. 13).

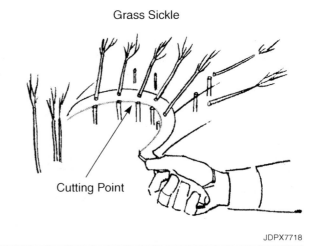

Grass Sickle

Cutting Point

JDPX7718

Fig. 12 — Cutting Points Are Created by Single Sharp Parts Moving with Enough Force to Cut Softer Material

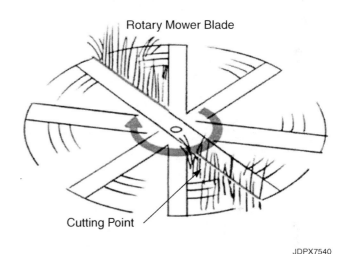

Rotary Mower Blade

Cutting Point

JDPX7540

Fig. 13 — Cutting Points on a Rotary Mower Blade

Devices Designed to Cut or Shear

Shearing and cutting devices on farm machines cut crops. The shearing and cutting parts may rotate (Fig. 12 and Fig. 13) or reciprocate (Fig. 14). Examples are sickle bar mowers, rotary shredders and cutters, and cutter heads of forage harvesters. Because they must cut crops at several tons per hour, they are aggressive. They do not know the difference between crops and fingers. And because they must cut crops on the move, they usually cannot be guarded to keep out hands and feet.

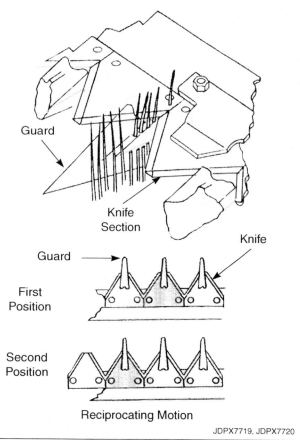

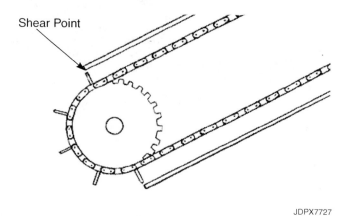

Fig. 16 — Chain and Paddle Conveyor

Fig. 14 — Some Shearing Parts Move in Reciprocating Motions

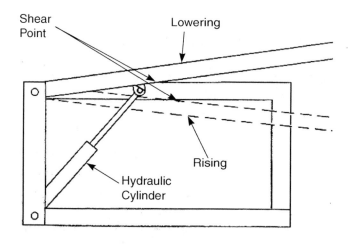

Fig. 17 — Implement Frame Rising or Lowering

Devices Not Designed to Cut or Shear

There are cutting or shear points on devices that are not actually designed to cut or shear. For example, grain augers (Fig. 15), chain and paddle conveyors (Fig. 16), and hinged implement frame members that move when an implement is raise or lowered (Fig. 17).

Avoid Shear Points

Because some shear points cannot be guarded, learn to recognize potential hazards and plan activities so you will not be injured.

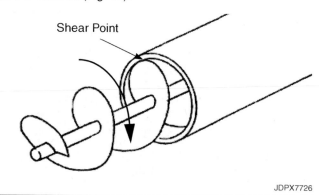

Fig. 15 — Auger in Tube

Crush Points

There are crush points between two objects moving toward each other or one object moving toward a stationary object (Fig. 18). Crush accidents often involve a second person. There are several dangerous crush points on hitches (Fig. 19).

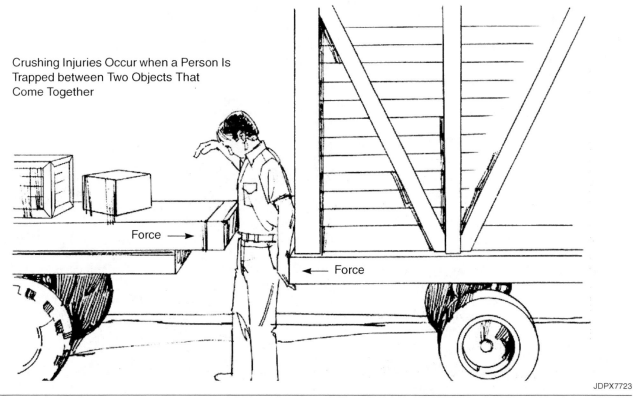

Crushing Injuries Occur when a Person Is Trapped between Two Objects That Come Together

Force →

← Force

JDPX7723

Fig. 18 — Crushing Injuries Occur when a Person Is Trapped between Two Objects Moving Together with Force or Speed.

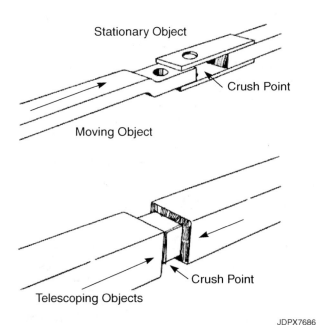

Stationary Object

Crush Point

Moving Object

Telescoping Objects

Crush Point

JDPX7686

Fig. 19 — Some Hitches' Telescoping Shafts Present Crush Points

A person working under a raised machine or other heavy object (Fig. 20) means a risk. Hoods and doors can crush or cut if a hand is in the way. Four-wheel drive articulated-steering tractors pivot in the middle and can turn and create a crush point at start-up. A person standing between the front and rear tires could be crushed between them. Keep in mind that where two objects move closer together, there is a potential crush point.

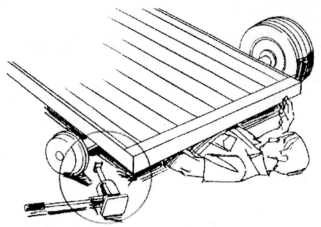

Working Under Heavy Objects That Aren't
Securely Blocked May Result in Crushing Injuries

JDPX7722

*Fig. 20 — Improperly Supported Objects Can Provide Extremely
Dangerous Crush Points*

Avoiding Crush Points

Do two things to avoid being crushed. First recognize all potential crush situations. Some of these are mentioned in examples above.

Second, stay clear of the hazards, This statement seems simple and obvious. But many people are killed and injured because they get into a crush point. Be constantly alert. Block all equipment securely if you must work under it. If an implement can roll, block the wheels so it will not roll. When two people hitch implements, make sure both are aware of the danger, and neither stands in a crush point. Know what the other person is doing at all times.

Pull-In Points

Pull-in accidents often happen when someone attempts to remove a corn stalk, hay, weeds, or something from feed rolls. People may be pulled into the moving parts of the machine and be seriously injured.

The real cause of the accident is attempting to remove material while the machine is running, thinking that the machine will pull in the plugged material and be more quickly cleared. It isn't worth the risk!

Another situation that leads to pull-in accidents is attempting to feed materials by hand into feed rolls, grinders, and forage harvesters to help them along (Fig. 21). Machines are faster and stronger than people. A machine can jerk in your hand and mangle it before you can react. You cannot pull free.

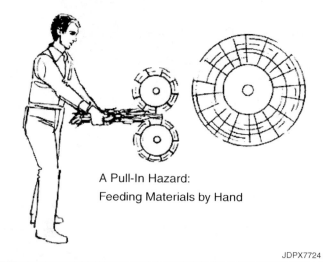

A Pull-In Hazard:

Feeding Materials by Hand

JDPX7724

Fig. 21 — A Pull-In Hazard — Feeding the Machine by Hand

Consider a person attempting to remove or feed corn stalks from corn picker rolls (Fig. 22). Feed rolls move the stalks at about 12 feet per second (3.7 meters per second). If the person grabs the corn stalk about two feet (0.6 meter) from the rolls, and the rolls suddenly take the stalk in at 12 feet per second (3.7 meters per second), the person's hand and arm will go in with the stalk. It takes nearly 3 seconds just to release the grip on the corn stalk. In that short time the corn stalk travels about 33 feet (10 meters) between the stalk rolls, and that's far beyond the hand.

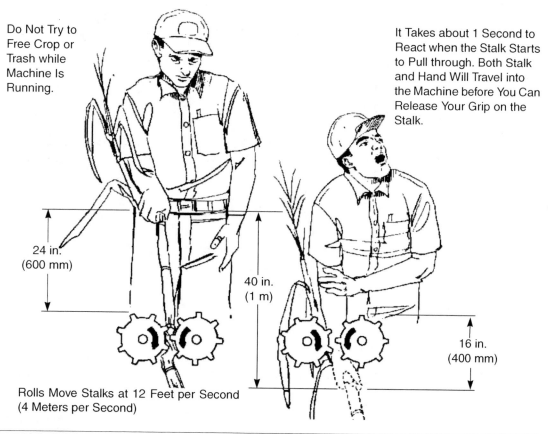

Do Not Try to Free Crop or Trash while Machine Is Running.

It Takes about 1 Second to React when the Stalk Starts to Pull through. Both Stalk and Hand Will Travel into the Machine before You Can Release Your Grip on the Stalk.

24 in. (600 mm)

40 in. (1 m)

16 in. (400 mm)

Rolls Move Stalks at 12 Feet per Second (4 Meters per Second)

JDPX7611

Fig. 22 — A Major Pull-In Hazard — attempting to Unplug a Machine while the Engine Is Running.

Other pull-in hazard situations:

Attempting to remove the stalk, vine, or piece of twine wrapped around a rotating shaft. As you pull, it may wrap more tightly and pull you into the rotating shaft instead of you removing the material (Fig. 23).

When Pulling Wrapped Material from a Rotating Shaft, the Material May Tighten Its Grip on the Shaft and Pull You into the Machine

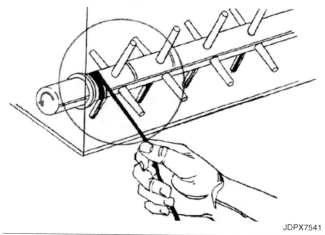

JDPX7541

Fig. 23 — A Pull-In Hazard—Pulling On A Wrapped Material To Remove It From A Rotating Shaft

Attempting to push an obstruction through a machine. For instance, kicking or pushing a piece of tree root through a cotton harvester or hay baler with your foot. When the root goes through, your foot follows.

Avoiding Pull-In Accidents

First, recognize the potential hazards. Second, realize that you can't win speed contests with farm machines. Third, remember it is impossible to design machines to do jobs and still be completely safe for people who misuse them.

CLEAN, LUBRICATE, OR UNPLUG A MACHINE ONLY WHEN IT IS SHUT DOWN. Always disengage power, shut off the engine, remove the key, and wait for all parts to stop moving before performing any of these operations unless the operator's manual instructs you do to otherwise. Some machines have powered reversers to clear headers.

Freewheeling Parts: Inertia

The heavier a part is and the faster it moves, the longer it will continue to move after the power is shut off. This is "freewheeling."

Many farm machines, especially those which cut or harvest, have components that continue to move after the power is shut off. The parts continue to move because of their own inertia (Fig. 24), or the inertia of other moving parts connected to them (Fig. 25).

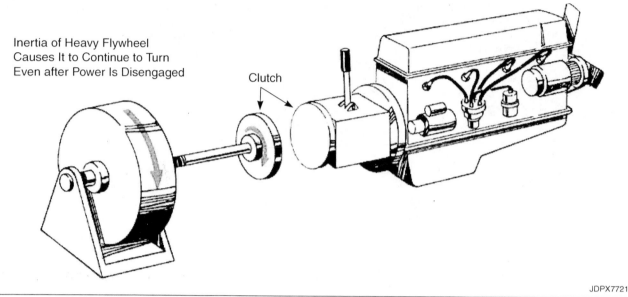

Inertia of Heavy Flywheel Causes It to Continue to Turn Even after Power Is Disengaged

Clutch

JDPX7721

Fig. 24 — Freewheeling Parts Continue to Turn after the Power That Drives Them Is Disengaged

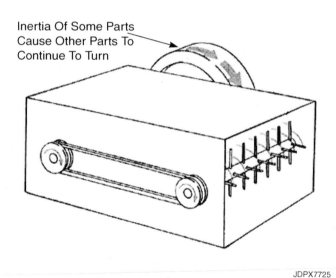

Inertia Of Some Parts Cause Other Parts To Continue To Turn

JDPX7725

Fig. 25 — Some Lightweight Parts May Continue Freewheeling Motion if Connected to Other High-Inertia Parts

Some freewheeling components on farm machines continue to rotate several minutes after power is disengaged. Though moving slowly, they're still dangerous (Fig. 26). Examples of freewheeling farm machinery components include:

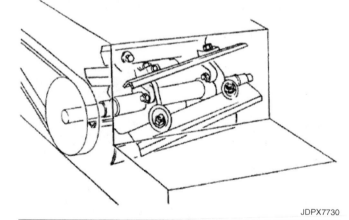

Some Parts Can Cause Serious Injury Even When Coasting Slowly To A Stop

JDPX7730

Fig. 26 — Even Parts That Are Moving Slowly Can Cause Serious Injury

- Cutter heads of forage harvesters

- Hammer mills of feed grinders

- Fans and blades on ensilage blower

- Flywheels on balers

How Injuries Occur

Why do people get injured by freewheeling mechanisms? There are several reasons. For instance, they may disengage the power, open an access door, and reach in to clean or unclog the machine before the parts stop moving. Perhaps they concentrate so much on servicing the machine they forget parts are still moving. Perhaps they get impatient and just can't wait for all motion to stop. Even though the parts may have almost come to a stop, their inertia may still be enough to cause serious injury.

Fatigue can cause people to do things they know are dangerous, and would never do if they were not tired.

Avoiding Freewheeling Parts

Listen! Almost any freewheeling part makes some sound — often whirring or humming sounds — especially at higher speeds. Also, clutches may make a fairly loud, distinctive clicking or clanking noise until freewheeling parts stop.

Watch for motion! Watch flywheel, pulleys, PTO shafts, and the ends of shafts to see if parts are still moving. Make sure you understand that freewheeling parts, even though moving very slowly, can still cause serious injury. Objects in motion stay in motion until they are stopped by some other force or object. Don't let your hand or foot be the object that stops a freewheeling part.

Thrown Object Hazards

A thrown object is a potential hazard, whether it is a stone from a slingshot, a baseball, or a pea from a peashooter (Fig. 28). If it strikes with enough force, it will cause serious injury.

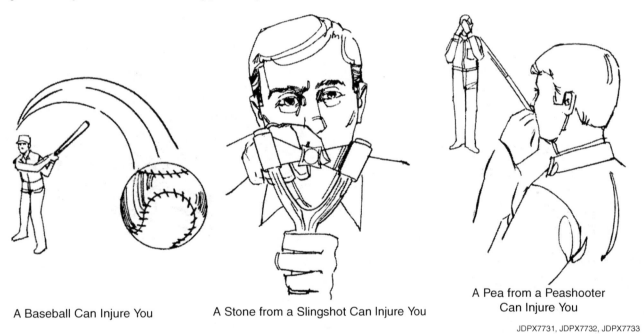

A Baseball Can Injure You

A Stone from a Slingshot Can Injure You

A Pea from a Peashooter Can Injure You

JDPX7731, JDPX7732, JDPX7733

Fig. 27 — Any Object Thrown with Enough Speed and Force Can Cause Serious Injury

Some farm machines can throw objects great distances with tremendous force (Fig. 28). Recognize those machines so you can avoid injury.

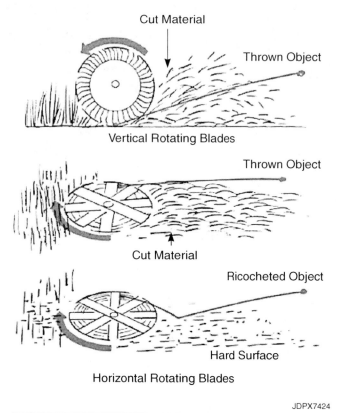

You Can Be Injured by Thrown Particles from Machines That Grind or Chop

JDPX7734

Fig. 29 — Particles of Ground or Chipped Grain May Be Thrown Hard Enough to Cause Injury

JDPX7424

Fig. 28 — Stones and Sticks Are Thrown Farther and Harder than Grass or Crops. They Can Also Ricochet.

Machines That Throw Objects

Perhaps the greatest potential for injury by thrown objects is around machines that have rapidly rotating parts to cut or chop crops out in the open field. In order to cut or shred, the machine must strike the crop with considerable energy. Examples of such machines are rotary mowers, cutters, and shredders.

Remember how you have thrown a flat stone so it skips on water? In a similar manner some machines, such as rotary cutters and mowers, can cause rocks or other objects to ricochet off the ground and upward toward a person. Be careful! Don't assume shields will always prevent all thrown object hazards. Stay a safe distance away and make sure others are not nearby when you operate such machines (Fig. 28).

There is another type of thrown object hazard around machines that chop or grind crops. Particles of crop may be thrown from the machine (Fig. 29). A straw chopper on a combine or a feed mill may fling pieces of corn cobs and kernels. They may not be thrown as far as a rock thrown by a rotary mower, but they may still have enough energy to cause injury.

Avoiding Thrown Object Hazards

1. Recognize which machines can throw objects.

2. Keep the machines properly shielded to reduce the possibility of thrown objects. Some operations may require removing or adjusting shields or deflectors to handle crops or feeds. Make sure you replace the shielding before you begin the next job. Follow the manufacturer's instructions for shielding to reduce the hazard of thrown objects.

3. Know how far and in what direction objects may be thrown, even with shielding in place (Fig. 30).

4. Stay a safe distance away from the likely path of thrown objects when you approach a machine. Make sure the machine has been turned off before you go closer.

5. When operating a machine that may throw objects, make sure the machine will not discharge toward people or animals.

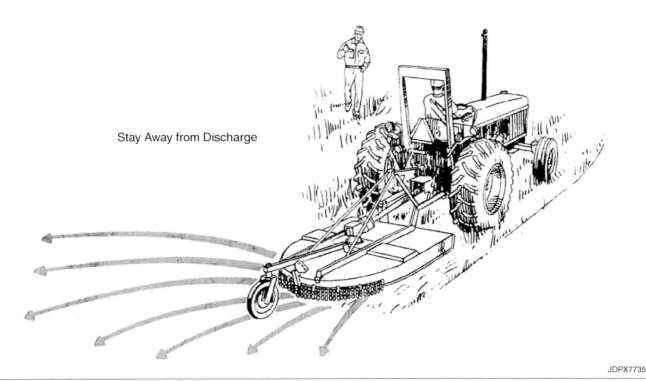

Stay Away from Discharge

JDPX7735

Fig. 30 — Know How Far and in What Direction a Machine May Throw Objects. Use Shields as Recommended by Manufacturer.

Stored Energy

Stored energy is energy confined and just waiting to be released. It is completely safe as long as it is confined. But if is released unexpectedly, stored energy can cause injury. Learn to recognize potential stored energy hazards and know how to handle them.

A simple slingshot made of a strong rubber band and a wooden handle (Fig. 31) uses stored energy. You slowly stretch the rubber band to "ready" position. Stretching uses some of your own energy. That energy is "stored" in the rubber band. It is harmless as long as you hold back the rubber band. But, when you release the slingshot you release all the energy you used to stretch the rubber band.

Stored energy can work for you, or it can be carelessly released and cause injury.

In many farm machines, energy is stored so it can be released at the right time, in the right way for you. Here are some systems and components that store energy. You should be alert for them as you use and service farm machinery:

- Springs
- Hydraulic Systems
- Compressed Air
- Electrical Systems

The Energy Is "Stored" in the Rubber Band until the Band Is Released

JDPX7543

Fig. 31 — Energy Is Stored In the Rubber Band of the Slingshot, Ready to Be Released

Springs

Springs are energy storing devices. They are used to help lift implements, to keep belts tight, and to absorb shock. Springs store energy in tension, like the slingshot, or in compression (Fig. 32 and Fig. 33).

No Energy

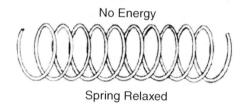

Spring Relaxed

Stored Energy

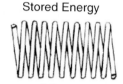

Spring Compressed Energy Attempts to Push Ends Outward

JDPX7544

Fig. 32 — A Spring Stores Energy in Compression

No Energy

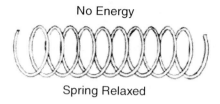

Spring Relaxed

Stored Energy

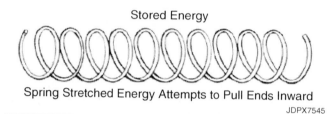

Spring Stretched Energy Attempts to Pull Ends Inward

JDPX7545

Fig. 33 — A Spring Stores Energy In Tension

When you remove any device connected to a spring, be sure you know what can happen. Know which direction the spring will move, and in what direction it will move other components when it is disconnected (Fig. 34). A compressed spring can propel an object or itself with tremendous force.

Make sure you and others will not be in the path of any part that will move when the spring moves. Plan exactly how far and where each part will move. Use proper tools to assist you in removing or replacing spring-loaded devices Even small springs can store a lot of energy.

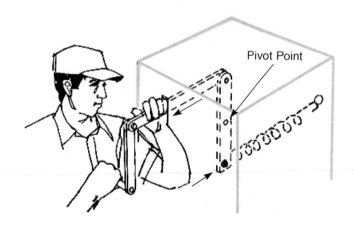

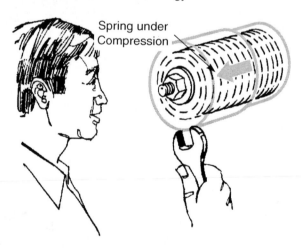

JDPX7685, JDPX7729

Fig. 34 — Be Sure You Know What Can Happen before Moving or Disconnecting Any Part Attached to a Spring

Hydraulic Systems

Hydraulic systems on farm machines also store energy. Hydraulic systems must confine fluid under high pressure, often higher than 2500 psi (17 200 kPa) (172 bar).

A lot of energy may be stored in a hydraulic system, and because there is often no visible motion, operators do not recognize it as a potential hazard. Carelessly servicing, adjusting, or replacing parts can result in serious injury. Fluid under pressure attempts to escape (Fig. 35). In doing that, it can do helpful work, or it can be harmful.

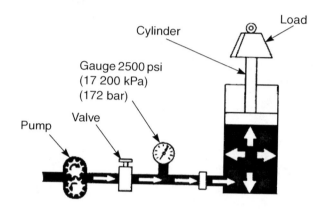

JDPX7612

Fig. 35 — Hydraulic Fluid under Pressure Attempts to Escape or Move to a Point of Lower Pressure

Servicing and Adjusting Systems under Pressure

Adjusting and removing components when hydraulic fluid is under pressure can be hazardous (Fig. 36). Imagine attempting to remove a faucet from your kitchen sink without relieving the water pressure. You'd get a face full of water! It is much more dangerous with hydraulic systems. Instead of just getting wet from water at 40 psi (275 kPa) (2.75 bar), you could be seriously injured by oil under the high pressures of farm equipment hydraulic systems. You could be injured by the hot, high-pressure spray of fluid, and by the part you are removing when it is thrown at you.

JDPX7429

Fig. 36 — Always Relieve Hydraulic Pressure before Adjusting Hydraulic Fittings. You Could Be Injured by a Hot, High-Pressure Spray of Hydraulic Fluid or by a Part Flung at You.

To avoid this hazard, always relieve the pressure in a hydraulic system before loosening, tightening, removing, or adjusting fittings end components. Keep all hydraulic fittings tight to prevent leaks. But do not tighten fittings without relieving the pressure. Also, if you over-tighten a coupling, it may crack and release a high-pressure stream of fluid. You could be injured by the fluid and the implement, which may move or drop to the ground.

Before attempting any service:

1. Shut off the engine which powers the hydraulic pump.

2. Lower implement to the ground or onto a solid support.

3. Move the hydraulic control lever back and forth several times to relieve pressure.

4. Follow instructions in operator's manuals. Specific procedures for servicing hydraulic systems are very important for your safety

Pinhole Leaks

If liquid under high pressure escapes through an extremely small opening, it comes out as a fine stream. The stream is called a pinhole leak. Pinhole leaks in hydraulic systems are hard to see, and they can be very dangerous. High-pressure streams from pinhole leaks penetrate human skin, flesh, and eyes. Hydraulic systems on many farm machines have pressures of 2500 psi (17 200 kPa) (172 bar) or higher. That's higher than the pressure in medical syringes used to give injections. Injury from pinhole leak injections comes from the fluid cutting through flesh and from complications such as injection and restricted circulation.

You may see only the symptoms of pinhole leaks from high-pressure systems. There may appear to be only a dripping of fluid, when actually it may be an accumulation of fluid from a high-pressure jet stream so fine it is invisible. Don't touch a wet hose or part with bare or even gloved hands to locate the leak. Wear safety glasses and pass a piece of cardboard or wood over the suspected area instead (Fig. 37). Relieve the pressure before correcting the leak.

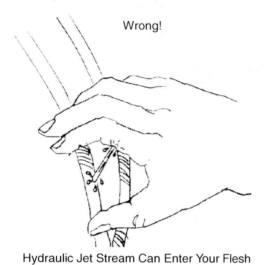

Wrong!

Hydraulic Jet Stream Can Enter Your Flesh

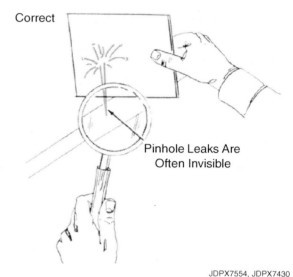

Correct

Pinhole Leaks Are Often Invisible

JDPX7554, JDPX7430

Fig. 37 — The Jet Stream or Mist from a Hydraulic Pinhole Leak Can Penetrate Flesh. Don't Touch It!

Diesel fuel injectors are designed to force fuel into engine cylinders under high pressure. Don't touch the jet stream from a diesel injector nozzle. It is just as dangerous as a pinhole leak.

If ANY fluid is injected into the skin, it must be surgically removed within a few hours by a doctor familiar with this type of injury or gangrene may result.

Gangrene often leads to amputation. If you ever have oil injected into your skin, get to a doctor immediately, and make sure the doctor understands how to treat the injury.

Trapped Oil

Hydraulic oil can be trapped in the hydraulic system even when the engine and hydraulic pump are stopped (Fig. 38). Trapped oil can be under tremendous pressure. You can be seriously injured by escaping fluid and moving machine parts if you loosen a fitting.

Trapped Hydraulic Oil Can Be under Tremendous Pressure

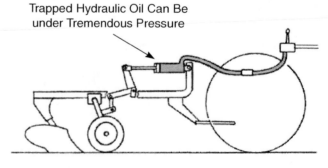

JDPX7736

Fig. 38 — Even Though the Engine and Hydraulic Pump are Stopped, Hydraulic Oil Can Be Trapped under Tremendous Pressure — A Potential Hazard

Another hazard with trapped oil (Fig. 39) is heat. Heat from the sun can expand the hydraulic oil and increase pressure. The pressure can blow seals and move parts of an implement or machine.

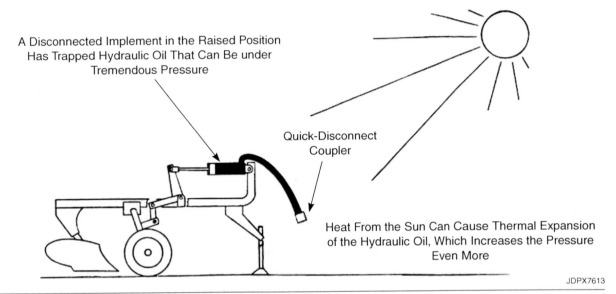

Fig. 39 — Trapped Oil Can Move an Implement and Blow Seals

Crossing hydraulic lines creates hazards. When lines are coupled to the proper port, you get the results you expect. But if lines are crossed, the implement may rise when you expect it to drop (Fig. 40). Serious injury could result. Make sure hydraulic lines are coupled exactly as specified in the machine operator's manual (Fig. 41). Color code lines with paint or tape. After you attach hydraulic lines, try the controls cautiously to see if you get the proper result.

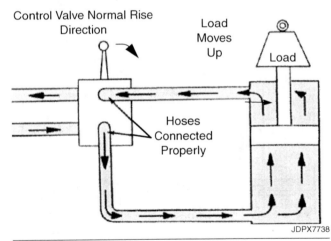

Fig. 41 — Properly Connected Hydraulic Hoses

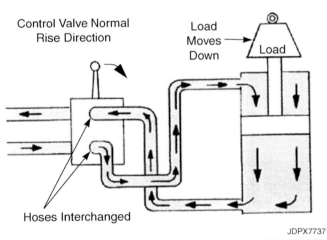

Fig. 40 — Crossing Hydraulic Lines Creates Hazards

Another hazard is coupling a high-pressure pump to a low-pressure system. Attaching tubing from a 2500 psi (17 200 kPa) (172 bar) system to an implement equipped with hoses, cylinders, and fittings designed for 1000 psi (6900 kPa) (69 bar) is inviting trouble (Fig. 42). The low-pressure system could burst or explode. Never improvise or adapt fittings, tubing, or hoses to attach a high-pressure system to a low-pressure system.

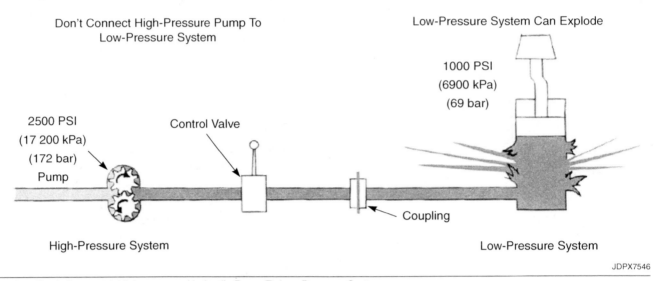

Fig. 42 — *Don't Connect A High-pressure Hydraulic Pump To Low Pressure System.*

When you replace hydraulic fittings, tubing, or hoses, be sure the new parts meet the working pressure requirements of the system. Inadequate parts could fail and cause machine breakdown and personal injury. Outside diameter is not a good indicator of strength or pressure rating, and steel tubing is not necessarily stronger than flexible hoses. Check with your equipment dealer to select proper replacement parts.

Hydraulic Accumulators

Some hydraulic systems have accumulators to store energy (Fig. 43). They may also be used to absorb shock loads and to maintain a constant pressure in the system (Fig. 44). Accumulators that store energy are often boosters for systems with fixed displacement pumps. Accumulators store oil pressure from pumps during slack periods and feed it back when it's needed (Fig. 45). Sometimes accumulators are used for backup devices to protect against loss of oil supply. For example, on some large machines an accumulator can be used to maintain pressure for emergency brakes or steering in case the engine stalls (Fig. 46).

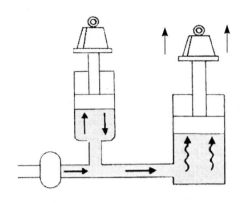

Fig. 46 — Accumulators Can Maintain Constant Pressure

Accumulators may be pneumatic (gas loaded), weight loaded, or spring loaded. They all store energy. That's their job. But because they store energy, they must be respected and properly serviced. Pneumatic accumulators use an inert gas, usually nitrogen, to provide pressure against the oil. (Inert gas is a gas that will not ignite.) Air or oxygen will explode if mixed with oil under high pressures. Gas in accumulators is separated from the oil by a flexible bladder (Fig. 47).

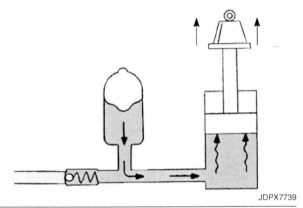

Fig. 43 — Accumulators Can Store Energy

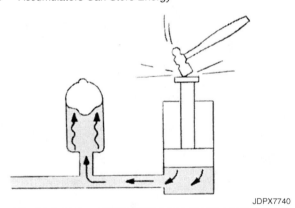

Fig. 44 — Accumulators Can Absorb Shock

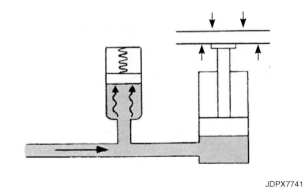

Fig. 45 — Accumulators Can Build Pressure Gradually

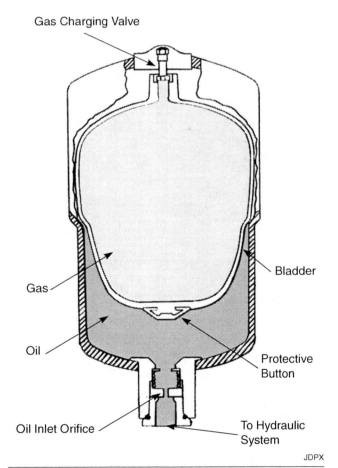

Fig. 47 — Construction for a Typical Hydraulic Accumulator

Recognize that the accumulators and the entire hydraulic system may have energy stored in them, if the pressure has not been relieved. When the hydraulic pump builds up pressure against the diaphragm or bladder, it compresses the nitrogen in the accumulator (Fig. 48). This energy, stored as compressed nitrogen, provides pressure against the oil. It may be released when the system calls for pressure, or absorb shock if that is its purpose. Even though the pump may be stopped, or an implement has been disconnected from the tractor, energy is stored in the accumulator unless the pressure was relieved before shutdown. The nitrogen is under pressure, so the hydraulic fluid is also under pressure.

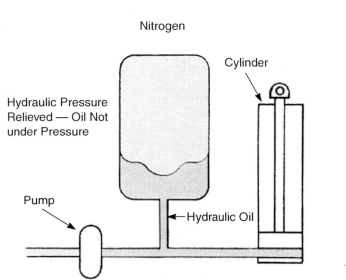
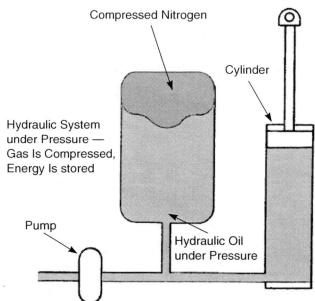

JDPX7744, JDPX7745

Fig. 48 — Hydraulic Accumulators Store Energy

Observe these basic safety considerations for hydraulic accumulators:

1. Recognize accumulators as sources of stored energy.

2. Relieve all hydraulic system pressure before adjusting or servicing any part of a system that has an accumulator.

3. Relieve all hydraulic pressure before leaving a machine unattended for the safety of others as well as for you.

4. Make sure pneumatic accumulators are properly charged with the proper inert gas (usually nitrogen). A pneumatic accumulator without gas is a potential "bomb" when charged only with oil.

5. Read and follow manufacturer's instructions for servicing accumulators.

Generally, manufacturers recommend only authorized dealers service gas charged accumulators. Read and follow the manufacturer's instructions thoroughly.

See John Deere's Fundamentals of Service Manual, HYDRAULICS, for more information.

Compressed Air

Compressed air is dangerous. When air is compressed, its volume is reduced and energy is stored (Fig. 49). It is like squeezing a rubber ball in your hand (Fig. 50). You can squeeze the ball until it is smaller. But it springs backs to normal size when you release it. The energy stored by compressed air can be tremendous, compared to squeezing the rubber ball. It can literally explode!

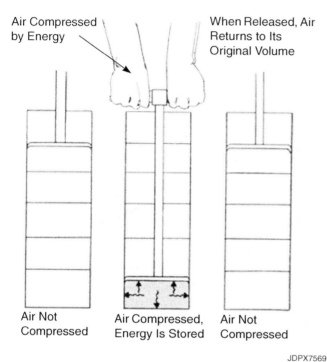

Air Compressed by Energy

When Released, Air Returns to Its Original Volume

Air Not Compressed

Air Compressed, Energy Is Stored

Air Not Compressed

JDPX7569

Fig. 49 — When Air Is Compressed, Energy Is Stored. The Air Attempts to Return to Its Original Volume.

When Compressed, a Rubber Ball Attempts to Return to Its Original Volume

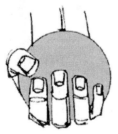

JDPX7746

Fig. 50 — Air, Like a Compressed Ball, Attempts to Return to Its Original Volume When Released

Most important for you to understand is that compressed air is a form of stored energy and that it can be very dangerous if not properly respected and controlled.

Probably the most common use of compressed air for farm machinery is for tire inflation. An inflated tire can be dangerous, under some conditions, especially the large ones, because they contain significant volumes of compressed air.

When a large volume of compressed air gets an opportunity to return to its original volume, it can occur with explosive force. That can happen when you are mounting and inflating a tire on a rim if you don't do it properly. The tire or the tire-to-rim seal could fail, or a multi-piece rim could suddenly break loose from the wheel with explosive force.

A tremendous amount of stored energy can be released from a tire in a split second. A 10.00-20 12PR truck tire inflated to 75 psi, (500 kPa), (5 bar) has 46 500 ft-lb, (63 000 joules) of stored energy. That is enough to raise a 3,000 lb, (1360 kg) car 15 feet, (4.6 meters). A 24.00-49 tire inflated to 75 psi, (500 kPa), (5 bar) has 354 000 ft-lb, (480 000 joules) stored energy which could lift a 134-lb, (60 kg) person one-half mile (0.8 kilometer) into the air! That could be you.

You should always follow recommendations of manufacturers of tires, rims, and wheels when demounting and mounting tires. Various types of tires and rims or wheels must be handled in some special way to avoid personal injury during demounting and mounting, especially with regard to the stored energy of the compressed air.

Don't exceed the recommended tire pressures, and stand to one side when inflating large tires (Fig. 51).

JDPX

Fig. 51 — Compressed Air Is Stored Energy. Always Stand to One Side when Inflating Tires and Never Over-Inflate Them.

For more information on tire safety, refer to a later chapter in this book, "Equipment Service And Maintenance".

See John Deere's Fundamentals of Service Manual, TIRES AND TRACKS, for more information.

Other sources of information on safe handling of pneumatic tires can be obtained from the following organizations:

Rubber Manufacturer's Association
1400 K Street, N.W.
Washington, D.C. 20005

National Wheel and Rim Association
4836 Victor Street
Jacksonville, FL 32207

Further safety information can be obtained from:

U.S. Department of Transportation

National Highway Traffic Safety Administration
400 Seventh St., S.W.
Washington, D.C. 20590

When working with compressed air in any situation, remember that stored energy can be very dangerous if it is not properly controlled. Always follow the recommendations of manufacturers of compressors and any equipment operated by compressed air. It is always a good idea to wear safety glasses when working with compressed air.

Electrical Systems

One of the most common forms of stored energy is electricity stored in 12-volt batteries. When properly used, it makes your work easier. If handled carelessly, it can cause serious injury. If you recognize the potential hazards of electricity, you'll be able to avoid serious accidents.

Fires

Electrical systems can cause fires if not properly maintained. The energy stored in the battery may be tapped to start the engine. But if a bare wire touches a metal part and becomes hot or sparks, it can start a fire in dust, chaff, and leaves. Most machinery fires do not result in personal injury, but every fire is a potential source of injury. Inspect electrical systems. Make sure wires are properly insulated and clean dust, chaff, leaves, and oil off wires.

Every self-propelled machine should have a multipurpose dry chemical fire extinguisher on board. Everyone involved with the machine should know how to use it, and it should be checked annually.

Short-Circuit Starting

If insulation on electrical wires is cracked or worn, a short circuit can occur. Electricity could flow to the cranking motor and start the engine when no one is around. If the positive and negative terminals of a cranking motor are accidentally contacted by another metal object, such as a wrench, the current will flow between the two terminals and accidentally start the engine.

By-Pass Starting

Tragically, some people misuse electrical energy by shorting across tractor starter terminals with screwdrivers or other devices. That bypasses the neutral start safety switch. If the tractor is in gear when the engine starts, it can suddenly lurch forward and crush you. Many people have died doing it. Don't try it!

Never bypass start a tractor or other machine, and never start it while standing on the ground. Start tractors and self-propelled machines only from the operator's station and with transmission in the neutral or park position. Bypass starting of tractors and other farm equipment is a serious safety concern. It is addressed in more detail in the chapter, "Tractors And Implements." (See "Neutral-Start Safety Switches" and also "Starting" in that chapter).

In summary, the energy in an electrical system is waiting to do something. If it does what was planned for it, at the right time, there's no problem. If it does what was planned for it at the wrong time, or does the wrong thing, injury and property damage result. Inspect electrical systems on all machines. Replace worn wiring, contacts, and switches.

Slips And Falls

According to reports from the National Safety Council and other organizations, slips and falls are among the most frequent personal injury accidents around the home, on farms and ranches, or other worksites.

Slipping and falling can put a person out of commission for hours, days, or years. Slips and falls can be prevented by recognizing the potential hazards and avoiding them.

On-Off Accidents

Farm machines are equipped with steps, handholds, ladders, and platforms to help you get on and off safely (Fig. 52). They are arranged so you can have three contact points on the machine at all times: two feet and one hand, or two hands and one foot. Always use steps, handholds, ladders, and platforms deliberately, facing the machine when getting on and off to avoid slipping. Also, keep them in good repair and uncluttered.

Fig. 52 — Use Steps, Ladders, and Handholds Deliberately to Get On and Off. Face the Machine, and Don't Jump.

Don't jump when getting on or off machines, and don't attempt to get on or off a machine when it is moving. That's asking for a much more severe injury than an ordinary slip or a fall. You could be run over.

Slippery Foot Surfaces on Machines

Mud, snow, ice, manure, and grease may build up on steps, platforms, and other surfaces. When it does, you can slip and fall (Fig. 53). Or you could slip and bump a control, causing the machine to lurch into action, injuring yourself or someone else. Also, you could fall into a moving part. Such falls can be fatal.

Fig. 53 — Mud, Snow, Manure, and Grease on Foot Surfaces of Machines Could Cause a Serious Fall

Take time to clean foot surfaces for your own safety and for others who will use them. Also, wear boots with non-skid soles.

Cluttered Steps and Operating Platforms

Chains and tools on operator platforms are accidents waiting to happen. You need to be able to move about without having to watch where you place your feet. Slipping on mud or snow may be somewhat excusable, but tripping on something you placed on a step isn't. Machine surfaces intended for your feet should be kept clear for your feet (Fig. 54).

Fig. 54 — Keep Machine Platforms Clear

Slippery Ground Surfaces

When there's snow, ice, or mud on the ground you can't do much to change it. But you can learn to recognize that those slippery surfaces can lead to accidents and try to avoid them.

The danger is usually slipping or tripping and falling against or into a machine that is running. For instance, when grinding feed in a muddy barnyard you could slip and fall into the hopper intake. When you must work under these conditions, the best practice is to slow down, step deliberately, and be on the lookout for slippery surfaces or objects that may cause you to lose your footing (Fig. 55). Also, wear boots with non-skid soles.

JDPX7747

Fig. 55 — Watch Where You Step

Slow-Moving Vehicle

Everyone knows that speed kills on the highway. Right? Then why consider slow-moving vehicles as major hazards on the highways? A study of slow-moving vehicle hazards may surprise you.

Recognizing the Hazards

Most motorists on public roads are not farmers. In fact the percentage of people who have farming backgrounds or those who even know farmers is becoming smaller each year. So, only a small percentage of the motoring public is likely to give much thought to the unique nature of a farm machine when they see it on the highway. They're traveling comfortably, visiting with passengers in the car, while meeting and overtaking other cars and trucks traveling at similar speeds.

When they approach a farm tractor whose speed is only 15 or 20 mph (24 or 32 km/h), they just don't realize how much slower it is traveling than they are. And even if they think about it, it may not occur to them that they must react differently to safely handle this situation. After all, they've met and overtaken hundreds or thousands of cars and trucks, but relatively few farm vehicles (Fig. 56). They haven't had the practice that provides the reflexes and judgment for proper action.

JDPX7615

Fig. 56 — Motorist Have Unpredictable Reactions to Slow-Moving Farm Equipment

Closure Time

Consider a motorist from a town or city travelling 55 mph (88 km/h). The car tops a hill, and the motorist sees another car 400 feet (122 m) ahead going just a little slower at 45 mph (72 km/h). The closing speed is only 10 mph (16 km/h), which is fairly normal for highway travel. At that closing speed the motorist has 27 seconds to recognize the slower speed of the car ahead, to react, and to take action to slow down.

Now picture that same non-farm motorist topping the same hill again. This time, 400 ft, (122 m) ahead is a tractor travelling 15 mph, (24 km/h). Now the closure speed is 40 mph (64 km/h). There are less than 7 seconds to recognize the slow speed of the tractor ahead, to react, and to slow the car (Fig. 57). That's only about one-fourth the time the motorist had to slow to a safe speed behind the auto described above. It may not be enough if the motorist doesn't immediately recognize the need to slow down.

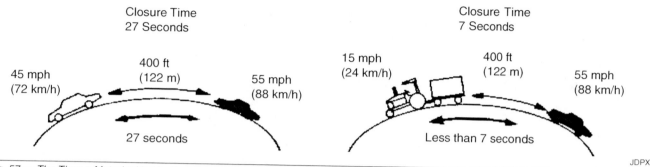

Closure Time
27 Seconds

45 mph
(72 km/h)

400 ft
(122 m)

55 mph
(88 km/h)

27 seconds

Closure Time
7 Seconds

15 mph
(24 km/h)

400 ft
(122 m)

55 mph
(88 km/h)

Less than 7 seconds

JDPX

Fig. 57 — The Time a Motorist Has to React to Avoid an Accident Is Much Less for Farm Machinery than for Automobiles

Situations similar to the one above lead to thousands of slow-moving vehicle collisions each year. The motorists aren't able to slow down or stop in time to avoid collisions, often because they don't quickly recognize the rapid closing speed involved in approaching farm machinery on the road. Motorists are often the most seriously injured victims. Many of the accidents are fatal.

Many of the collisions with slow-moving farm machines also involve other rural motorists, not just city folks. Every motorist must be alert to slow-moving vehicles on public roads to avoid collisions.

Avoid Slow-Moving Vehicle Accidents

As an operator of slow moving farm machinery on public roads, you have a major responsibility for helping motorists avoid hitting you from the rear. It's for your safety and for theirs.

1. Use a Slow-Moving Vehicle (SMV) Emblem. Identify your farm machinery as slow-moving vehicles whenever you travel on a public road, even for short distances. The triangular SMV Emblem, visible from the rear (Fig. 58), is the universal symbol to tell everyone a vehicle travels 25 mph (40 km/h) or less. It is required by state law throughout most of the United States. Also, federal regulations (OSHA) require SMV emblems on machines operated by employees. It should remind other drivers that the closing speed between a car and a slow-moving vehicle is much faster than between cars.

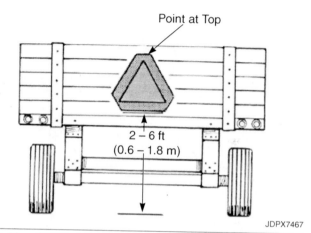

Point at Top

2 – 6 ft
(0.6 – 1.8 m)

JDPX7467

Fig. 58 — SMV Emblems on Your Farm Machinery Help Avoid Collisions on Public Roads

2. Keep the emblem surface clean and in good repair for both day and night identification (Fig. 59). When the reflective red border or fluorescent orange center loses its brilliance, replace the emblem. If motorists can't see them clearly, they'll not give you any protection. Mount the emblem securely, and always with a point upward. If a point is downward, the emblem does not look like a universal symbol as intended. It should always give every motorist the same message.

JDPX7748

Fig. 59 — Keep SMV Emblems Clean and Visible. Replace Them When the Colors Fade.

3. Keep lights and reflectors in good working order for farm machines traveling on public roads. Operate flashing lights both day and night so you can be recognized as a slow-moving vehicle.

4. Anticipate problems motorists may have when your machinery is on the road. Drive with others in mind.

Use care, courtesy, and common sense. For example:

• Move over to the shoulder so motorists won't be in a tight spot. Stop if necessary.

• Signal clearly for left turns, and be sure motorists recognize that you intend to turn.

• When entering the road from a field or farmstead, allow plenty of time and distance for approaching motorists. Remember, your machinery moves slowly.

Do whatever you can to avoid slow-moving vehicle collisions, even if the motorist is in error. If there's an accident, it really doesn't matter who was "right." For more details on highway safety equipment and how to avoid slow-moving vehicle accidents, see Chapter 7, "Tractors and Implements".

5. When you're the motorist, be alert for slow-moving vehicles. Remember that closing speeds are relatively fast. Act quickly.

Second-Party Hazards

The rear wheel of a tractor backed over a young farm boy because the father didn't know his son was behind the tractor. A farm operator had his hand tangled in an unshielded V-belt drive when his helper started the combine. A child passenger on a tractor was bounced right out the door and run over by a rotary cutter.

These three farm accidents killed a second party. We call them "second-party" accidents. They can be divided into two categories: (1) Necessary second-party accidents involving other persons who are needed to help with machinery operation, and (2) Unnecessary second-party accidents involving other people who are not needed for machinery operation. Both types of accidents can be avoided.

Accidents to Necessary Second Parties

Many farm jobs need two people. You may need someone to hold a part in place while you adjust it. Sometimes you need help to get something done sooner. And some jobs are safer if you have help.

Any time there's more than one person involved, there's a chance that what one person does may cause injury to the other. Second-party accidents can be avoided if each person knows exactly where the other person is and is going to do. For example, when hitching, a helper stands aside while the driver backs up and aligns the hitch. The driver puts the tractor in the park position or forward gear. Only then does the assistant step in to make the hitch. Final adjustments are made by moving the tractor forward, never backward toward the helper. Clear communication is essential. There are different ways of communicating, and certainly the most effective and convenient method is by voice. When there's too much noise for you to be heard, however, use standard hand signals (Chapter 1, "Safe Farm Machinery Operation"). Whatever communications you use, make sure the other person understands you. Here's one good overall rule to follow when two or more persons are working around machines:

Before starting or moving a machine, tell people to stand where you can see them. Some operators make a habit of sounding the horn before starting the engine.

To avoid injuries to yourself when you are the second party, try to anticipate actions and errors of the other guy. Be defensive, like a defensive driver. Expect the unexpected. When you're helping hitch an implement to a tractor, the driver's foot might slip off the clutch or brake. If you think of that possibility, you'll stand aside while the operator backs up to the implement. Then, when the tractor is in the park position, you will step in to insert the hitch pin (Fig. 60).

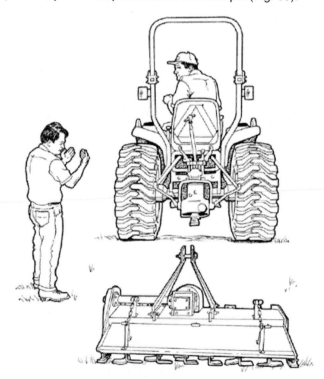

JDPX7434

Fig. 60 — Anticipate Actions and Possible Errors by Other Persons

If you want to talk to a tractor driver operating a rotary cutter in a stubble field, recognize that the cutter could pick up a rock and throw it at you. Move to a safe place, signal the operator to stop, and then approach the machine after the moving parts stop.

Accidents to necessary second parties around farm machinery could practically be eliminated if these precautions were followed:

1. Know what other people plan to do.

2. Anticipate the actions and errors of other people.

Accidents to Unnecessary Second Parties

The unnecessary second parties in farm machinery accidents shouldn't be there. Usually they just want to watch, or want to be with the person operating the machine. Although some are adults, the highest percentage are children under 15 years of age. Many of them are just riding along for the fun of it when they are injured (Fig. 61). They can interfere with the operator and cause injury to him or her, too. The second party doesn't anticipate what could happen because he or she doesn't understand, or is not alert to, all that is going on. The results are often tragic,

JDPX7682

Fig. 61 — These Children Are Riding in a Dangerous Place. Do Not Allow It.

Keep people away from machines. Don't give them rides! Even in tractors with cabs, riders can be bounced around by unexpected bumps, accidentally unlatch the door, and fall out.

Look for the unexpected. Check all around your machine before you move it in any direction (Fig. 62).

JDPX7627

Fig. 62 — Make Sure Everyone Is Clear before Starting Any Machine

Test Yourself

Questions

1. An error that results in accident or injury could be caused by:

   a. taking a "shortcut" in order to get a job done faster.

   b. ignoring a safety warning.

   c. failing to recognize a hazard.

   d. all of the above.

2. A very effective way to eliminate many common machine hazards is to:

   a. paste safety stickers in obvious places on a machine.

   b. always work with another person so there will be two people to recognize potential hazards.

   c. disengage the power to the machine, shut off the engine, take the key, and wait for all parts to stop moving before working on or near a machine.

   d. all of the above.

3. You must learn to _____ and _____ hazards in order to prevent accidents.

   a. recognize / understand

   b. see / smell

   c. read / observe

   d. none of the above

4. (T/F) Smooth rotating shafts are harmless.

5. Which of the following would be most likely to have a pinch point hazard?

   a. belt or pulley drive

   b. hydraulic hose

   c. 12-volt battery

   d. spring

6. (T/F) Loose or frayed clothing is often a contributing factor in "wrap point" accidents.

7. A grain auger would be likely to present which of the following types of hazards?

   a. pull-in points

   b. shear points

   c. all of the above

   d. none of the above

8. A knife blade is an example of a _____ point, while a sickle bar mower blade is an example of a _____ point.

   a. cutting / shear

   b. wrap / cutting

   c. pull-in / cutting

   d. wrap / shear

9. If you are working under a piece of equipment, what is the most effective way to prevent the machine from collapsing and crushing you?

   a. lock the brakes.

   b. use a hydraulic jack that is in good condition.

   c. have a partner warn you if the equipment starts to shift.

   d. secure equipment with jack stands.

10. (T/F) A baler flywheel may continue to rotate or "freewheel" for several minutes after the power to the machine is disengaged.

11. Springs, batteries, hydraulic systems, and compressed air are all examples of:

    a. freewheeling parts

    b. stored energy

    c. thrown object hazards

    d. all of the above

12. (T/F) The hydraulic system of an implement disconnected from a tractor may contain oil under high pressure.

13. One point of a slow moving vehicle sign should point what direction?

    a. straight down

    b. left

    c. right

    d. straight up

14. (T/F) "Second-party" machinery-related injuries can happen to bystanders or children who are not even working on the farm.

15. (T/F) You can assume that motorists in rural areas are used to sharing the roads with tractors and farm equipment, and will react accordingly.

References

1. Slow-Moving Vehicle Identification Emblem (ANSI/ ASAE S276.3/SAE J943). 1991. American Society of Agricultural Engineers, St. Joseph, Michigan 49085.

2. Lighting and Marking of Agricultural Field Equipment on Public Roads (ANSI/ASAE S279.9/ SAE J137). 1990. American Society of Agricultural Engineers, St. Joseph, Michigan 49085.

Managing Farm/Ranch Safety

4

Fig. 1 — Safety Should Be Managed on Farms and Ranches of All Sizes, Whether Family or Corporate Operations

Introduction

Productivity and profitability in farming and ranching don't just happen because people work hard. They require analysis, planning, and management. Management methods will also help ensure a reasonably safe and healthy workplace for families and employees (Fig. 1). This chapter briefly describes a management approach to farm safety.

A safety management program may at first seem impractical for a small operation. However, without some deliberate approach, safety will not likely be improved. Safety management principles can be applied to family farms or larger corporate operations. This chapter addresses the following safety management topics:

- Organization and safety committee

- Family safety management team

- Safety status assessment

- Safety policy

- Objectives to support policy

- Activities to meet objectives

- Safety checklist

- Monitoring progress

- Minimum safety program

- Plan for emergencies

- Youth and children

- Older adults

- Employees and regulations

Organization

You can't have a good safety program by merely agreeing, "We all need to be more safety conscious." There must be a plan, and someone must be responsible for specific plans and actions and for making sure they get done. Someone must also have ultimate responsibility for safety decisions. Ultimate responsibility would generally rest with the owner or operator, or whoever makes the final decisions on other business matters.

Safety Committee

Some large operations appoint a safety director to oversee the safety program. That person generally reports directly to the owner or top management. Regardless of farm size, a safety committee is recommended (Fig. 2).

Fig. 2 — Safety Committees Work for Family or Corporate Operations. Mom Could Be Chairperson

Why a committee? A committee provides for a sharing of ideas and judgment, and it helps spread responsibility and continued interest. A safety committee can develop plans and guide their implementation.

Committee members could include two or three people or more. On a family farm, it could include the entire farm family, perhaps with the mother as chairperson. If there are employees, some of them should also be included. Whether there is a committee or only one or two people with safety responsibility, the responsibility should be assigned and understood.

The committee should meet several times a year, and also whenever needed for special issues. It could meet several weeks before each primary season, such as before planting and harvesting. That allows time for corrective action before the work pressure begins in those busy seasons. Meetings might be at the breakfast table on a rainy day. They would be more formal for large operations. However, meetings should be scheduled to make sure the safety work is planned and completed.

The safety committee determines needs, proposes action, and follows up on approved activities to ensure completion. The committee also submits recommendations and progress reports to the person who is ultimately responsible for operations, seeking that person's approval and support as necessary.

Family Safety Management Team

Not only do the adults, children, and youth of farm and ranch families live together; they also work together. And, for the most part, they work where they live. They depend on each other in their work, and they care for each other. Thus, the family is uniquely postured to work as a team to provide for each other's safety (Fig. 3).

Fig. 3 — Farm and Ranch Families Should Work Together as a Team to Manage Safety and Health

Developing a Family Safety Management Plan

The family can develop a safety management plan as described in this chapter. In fact, the planning and managing of safety activities through the safety committee will be good business experience for the children and youth.

Much of the information you need to begin developing a family safety management plan is in this book, especially in the early chapters. There are other sources of information, including these:

- Farm and ranch magazines and newspapers

- Cooperative Extension Service

- Safety clinics, video tapes, DVD's, and information from farm equipment and chemical dealers

- Safety exhibits at state and county fairs

- Safety projects and programs of 4-H, FFA, Farm Bureau' and other farm organizations

Arrange for the family to attend safety activities together (Fig. 4). Also, discuss safety information as a family, so all can agree on the need to manage family safety. Then, make the commitment to do it, together. Most important: Keep all family members involved in managing safety.

JDPX8067

Fig. 4 — Arrange for Family Members to Participate in Safety Activities Together

Safety Status Assessment

A farm or ranch safety program should begin with an assessment of the current safety status and needs of all areas of the operation, such as these:

- Tractors and implements
- Field operations
- Shop
- Buildings (including residence)
- Livestock handling
- Pesticide handling, storage, and application
- Personal protective equipment and its use
- Safety training
- Emergency preparedness
- Compliance with safety and health regulations
- Liability and workers' compensation insurance
- Records system for accidents, pesticides, regulatory compliance, insurance and others

The safety committee should assess the safety status and needs of all applicable areas such as, but not limited to, those above. The written assessment should briefly describe the present status of safety, whether or not it is adequate, and what changes should be made (Table 1).

SAFETY STATUS ASSESSMENT
1. Describe Present Status Of Safety.
2. Is Present Status Adequate? Yes/No
3. If No, List The Changes Needed.

Table 1 — Make a Safety Status Assessment of All Farm or Ranch Operations

The safety status assessment can be a basis for developing a safety policy, as well as objectives and activities to improve safety.

Safety Policy

Develop a safety policy, regardless of the size of the operation. Most successful businesses are guided by statements of policy, which help entire organizations work together toward common goals. The act of developing policy actually helps the people involved to thoroughly understand the needs and commit to meeting them.

The policy statement becomes the safety guideline for everyone involved in the farming operations (Fig. 5). The safety policy should state what the owners or operators and management expect of the farm family and employees. It should also indicate what family and employees are entitled to. The policy should also assign safety responsibility.

JDPX7439

Fig. 5 — A Safety Policy Outlines What Everyone Is Expected to Do for Safety

Here is an example of a farm safety policy:

The Shady Maple Farm considers safety an integral part of its operation. We intend that our equipment, facilities, and practices will not present an unreasonable risk of injury or health to any family member, employee, or visitor. No task or operation is to be considered more important than personal safety.

We intend to comply with safety and health regulations.

All family members and employees are expected to work together to provide a reasonably safe and healthy workplace, and to continually improve it.

The safety committee is responsible for proposing, promoting, and monitoring specific safety plans and activities to support this policy.

Since owners or management must support the safety program, they must be involved in developing the policy, and must approve it.

Objectives to Support Safety Policy

The safety committee should develop specific objectives that must be accomplished to fulfill the safety policy (Fig. 6).

JDPX7616

Fig. 6 — Safety Committees Develop Specific Objectives to Support Safety Policy

Here are some examples of objectives that could support the sample safety policy presented earlier:

1. All safety features on tractors and implements in place and in good working condition. All safety signs in place and legible.

2. All operations, equipment, and facilities in compliance with applicable safety and health regulations.

3. All persons aware of and understand potential hazards related to their work and how to prevent them.

4. All persons adequately trained before operating machinery.

5. Periodic safety inspection and maintenance of equipment and facilities.

6. Adequate plans, procedures, and training for emergencies.

Develop objectives in support of your safety policy, for all primary functions or areas of operation.

Refer to Cooperative Extension Service information and checklists, to other chapters in this book, and to equipment operator's manuals to help identify safety objectives and activities for your farm and ranch operations.

Activities to Meet Objectives

After developing safety objectives, develop specific activities to carry out each objective. Just giving people a list of objectives won't get the job done. Consider, for example, Objective No. 1: "All safety features on tractors and implements in place and in good working condition. All safety signs in place and legible."
Here are sample activities to meet Objective No. 1:

a. Inspect safety features and safety signs on all tractors and implements. If damaged or missing, replace or repair them immediately.

b. Inform all persons that in the future any damaged or missing safety features and signs are to be replaced or repaired immediately.

c. Conduct safety inspections of all equipment annually or before each major use season. Replace or repair damaged and missing safety features and signs before using machines.

Assign Responsibility and Dates for Action

For each activity, assign responsibility to a person and indicate when it should be completed. Examples for activity a, above:

• Inspect safety features and safety signs on all tractors. Replace or repair as needed. (Jerry and Tanya, to complete by June 20).

• Safety committee review inspection results and corrective action (June 25).

Similarly, responsibilities and completion dates should be designated for all listed activities. The safety committee should report all planned actions to the person who has final responsibility for safety. Get formal approval if needed.

Safety Checklist

An excellent way to inspect equipment and facilities is to use safety checklists (Fig. 7). They list features and practices necessary for safety; and you check each item on the list, indicating whether it is adequate or needs improvement. To assess safety needs of your tractors and implements, for example, you can use a farm machinery safety checklist for each machine.

FARM SAFETY
CHECKLIST

Date checklist completed: /   /

Date checklist to be reviewed (annually or when there is a change or addition to tasks on the farm):   /   /

Name of person(s) who completed checklist: Initial:

Business Name:

This checklist covers hazards found in typical rural workplace situations. It is not designed to cover all of the risks on a farm but to help you to get started on the process of identifying the hazards around you. You should involve your employees in filling out this checklist.

An important part of managing your farm is to ensure the health and safety of your employees and other people, such as customers, visitors or tradespeople who visit your workplace. In fact, the *Occupational Health and Safety Act 2000* requires you to ensure your workplace is safe.

To ensure you fulfil your obligation for a safe workplace, you need to become aware of what can cause harm and then take action to ensure no one is at risk while they are in your workplace. The following questions will help you evaluate how well you are currently managing safety on your farm.

Do you talk to your employees about safety issues?   Yes ☐   No ☐
Do you encourage your employees to report safety problems?   Yes ☐   No ☐
Do you regularly inspect your farm to identify safety problems?   Yes ☐   No ☐
Do you fix identified problems?   Yes ☐   No ☐

JDPX8068

Fig. 7 — State Cooperative Extension Offices Can Provide Farm and Ranch Safety Checklists

Checklists include items such as these:

- Do you talk to your employees about safety issues?

- Are PTO shields in place and in good condition?

- Do you regularly inspect your farm to identify safety problems?

The safety committee should review completed safety check lists and assign action to correct any listed problems.

To get safety inspection checklists, contact your state or local Cooperative Extension Service. Some offices have safety checklists for nearly every major area on a farm or ranch, including machinery, shop, livestock handling, pesticides, and others. Ask for a complete set and also ask for other safety management materials. (You could also develop checklists based on recommendations throughout this book.)

Monitor Progress

Safety management isn't a one-time effort. Just as all farmers review their finances at least annually for tax reporting, safety program effectiveness must be reviewed as follows:

1. Follow up on each activity to make sure work gets done on time. The safety committee chairperson could be the one to check up on the activity. The committee should review progress on each activity at future meetings.

2. Make periodic safety inspections. Safety checklists are useful for this. Inspections can be done by one person, but it's good to involve two or more persons, or the entire safety committee in some instances (Fig. 8).

JDPX7750

Fig. 8 — Periodically Inspect Each Machine for Safety. Use a Safety Checklist

3. Review results of each periodic and special inspection and follow up on each one. The safety committee should review the inspection findings and determine necessary plans, actions, and follow-up needed. The safety committee should also seek approval of the operation's ultimate decision-maker.

4. Evaluate the entire safety program each year. Don't rely only on inspections. Review the entire program annually to be sure it keeps up with changing needs, including regulations, training, and general awareness.

5. Keep orderly records of policy, objectives, activities, inspections, decisions, assignments, and accidents. Records will help you make sure plans and assignments are carried out. Try a 3-ring binder or folders in a filing cabinet (Fig. 9).

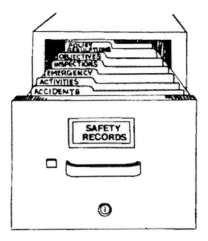

JDPX7610

Fig. 9 — Keep Good Records to Help You Manage Safety

Minimum Safety Program

Ideally, each farm and ranch should have the elements of a managed safety program outlined in this chapter. When that is not possible, the principles can be applied less formally.

Probably the two most important activities for any farm or ranch operation are these:

1. Get everyone involved in safety.

2. Use safety checklists or inspection sheets, especially for farm equipment; and be prepared for emergencies.

If all family members and employees are committed, and if you conduct and follow up on safety inspections each year, you can significantly improve safety for all persons in the operation.

Plan for Emergencies

Even with the best farm or ranch safety program and careful people, an emergency may someday threaten someone's life. You should have procedures in place and people trained to handle these and other emergencies:

*   Machinery and shop accidents

*   Fires

*   Electrical accidents

*   Pesticide and chemical accidents

*   Drowning (water and grain)

First Aid and CPR

All persons should have first aid and CPR (cardiopulmonary resuscitation) training (Fig. 10). Contact your local Red Cross, hospital, fire department, or Cooperative Extension Service to find out where and when to enroll in training and refresher courses.

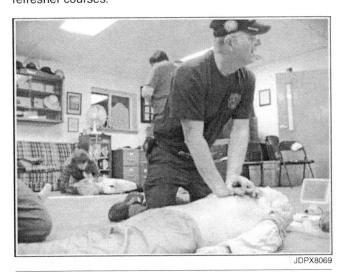

JDPX8069

Fig. 10 — CPR and First Aid Training Help Save Lives in Emergencies

Provide and maintain first aid kits and supplies. First aid kits are available in retail outlets. Provide a small one for each major machine. Also, display larger kits in the shop, house, and other buildings. Be sure you have proper supplies and equipment near pesticide sites as well as near grain and liquid manure storage facilities.

Emergency Rescue

Rural accident victims are often not found immediately. Professional help may be too far away for prompt response. So, proper action by the person who discovers the accident can be the difference between life and death.

All farm and ranch persons should know and be trained for the following:

*   What to do first for the victim when you arrive

*   How to avoid becoming a victim yourself

*   Whom to call or where to go for help

*   What information, including location and directions, to give to emergency professionals

*   What to do until professional help arrives

As a minimum, you should list Emergency Medical Services to contact for help. Display telephone numbers and calling instructions near each phone (Fig. 11). Include these: Emergency 911, if available; fire department, police or sheriff; ambulance; poison control center; electric/gas supply; and 1-800-424-9300 CHEMTREC for chemical spills.

Fig. 11 — Display Emergency Phone Numbers and Calling Instructions Where They're Easy to Read and Use

Train all persons in emergency rescue. Proper action at the scene of an accident includes the following: Assess the situation; remove victim from hazard area if necessary and possible; check breathing and circulation, and control bleeding; get professional help and transportation. If you are properly trained, check the victim for shock and spinal injury, and apply splints for fractures. In the case of machinery accidents, proper action in extricating the victim is especially critical. Training is important for each step.

Your farm or ranch safety program is not complete unless all persons are trained in CPR, first aid, and emergency rescue. The shock of an accident often interferes with an untrained person's ability to think and act properly. That could cost a life. Contact your Cooperative Extension Service, fire department, or community hospital to locate a training course such as "first-on-the-scene," "first care" or others.

Youth and Children

The safety of youth and children calls for special understanding and effort in managing safety. Many youth and children, including infants, are unnecessarily exposed to hazards and are seriously injured and killed each year in farm work-related accidents. (See Chapter 1, "Safe Farm Machinery Operation").

An Indiana-Wisconsin study indicated that 70% of farm work-related fatalities of children and adolescents involved tractors and implements. Other studies reveal similar conclusions.

One of the most common accidents involves children being run over by tractors and implements (Fig. 12). Some are entangled in unshielded PTOs, augers, chains, belts, and gears. Others are victims of tractor overturns, machines falling on them, grain suffocation in wagons, and chemical poisoning.

Fig. 12 — Many Children and Youth Are Injured and Killed in Farm Machinery Accidents

Why So Many Children's Machinery Accidents?

There are many reasons youth and children are injured and killed in farm equipment accidents, but they can be generally summarized as follows:

Children and youth are often exposed to hazards they aren't prepared or able to recognize, understand, or avoid.

How to Prevent Injuries to Children

Children should not be subjected to hazards they don't understand and aren't prepared to avoid. Follow these recommendations:

1. Start early to teach children that farm machines are for adult work and that they can injure and kill people who aren't experienced and trained to use them. They are not toys.

2. Don't let children play on or near machinery, even when it is parked or stored. Reinforce the idea that machines aren't toys.

3. Don't allow children near machinery when it is operating or being repaired unless an adult is closely supervising them.

4. Enforce the "No Rider" rule. Don't allow children to ride on farm equipment with you or others. Hire someone to care for the children when needed.

5. Set a good example. Make it obvious to children that safety is important. Enforce the "No Rider" rule for adults, too. Include children in safety discussions (Fig. 13).

JDPX8071

Fig. 13 — Always Set a Good Safety Example for Children. Include Them in Safety Discussions

6. Have children and youth participate in summer safety camps, safety activities of 4-H, FFA, and other youth and safety organizations such as Safe Kids Worldwide, or Farm Safety 4 Just Kids.

7. Provide training for each type of machine they will operate, but only when you're confident they are mature enough to do it safely. State or county Cooperative Extension Services, 4-H clubs, high school vocational agriculture instructors, and farm equipment and chemical dealers can help locate safety training clinics.

8. Supervise all persons around machinery until you're sure they understand the hazards and can safely deal with them.

When Are Children Old Enough?

There is no specific age when youth or children can safely be around or operate farm equipment, but many start too young. Each child is unique in temperament, physical and emotional maturity, and experience. When considering whether people are capable of safely operating machinery, consider these questions:

• Are they large enough and strong enough?

• Are they mentally mature enough?

• Can they read and understand the safety instructions in operator's manuals and machinery safety signs? Do they understand the machinery controls, gauges, and symbols?

• Are they adequately trained and experienced?

• Do they recognize and understand the potential hazards and how to avoid them?

• Will they follow the safe practices? Can they safely cope with emergencies?

Suggestion: The safety committee could apply the above questions in deciding whether people of any age, including children, youth, older adults, or employees, should be allowed to operate farm equipment.

Older Adults

Some studies indicate that persons over age 60 are involved in up to one-third of the fatal farm accidents. As people age, their strength, vision, hearing, and reflexes generally decline. Also, they may be taking medication that affects their ability to operate machinery. When deciding when persons should no longer operate machines, consider the same questions listed above for youth and children.

Employees and Regulations

When hiring anyone to work on a farm or ranch with farm equipment or other operations, consider the same questions outlined above for youth and children. Also, follow the federal, state, and local safety and health regulations for employees.

U.S. Public Law 91-596 (OSHA) includes legal safety and health requirements for employee equipment and workplaces. Also, the Hazardous Occupations Order in Agriculture (HOOA) makes it unlawful to hire youth under age 16 for certain work activities, including machinery operation. It allows hiring 14- and 15-year-olds for some tasks only if they have special training. (See Glossary, Laws, And Resources at this end of the book.)

Test Yourself

Questions

1. (T/F) A safety committee is more appropriate for a large industrial corporation than for an average farm or ranch.

2. What should be the first step in an effective ranch safety program?

    a. Perform a safety status assessment

    b. Replace all missing machinery shields

    c. Have all family members or workers complete a chemical safety program

    d. Designate one person to handle all safety-related concerns

3. Who should be involved in a typical family farm safety committee?

    a. The farm operator

    b. Spouses and children

    c. Hired labor

    d. All of the above

4. (T/F) In large corporate farming operations, management should not be involved in the development of safety policy.

5. (T/F) A safety committee determines needs, proposes action, and follows up on proposals to ensure completion.

6. A farm or ranch safety program should be thoroughly reviewed:

    a. Continually

    b. At least annually

    c. Every five years

    d. An effective program should not need altering once put into place

7. (T/F) Local emergency services are usually close enough that farm family members have no real need for first aid or CPR training.

8. Young people can be trusted to operate equipment safely by what age?

    a. 12

    b. 14

    c. 16

    d. There is no specific age recommendation, because all children are different

9. (T/F) There are no government regulations concerning safety and health of farm workers.

10. Which law prohibits the hiring of youth under age 16 for certain farm work activities?

    a. Hazardous Occupations Order in Agriculture

    b. Occupational Safety and Health Act

    c. Environmental Protection Act

    d. Safe Youth Farm Workers Act

11. Farm safety recommendations and information can be obtained from:

    a. Cooperative Extension Service

    b. Farm magazines

    c. FFA programs

    d. All of the above

12. (T/F) Farm safety and health can be improved, just as productivity or profitability, through sound management and planning.

References

1. Farm Safety Walkabout Handbook. 1992. Iowa Department of Public Health, Lucas State Office Building, Des Moines, Iowa 50319.

2. Farm safety checklists. Cooperative Extension Service, University of Arizona, Dept. of Education, Tucson, Arizona 85719.

3. First On The Scene. NRAES-12. 1989. Northeast Regional Agricultural Engineering Service, 152 RileyRobb Hall, Cornell University, Ithaca, New York 14853.

4. Responding to Farm Accidents. 1992. University Extension, University of Missouri, Columbia, Missouri 65211.

5. Injury in the Agricultural Workplace. 1987. W. Field and M. Purshchwitz. Dept. of Agricultural Engineering, Purdue University.

6. Pediatric Exposure to Agricultural Machinery: Implications for Primary Prevention. 1993. Hawk, Denham, and Gay. Inst. of Agricultural Medicine, University of Iowa.

7. Review and Analysis of Fatal and Nonfatal Farm Work-Related Injuries Involving Children and Adolescents Through Age 17. 1992. E. Sheldon, M.S. Thesis. Agricultural Engineering Dept., Purdue University.

8. Working Together for Kids Safety. Notebook/kit. Farm Safety 4 Just Kids, 130 East First Street, Earlham, Iowa 50072.

Equipment Service
and Maintenance

5

JDPX8059

Introduction

Zero In on These Targets

 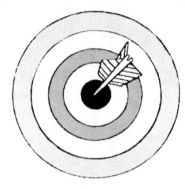 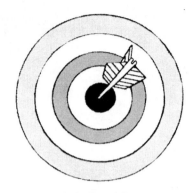

Safe Operating Procedures   Safe Operating Condition   Safe Servicing

JDPX7752

Fig. 1 — Hit Each Target for Farm Equipment Safety

Now that you have learned to recognize common hazards, you should be better prepared to avoid personal injury in operating tractors and implements in farm or ranch work. This chapter addresses specific safety hazards and safe practices related to servicing and maintaining farm equipment. In addition, you should follow the recommendations and instructions in machinery operator's manuals and service manuals to help you avoid damage to the machines as well as to avoid personal injury.

The safe farm equipment operator zeros in on three targets (Fig. 1):

- Safe operating procedures and practices

- Machinery in safe operating condition

- Safe servicing and maintenance procedures

Hit each safety bull's eye, and follow the procedures in machinery operator's and service manuals (Fig. 2), and your work with farm equipment can be more efficient and accident free.

JDPX7753

Fig. 2 — Follow Recommended Service Procedures to Protect You and Your Equipment

Keeping Equipment In Safe Operating Condition

Keep all equipment in top-notch condition. See your dealer for repairs. Accidents are most likely to happen when:

- Machines are out of adjustment.

- Worn or broken parts are not replaced.

- Cutting edges are dull.

- Shields and other safety devices are not in place or working properly.

- Safety procedures aren't followed during maintenance.

- Safety signs are missing or not legible.

How can you prevent accidents? Here are three recommendations:

1. Check each machine before you use it, especially the safety features.

2. Follow the maintenance and service schedules the manufacturer provides for each machine.

3. Be alert to changes in machine performance and operation.

Here is a closer look.

Operation Checks Before Use

Make it a habit to check all components and systems that affect the safety and performance of machine operation before and during each use. These checks are explained in detail in other chapters.

Pay special attention to these:

- Steering

- Brakes

- Hydraulics

- Fuel system

- Warning lights, reflectors, turn signals, SMV emblem

- Tires

- Controls

- Safety switches

- Rollover protective structures

- Shields, guards, other safety features

When you discover problems, take immediate action to make the necessary adjustments or repairs. Follow the procedures outlined in your operator's or service manual, or take the equipment to your dealer for service.

Service and Maintenance Schedules

Follow the service and maintenance schedule the manufacturer provides for each of your machines (Fig. 3). These schedules, along with the instructions found in your operator's manual, tell you when and how:

- To lubricate and adjust moving parts to prolong machine life.

- To replace parts that deteriorate with use and age.

- To maintain clearances to compensate for wear.

- To sharpen cutting edges to maintain efficient operation.

- To tighten bolts and nuts to protect the safety of the machine and the operator.

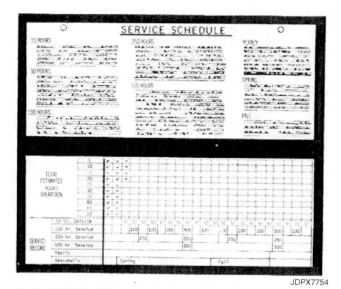

JDPX7754

Fig. 3 — Following the Manufacturer's Service Schedule Helps Keep Each Machine in Safe Operating Condition

Most operators agree that the small expense it takes in time and money to follow maintenance recommendations pays big dividends by maintaining machine performance and efficiency, prolonging machine life, and keeping the machine's safety devices working properly.

In addition to scheduled maintenance, the machine must be kept clean to operate safely. Pay special attention to steps, handholds, and the working or walking surfaces.

Changes in Tractor and Machine Performance

A good way to spot problems early is to be alert to changes in the operating characteristics and performance of your equipment. To detect these changes, use your senses of hearing, sight, touch, and smell. Be alert to these warning signals:

- Unusual noises

- Increased vibration

- Indications that moving parts (bearings. belts, etc.) are too hot

- Lack of response to controls

- Increased power requirements for engine-driven machines

- Changes in operating speed

- Changes in engine exhaust characteristics

- Operation of warning lights or horns

- Instrument panel gauge readings too high, too low, or flashing "Service Alert" or "Stop Engine" (Fig. 4)

JDPX8036

Fig. 4 — Watch Your Instrument Panel for Trouble Signs

Stop, investigate, and make necessary repairs or adjustments when you receive any of thee warning signals. Many times, safety is involved. Also, you may be able to avoid extensive repairs by catching a minor problem early.

NOTE: Lower all equipment to the ground, turn off the engine, remove the key, shift to the park position or set the parking brake, and disengage the PTO before dismounting from a tractor or a self-propelled machine. Do not attempt to unplug, clean, or lubricate any machine until all moving parts have stopped.

Who Will Do the Job?

When service and repairs are needed, you have to decide who will do the job. You can do it yourself, or you can take the work to your farm equipment dealer. Ask yourself these questions to help you decide:

• Do I have enough time?

• Do I know how?

• Do I have the necessary parts, tools, and equipment?

• Can I do the job without getting hurt?

• Do I have a safe place to work?

Let's review each of these factors.

Time

Try to do your maintenance and repair work during slack periods in your work schedule, and during periods of time when the equipment is not being used. Time is available then to work carefully without the pressures of other jobs

Keep in mind many service operations will take longer than you think. In many situations, you'll save time in the long run if you go ahead with other jobs and let your farm equipment dealer do your service work for you.

Know-How

Your operator's manual is full of maintenance, lubrication, and adjustment procedures. It tells you how to service equipment safely. Always follow those procedures.

What about jobs not described in your operator's manual? In some cases, the best answer is this: unless you have experience, specialized training, and the technical service manual for the machine, take the work to a qualified dealer (Fig. 5). Service manuals are essential for finding safety recommendations, clearances for close-fitting parts, disassembly and assembly information, and proper adjustment procedures. If you don't have previous experience servicing equipment, you may be faced with unexpected problems that you won't be able to solve. Also, you may be exposing yourself to safety hazards unknown to you.

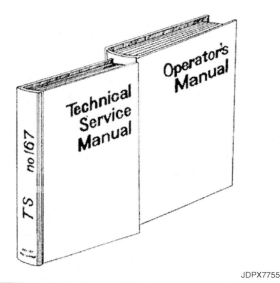

JDPX7755

Fig. 5 — Technical Information Is Needed to Perform Service Work Properly and Safely

Tools and Service Equipment

Your farm equipment dealer has a big investment in specialized tools and service equipment (Fig. 6). These are necessary to support equipment safely during repair, and for fitting close tolerances, adjustment, and testing. Without these tools, you may not be able to do the job properly and safely. Before you tackle any job, read through the service procedure to see if you have the necessary tools and equipment.

JDPX7626

Fig. 6 — Specialized Tools Are Required for Many Service Jobs

Safety

Time, availability of the proper tools, and "know-how" are essential for safe service work. There are other factors, too.

The safe equipment operator:

- Maintains a hazard-free shop.

- Uses service tools and equipment safely.

- Guards against the hazards of service work.

Let's look at each of these factors.

Maintaining a Hazard-Free Shop

A safe farm shop meets three important requirements (Fig. 7):

- It is planned to be as free from hazards as possible.

- It is managed to keep it that way.

- It is equipped to handle emergency situations.

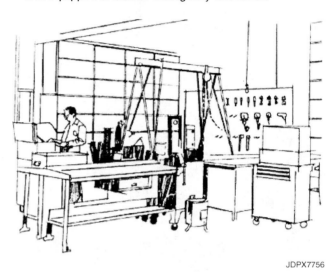

JDPX7756

Fig. 7 — Planning and Good Shop Management Are Necessary for Safety

Shop Planning

An efficient and hazard-free shop results from careful planning. When a new shop is planned, potential hazards are recognized in advance, and are eliminated or minimized as much as possible as the plans for the shop are worked out.

Let's look at some important recommendations for planning a new shop. Even if you don't have the opportunity to plan a new shop, you can use these same recommendations to evaluate and modify, if necessary, the shop facilities you already have.

Location

Locate the shop in a convenient and accessible location for farm service work, and for the storage of spare parts, tools, and supplies (Fig. 8). Provide adequate drainage to keep the shop floor dry at all times. Give yourself enough space around the building and a service door large enough to maneuver equipment easily and safely. Provide a concrete apron outside for cleaning equipment, for welding, and to provide a solid foundation for hydraulic jacks and support stands. Build a fireproof wall between shop and storage.

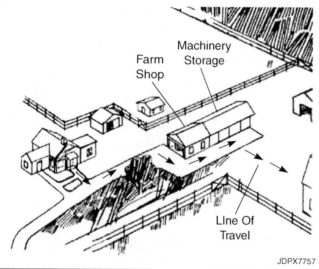

JDPX7757

Fig. 8 — Locate Your Shop for Accessibility and Convenience

Size and Arrangement

To determine size and arrangement, make a list of the jobs and activities you'll be doing in the shop, and the sizes of the equipment you'll be working on inside the shop facility. Then sketch floor plans until you are satisfied with the arrangement and location of work and storage areas, benches, and stationary tools (Fig. 9). Keep these points in mind:

1. Provide adequate and uncluttered work space around machines you plan to service.

2. Locate benches, stationary tools, bins, racks, and tool panels in work areas according to use and function.

3. Keep combustible materials out of the welding area, away from heating units, and in proper containers.

4. Locate welding, metalwork, and other dirty work areas away from woodworking and other areas you want to keep clean.

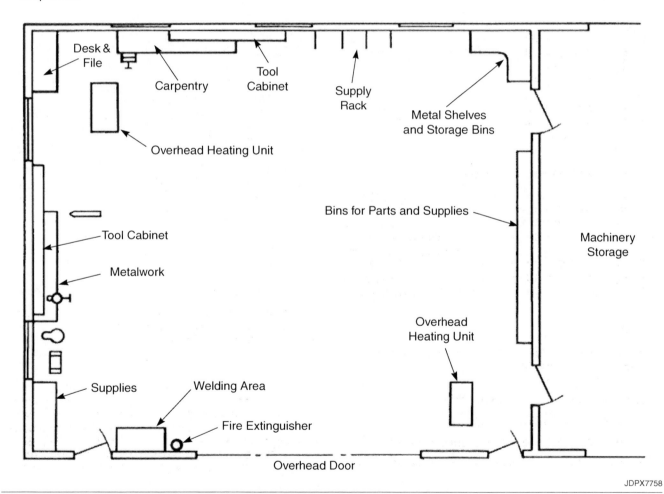

JDPX7758

Fig. 9 — Draw a Plan to Arrange Your Shop for Convenience And Safety

Lighting

Provide a sufficient number of windows, skylights, and overhead lights to ensure good general lighting. Place additional lights over benches, stationary power tools, and main work areas (Fig. 10). Use portable lights when necessary to eliminate shadows while servicing equipment. Keep windows clean. Apply light-colored paint to the walls. This will utilize existing light more effectively than darker walls, reduce shadows, and make potential hazards more visible by having contrasts in color.

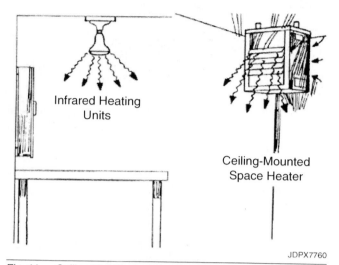

Fig. 11 — Ceiling Unit Heaters Are Good, They Leave Clear Working Space Below

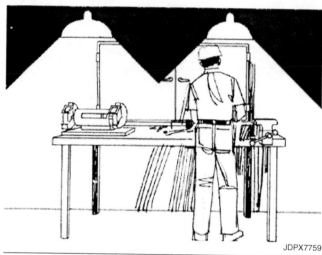

Fig. 10 — Provide Adequate Lighting Over Workbenches, Stationary Tools, And Main Work Areas

Heating and Cooling

You'll need some source of heat for cold weather work, and in warm climates air conditioning may be needed. Before selecting and installing heating and air conditioning units, get the services of a heating and air conditioning specialist. That expertise will help you get the most effective installation. Here are key points to keep in mind when planning:

1. Put units where they will distribute heat and cold efficiently. Do not place units where they can be struck by moving equipment (Fig. 11), or near areas where combustible or flammable materials are used or stored.

2. Make sure the installation is safe:
   • Adequate ventilation provided for the units inside
   • Vented safely outside
   • Equipped with overheating shutoff devices
   • Installed with emergency shutoffs for fuel and electric power.

Ventilation

Put enough doors and windows in your shop to ventilate smoke, fumes, and vapors outdoors. Ideally, window area should equal 25 percent of the shop's floor area. Be sure to open windows and doors and work under ventilating hoods when running engines inside, when welding, and when handling chemicals with poisonous dusts or fumes. When working under these conditions, gases, fumes, and toxic substances can rapidly build up to dangerous concentrations.

Use flexible metal tubing to carry engine exhaust fumes outside. If you don't have vent hoods, use exhaust fans to clear smoke and vapors away from welding, cleaning, and painting areas to the outdoors. Hoods equipped with exhaust fans are best for removing fouled air from specific work areas (Fig. 12). Remember that exhaust fans are only efficient if fresh air is available from an open window, or from a ventilator installed to provide an incoming source of fresh air.

JDPX7761

Fig. 12 — Adequate Ventilation: Large Doors for Summer, an Exhaust Hood for Winter

Wiring

Get assistance from a qualified electrician when planning the wiring system. In general, the system should meet five requirements:

1. Have adequate capacity to handle lighting, heating, and power tool requirements (Fig. 13).

2. Have enough outlets so extension cords aren't needed.

3. Have three-wire, grounding-type, 120-volt circuits to prevent electric shock while you're using power tools.

4. Have 240-volt circuits for welders and motors over 1/2 horsepower.

5. Be able to expand for future needs.

6. Have properly grounded and protected circuits, including GFIs (ground-fault interrupters), to prevent electrocution in case of shorted circuits.

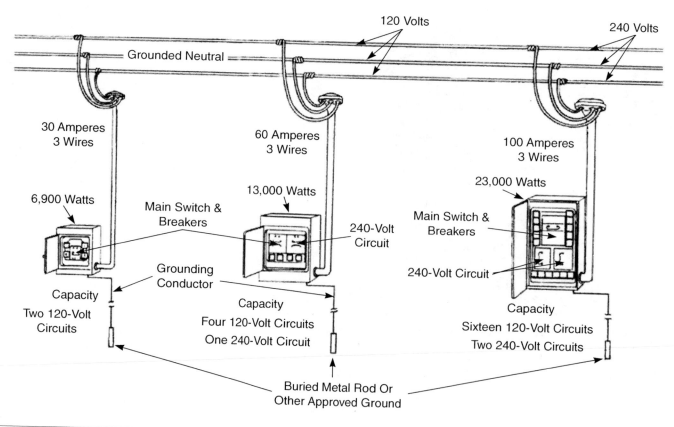

Fig. 13 — *Install Service Entrance Equipment with Adequate Capacity for Present and Future Needs*

JDPX7762

Storage

Develop a system of racks, bins, and tool boards so finding the right tool is easy and quick (Fig. 14). Don't store tools, supplies, or spare parts in aisles or on the work floor where someone will trip over them. Keep grease, oil, paint, solvents, lumber, and other flammable materials away from heaters and welding areas to prevent fire. Store oil, grease, paint, and solvents in closed containers in metal cabinets or on metal shelves. Stored in this manner they are protected from sparks and flame, and spills don't soak into shelves. Wooden shelves soaked with flammable liquids are fire hazards.

Fig. 14 — Finding Parts Is Quick and Easy with Labeled Storage Bins

NOTE: *Do not store gasoline or other fuels in the shop. If it is absolutely necessary to keep small quantities in the shop, use approved safety cans, and keep these cans in a ventilated area away from sparks and flame. Label the containers so others will know what is in them and can avoid accidental misuse.*

Properly store and dispose of hazardous wastes such as battery acid, cleaning solvents, and other harmful chemicals.

Use steel drums for storing trash. Use one drum for non-combustibles and another for materials that will burn. Also, provide a metal container with a self-closing cover for oily rags and oil-soaked filters that sparks, flame, spontaneous combustion, or a careless smoker could set on fire (Fig. 15). Empty trash containers frequently.

Self-Closing Lid

Fig. 15 — Store Oily Refuse in a Metal Container with a Self-Closing Lid

Shop Management

After a shop has been planned to be as hazard-free as possible, it must be managed to keep it that way. Consider these key management procedures:

Make sure someone knows you are working in the shop and will check on you and render aid if you are injured.

1. Keep all tools and service equipment in good condition.

2. Use personal protective equipment: goggles, face shields, gloves, and respirators. Display them near equipment with which they should be used to encourage their use.

3. Keep floors and benches clean to reduce fire and tripping hazards.

4. Clean up as you go while doing a job, and clean the area completely after the job is done.

5. Empty trash containers regularly.

6. Keep lighting, wiring, heating, and ventilation systems in good shape.

7. Lock your shop to prevent accidents. A shop is an attraction to a child.

8. Don't let anyone use tools or service equipment unless they've had adequate instruction.

9. Keep guards and other safety devices in place and functioning.

10. Use tools and service equipment for the jobs they were designed to do.

11. Supervise children carefully when they are in the shop.

12. Keep fire extinguishers serviced, and the first aid kit replenished with supplies.

Emergency Situations

Every farm shop should be equipped to handle emergency situations. The most common types of emergencies are fires and personal injury.

Fire

Your best protection against fire is prevention. But if a fire starts, you should be ready to take immediate action. Try to judge the situation quickly without panic. If you think you can extinguish the fire easily, do it; if not, immediately call the nearest fire department (always keep its number by the phone).

There are three types of fire: Class: A, B, and C (Fig. 16):

- Class A — Combustibles like paper and wood

- Class B — Gasoline, diesel fuel, grease, and solvents

- Class C — Electrical equipment fires

Class A Fires	Paper, Wood, Cloth, Excelsior, Rubbish, Etc., Where Quenching And Cooling Effect Of Water Is Needed	
Class B Fires	Burning Liquids (Gasoline, Oils, Paints, Etc.), Where Smothering Effect Is Required.	
Class C Fires	Fires in Live Electrical Equipment (Motors, Switches, Heaters, Etc.), Where a Non-Conducting Extinguishing Agent Is Required.	

JDPX7548

Fig. 16 — Be Prepared to Fight These Types of Fires

Your fire extinguisher must be effective and safe for fighting each type of fire. To be sure you have the proper extinguishers properly located, check with your insurance company. Study the following minimum recommendations for farm shop fire extinguishers.

For Class A fires (ordinary combustibles), provide at least one of the following methods for fighting fire:

1. One or more 20-pound (9-kg), multi-purpose, dry chemical pressurized fire extinguishers specifically approved for fighting Class A fires (Fig. 17). (Not all types of dry chemical extinguishers are approved for Class A fires.)

JDPX7763

Fig. 17 — Equip Your Shop With At Least One Multi-Purpose Dry Chemical Extinguisher

2. A garden hose attached to one of the water faucets in the shop, kept ready to use, and long enough to reach all areas of the shop (Fig. 18).

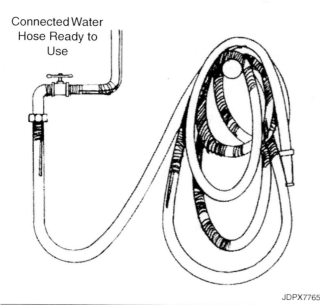

Connected Water Hose Ready to Use

JDPX7765

Fig. 18 — Have Water Available to Fight Class A Fires if Type A Commercial Extinguishers Are Not Available

3. Several pails submerged in a barrel of water (Fig. 19). These pails can be pulled quickly from the barrel, full of water, and ready to use in an emergency. Add antifreeze if the water could freeze in winter months, but not a flammable alcohol antifreeze. Calcium chloride, available from your farm equipment dealer, will not burn and is economical to use.

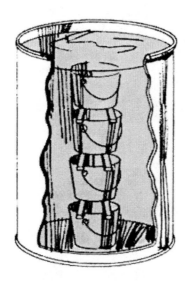

JDPX7764

Fig. 19 — Have Water Available to Fight Class A Fires if Type A Commercial Extinguishers Are Not Available

NOTE: Don't use water on Class B (burning liquids) or C (electrical equipment) fires. Water spreads Class B fires, and water may conduct electricity to give you a severe shock if used on Class C fires.

For Class B fires (burning liquids) and Class C fires (those involving electrical equipment), provide at least one pressurized dry chemical fire extinguisher of 20-pound (9 - kg) capacity, or follow the recommendations of your insurance company and state extension office.

Keep extinguishers in a convenient place, and close to, but not in, the fire hazard area. Make sure they are protected from damage and always within easy reach and easy for anyone to see (Fig. 20). Some, but not all, dry chemical extinguishers are effective against Class A fires. When buying extinguishers, select those rated for all classes of fires — A, B, and C.

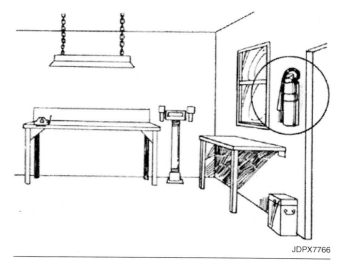

JDPX7766

Fig. 20 — Always Keep Extinguishers Where They Are Easy to Reach

Be sure you and others in your operation know, in advance, how to use the fire extinguishing equipment and how to extinguish fires. Don't wait until there's a fire to learn.

Personal Injuries

Be prepared to take care of injuries. Advance preparations are essential for coping with serious emergencies and preventing small injuries from becoming serious medical problems. Follow these recommendations:

1. Learn the basic rules of first aid: At least one member of your family or work force should take a basic course in first aid, such as those given by the American Red Cross, safety extension specialists, fire departments and other agencies. Emergency response time to farms is four to six times longer than urban places. Farm families should be able to sustain an injured person's life until professional help arrives. Keep a basic first aid book in a handy location known to all family members (Fig. 21).

JDPX7438

Fig. 21 — Keep a First Aid Book and a First Aid Kit Available for Immediate Use

2. Know who to call for help: Keep emergency numbers for doctors, ambulance service, hospital, and fire department near your telephone. You should have a telephone in your shop.

3. Apply immediate first aid to all injuries: If an injury appears severe, don't move the victim. Call a doctor and follow the doctor's instructions. If the injury is not severe, administer first aid and take the injured person to a doctor for further treatment.

4. Keep a small, industrial quality first aid kit in your shop: This kit should be backed up with more extensive first aid supplies in your home or office. You may purchase a kit or assemble your own. To assemble a first aid kit:

   a. Find a small metal or plastic box that will seal out dust and moisture.

   b. Wash the box and rinse it with boiling water.

   c. Include these items (and others you may wish to add):
   - Sterile gauze pads, individually packaged, 3 and 4 inches square
   - Rolls of sterilized gauze, 2 and 3 inches wide
   - Band-aid® adhesive bandages in assorted sizes
   - Sterile absorbent cotton
   - Roll of adhesive tape
   - Large triangular bandages made from 40-inch square cotton sheeting, cut diagonally
   - Scissors with rounded tips
   - Tweezers
   - Safety pins
   - An antiseptic
   - Cold pack

NOTE: Antiseptics are not necessary if the victim will soon receive medical care. However, they should be available for use when needed. Ointments should only be used on minor burns. Cold is best!

Using Service Tools and Equipment Safely

Don't take the use of hand tools and service equipment for granted. You're more likely to be injured when servicing equipment than when you're operating it. Small hand tools can inflict great injury.

Consider the safe use of these service tools:

- Hand tools
- Power tools
- Welding equipment
- Hoists and jacks
- Cleaning equipment

Hand Tools

You can avoid hand tool injuries if you follow four basic rules:

1. Select the right tool for the job.

2. Use it in the right way.

3. Keep it in good condition.

4. Store it safely when it's not in use.

To follow all of these rules, you need more information than can be given here. Study several good publications, available from book stores or libraries, that describe the proper use and care of hand tools, and follow the recommendations given for their use. Let's look at some of the important principles relating to personal safety for the hand tools most frequently used for service work.

Chisels and Punches

1. Wear eye protection: The hardened face of the hammer, the end of the tool, or other metal you strike with a hammer may chip or shatter and send metal fragments flying through the air (Fig. 22).

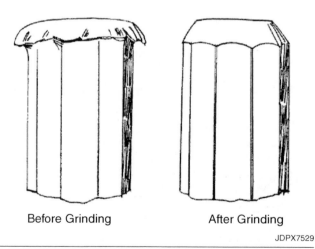

Fig. 22 — Wear Eye Protection and Hold Chisels and Punches Near the Head of the Tool

2. Grind off mushroom heads: The sharp edges can tear your skin if the tool slips. And when the tool is struck, chips could break off the mushroomed head and fly into your eyes or could strike another person near you. Keep a smooth bevel ground on the heads of all punches and chisels (Fig. 23).

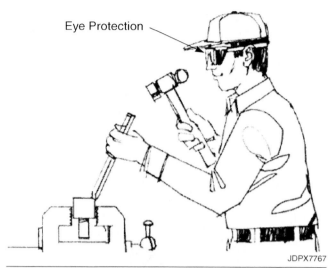

Before Grinding   After Grinding

Fig. 23 — Grind Mushroom Heads from Chisels and Punches

3. Don't use chisels and punches for prying: They are hard and brittle, and excessive force could break them with a snap.

4. Hold the tool steadily but loosely: The best place to hold it is just below the head. If you miss and strike your hand, your hand will not be caught between the hammer and the work piece.

5. Select the proper sized tool for the job: Heavy pounding on tools too small for the job increases the risk of injury from tool breakage. Tools too large for the job may not be safe either. The full cutting edge of a chisel, for example, should be used. Using only a section of the cutting edge of a larger chisel could result in breaking off the corners from the cutting edge.

Files

Keep a handle on every file: This will keep the tang from piercing your palm or wrist if the file should slip or catch. Also, it makes the tool much easier to use effectively (Fig. 24).

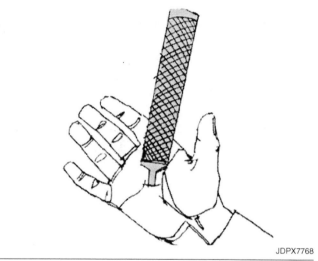

Fig. 24 — Files without Handles Can Pierce Your Hand or Wrist

Knives

1. Keep blades sharp: The greater the force you have to apply, the less control you have over the cutting action of the knife. The safest knife usually has the sharpest edge. Keep all of your knives uniformly sharp.

2. Cut away from the body: Your hands and fingers should always be behind the cutting edge. Keep knife handles clean and dry to keep your hand from slipping onto the blade.

3. Never pry with a knife: Blades are hardened and can break with a snap.

4. Store knives safely: Keep knives in their own box or scabbard when not in use. An unguarded blade could cut you severely.

Screwdrivers

1. Use screwdrivers only for driving screws: Using them for punches or pry bars breaks handles, bends shanks, and dulls and twists the tips. This makes them unfit to tighten or loosen screws safely.

2. Sharpen screwdrivers properly: File or grind worn or damaged tips to fit the slot of the screw (Fig. 25). A sharp, square-edged tip won't slip as easily as a dull one, and less pressure will be required to hold the tip in the slot. Keep an assortment of screwdrivers on hand to fit different sizes and types of screw heads.

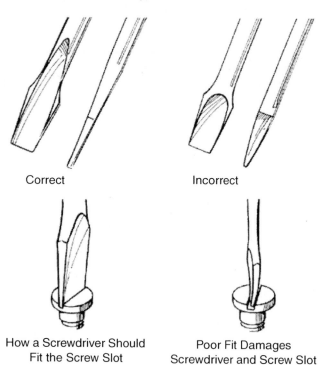

Correct   Incorrect

How a Screwdriver Should Fit the Screw Slot   Poor Fit Damages Screwdriver and Screw Slot

JDPX7656

Fig. 25 — File or Grind Screwdriver Tips to Fit the Slot of the Screw

3. Don't hold parts in your hand: Put the work on a bench or in a vise to avoid the possibility of piercing your hand with the screwdriver tip (Fig. 26).

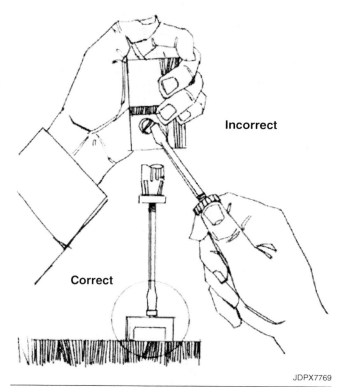

Incorrect

Correct

JDPX7769

Fig. 26 — Working on a Bench Will Keep the Screwdriver from Piercing Your Hand.

4. Use screwdrivers with insulated handles for electrical work: If the blade or a rivet extends through the handle while you are holding it, an electrical current could give you a serious shock.

Hammers

1. Wear eye protection: Always wear goggles when striking hardened tools and hardened metal surfaces. This will protect your eyes from flying chips. Whenever possible, use soft-faced hammers (plastic, wood, or rawhide) when striking hardened surfaces.

2. Check the fit and condition of the handle: Keep handles tightly wedged in hammerheads to prevent injury to yourself and others nearby. Replace cracked or splintered handles. And don't use the handle for prying or bumping. Handles are easily damaged and broken this way (Fig. 27).

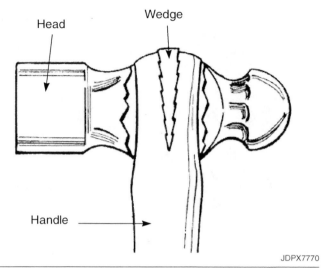

Head

Wedge

Handle

JDPX7770

Fig. 27 — Keep Handles Tightly Wedged to Prevent Injury to Yourself or Others Nearby

3. Select the right size for the job: A light hammer bounces off the work. One that's too heavy is hard to control.

4. Grip the handle close to the end: This increases leverage for harder, less tiresome blows. It also reduces the possibility of crushing your fingers between the handle and the projecting parts and edges of the workpiece if you should miss (Fig. 28).

Incorrect   **Correct**

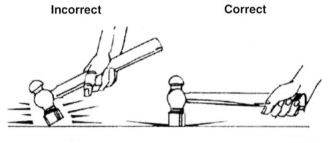

JDPX7771

Fig. 28 — Grip the Hammer Near the End and Strike Squarely with The Surface

5. Prevent injuries to others: Swing in a direction that won't let your hammer strike someone if it slips from your hand. Keep the handle dry and free of grease and oil.

6. Keep the hammer face parallel with your work: Force is then distributed over the entire hammer face, reducing the tendency of the edges of the hammer head to chip or to slip off the object being struck.

Wrenches

1. Use wrenches that fit: Wrenches that slip damage bolt heads and nuts, skin knuckles, and lead to falls. Don't try to make wrenches fit by using shims (Fig. 29).

   Avoid using metric wrenches on inch-sized bolts and vice versa: Use the exact sizes needed for your machinery. Most sizes are not interchangeable.

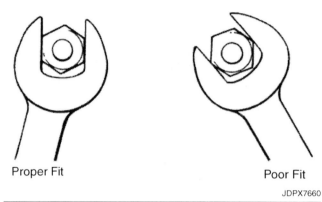

Proper Fit   Poor Fit

JDPX7660

Fig. 29 — Use Wrenches that Fit

2. Don't extend the length of a wrench: Do not use pipe to increase the leverage of the wrench. The handle was made long enough for the maximum safe force to be applied. Excessive force may break the wrench or bolt unexpectedly, or the wrench may slip, rounding off the corners of the bolt head or nut. Skinned knuckles, a fall, or a broken wrench may result. Don't hammer on wrenches unless they're designed for this type of use.

3. Pull on the wrench: This isn't always possible, but if you push, and if the wrench slips or the nut suddenly breaks loose, you may skin your knuckles or cut yourself on a sharp edge. Use the open palm of your hand to push on a wrench when you can't pull it toward you (Fig. 30).

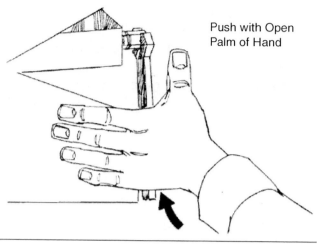

Push with Open
Palm of Hand

Pull the Wrench
Toward You

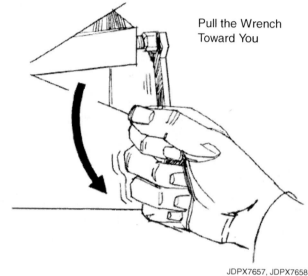

JDPX7657, JDPX7658

Fig. 30 — Pull on The Wrench or Push with the Open Palm of Your Hand to Avoid Injury

4. Replace damaged wrenches: Straightening a bent wrench weakens it. Cracked and worn wrenches are too dangerous to use, as they could break or slip at any time.

5. Keep the open jaws of adjustable wrenches facing you: Have the open jaws toward you when placing adjustable wrenches on bolt heads and nuts. Then pull on the wrench. This forces the movable jaw onto the nut, reduces its tendency to slip, and places most of the pressure on the solid, stronger jaw (Fig. 31), Adjust these wrenches to fit bolt heads and nuts to a snug fit.

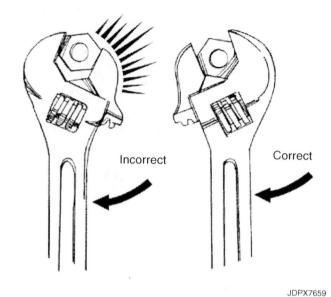

Incorrect   Correct

JDPX7659

Fig. 31 — Apply Most of the Force to the Solid Jaw by Keeping the Open Jaw Toward You

6. Use pipe wrenches only for pipe or round stock: If you tighten bolt heads and nuts, their sharp corners may slip and break the hardened teeth in the jaws. Then, the wrench may slip when used on pipe, making the wrench unsafe to use.

Pliers and Cutters

1. Do not use pliers as a wrench: They do not hold the work securely and can damage bolt heads and nuts.

2. Guard against eye injuries when cutting with pliers or cutters: Short and long ends of wire often fly or whip through the air when cut. Wear eye protection, or cup your hand over the pliers to guard your eyes.

3. Wear eye protection when cutting with bolt cutters: Chips of metal may break away from the cutting edges and be flung into your eyes. To help prevent chipping or breaking the cutting edges, observe these precautions:

   • Select a cutter big enough for the job.

   • Keep the blades at right angles to the stock.

   • Don't rock the cutter to get a faster cut.

   • Adjust the cutters to maintain a small clearance between the blades. This prevents the hardened blades from striking each other when the handles are closed.

Power Tools

Speed and reducing the physical effort required are the main reasons that workers use power tools. In your attempt to get a job done quickly, however, take the necessary precautions to get it done safely:

1. Read the operator's manual and observe all precautions.

2. Protect yourself from electric shock (refer to next section).

3. Keep guards and shields in place.

4. Keep the work area clean.

5. Give your job full attention.

6. Let each tool work at its own speed without forcing it.

7. Wear snug-fitting clothes to prevent entanglement of clothing.

8. Wear eye protection when recommended.

9. Maintain secure footing and balance at all times.

10. Keep your tools clean and sharp.

11. Be sure the switch is off before plugging in the power cord.

12. Keep your fingers away from the switch when carrying a portable tool.

13. Turn the switch off immediately if the tool stalls or jams.

14. Use portable tools only in areas completely free of flammable vapors and liquids. Sparks could cause a fire or explosion.

15. Before making adjustments or changing bits or cutters, disconnect the power cord or you could accidently touch the switch and be injured when the tool starts.

16. Repair or replace damaged extension cords and plugs.

17. Use clamps or a vise to hold your work.

18. Use power tools only for their intended functions.

19. Remove adjusting keys and wrenches before operating a tool.

20. Use three-wire cords and GFI (ground-fault interrupters) for shock protection.

21. Store idle tools safely to prevent damage to the tool and cord, and to prevent unauthorized use.

22. Provide enough light so you can see what you're doing.

23. Use double-insulated electrical tools.

24. Use ground-fault-interrupted circuits.

Electric Shock

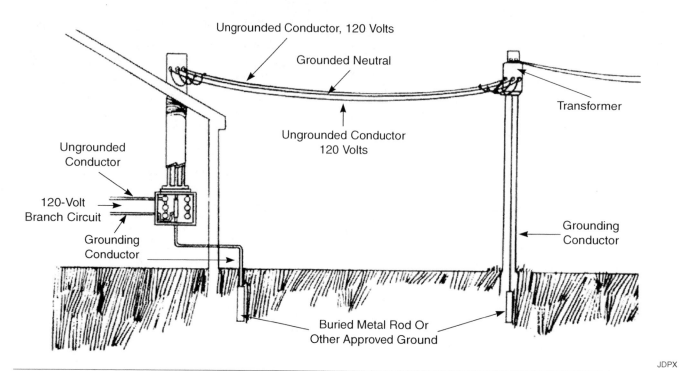

Fig. 32 — 240-Volt Electrical Service Showing a Two-Conductor Branch Circuit

Three conductors supply electrical power to your farm shop if it's equipped with 210-volt single-phase service (Fig. 32). One of those conductors is the grounded neutral. It is connected to a metal rod buried in the ground at the transformer, and again at the service entrance box that provides the branch circuits for the shop. The grounded neutral may or may not be insulated. If it is, the insulation is always white if the circuit has been wired according to the specifications of the National Electric Code of the United States.

The other two conductors are ungrounded. Each carries a voltage of 120 volts with respect to the grounded neutral. Because the grounded neutral is connected directly to ground, each of the ungrounded conductors has an electrical pressure of 120 volts above the earth's "neutral" charge. Ungrounded conductors may be covered with insulation of any color except white or green. (White insulation identifies grounded neutral conductors. Green insulation identifies grounding conductors.)

To illustrate how a person can receive an electric shock when using a power tool, let's look at the operation of a portable drill (Fig. 33). The drill is connected to a 120-volt branch circuit consisting of two conductors: an ungrounded conductor and the grounded neutral.

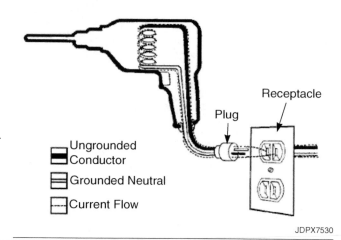

Fig. 33 — Safe Current flow Through an Electric Drill

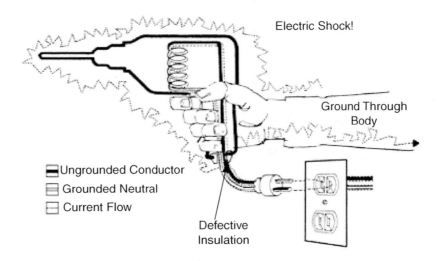

Electric Shock!

Ground Through Body

■ Ungrounded Conductor
▤ Grounded Neutral
▫ Current Flow

Defective Insulation

JDPX7530

Fig. 34 — Two-Wire, 120-Volt Circuits Do Not Provide Shock Protection from Defective or Inadequate Insulation

When the drill is turned on, current flows from the ungrounded conductor, through the insulated conductors in the drill, back through the grounded neutral, then to ground, If the insulation in the drill is in good condition, all of the current entering the drill will return through the grounded neutral. But if the drill's insulation is defective, current may also flow from the drill housing through the body of the operator to ground (Fig. 34). This causes electric shock.

The amount of current that flows through the operator's body is determined by the condition of the insulation in the tool, and the electrical resistance of the skin of the operator's body. His resistance is lowest when his skin is damp from perspiration, or when he's working in a damp location. These situations are dangerous. A current of only 0.006 ampere can electrocute a healthy man in less than a second.

In new power tools, the insulation is usually adequate to prevent current leakage from the insulated conductors within the tool to its frame or housing. Age and abuse, however, can deteriorate insulation, and so a shock hazard always exists when a power tool is used.

There are three ways to prevent electric shock:

• Using three-conductor, grounding-type circuits

• Using tools equipped with double insulation

• Installing ground-fault interrupters

Here is a closer look at these.

Three-Conductor, Grounding-Type Circuits

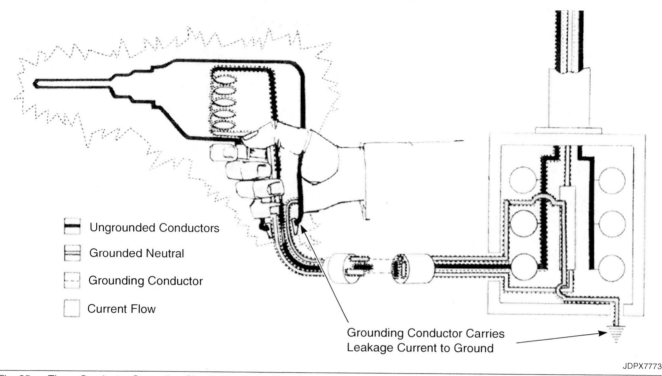

- ▭ Ungrounded Conductors
- ▤ Grounded Neutral
- ▦ Grounding Conductor
- ▯ Current Flow

Grounding Conductor Carries
Leakage Current to Ground

JDPX7773

Fig. 35 — Three-Conductor Grounding Circuit. The Grounding Conductor Carries Current Leakage Safely to Ground.

Grounding-type, 120-volt circuits use three conductors. The third conductor is called the grounding conductor. It's connected to the grounded neutral in the shop's service entrance box, and to the center terminals of the grounding-type outlets installed in the circuit (Fig. 35).

Power tools designed for use on grounding-type, circuits are equipped with three-wire cords. At one end, the grounding conductor is connected to the tool housing. The other end is connected to the center blade of a three-prong plug. (The insulation on this conductor is green.) When the tool is plugged in, the grounding conductor in the power cord and the branch circuit provides a continuous electrical path from the tool housing directly to ground. Should any current leakage occur from the insulated conductors, the grounding conductor carries the current directly and safely to ground.

Double-Insulated Tools

Because many power tool users don't have three-conductor, grounding-type circuits, manufacturers make power tools equipped with two layers of insulation. If one layer becomes defective, the second layer provides the necessary protection from shock.

Tools equipped with double insulation can be safely used on two-conductor circuits, since there's no need for the grounding conductor. These tools can be identified by the words "Double Insulation" or the symbol El marked permanently on the tool housing or nameplate (Fig. 35). Double-insulated tools are frequently encased in a plastic covering.

Look for Words or the Symbol

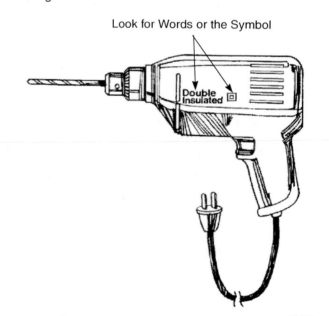

JDPX7534

Fig. 36 — Double-Insulated Tools Are Clearly Marked on Case or Nameplate

Double-insulated tools provide good protection from electrical shock if dry. But there is a danger of electrical shock if you are working in damp conditions. Three-wire grounded tools provide almost complete protection.

Ground-Fault Interrupters

All of the electrical current that enters a power tool on an ungrounded conductor should flow back again from the tool through the grounded neutral to ground. If it doesn't, a ground fault exists. This means that defective insulation is allowing some of the current to flow to ground by some other path. If this path is the operator's body, there is the danger of severe shock or electrocution.

A ground-fault interrupter is a device that compares the amount of current flowing to a power tool in the ungrounded conductor with the amount returning in the grounded neutral. If the ground-fault interrupter (often called a GFI) senses a difference as low as 0.005 ampere, the GFI snaps off the current by opening the circuit. In this way, the GFI protects the operator from shock.

Two types of GFIs are available. One type is permanently wired into the branch circuit at the service entrance box. The other is a portable unit that plugs into standard 120-volt outlets (Fig. 37). The plug of the power tool is then plugged into the GFI. Installation of GFIs in the service entrance panel is a job for a qualified electrician.

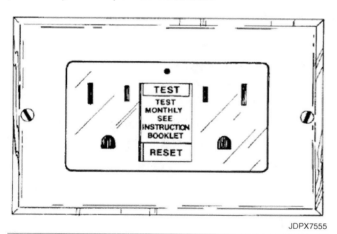

JDPX7555

Fig. 37 — The Ground Fault Interrupter Protects the Electric Circuit. The Power Tool Plugs Into the GFI. Portable and In-Outlet Ground Fault Interrupter Plugs Also Are Available.

Protect yourself from electrical shock by following these recommendations:

1. Select a shock protection system: If you have two-conductor circuits and a variety of tools — some with two-wire cords and plugs and some with three-wire cords and plugs — you have four alternatives:

   • Have an electrician install ground-fault interrupters permanently in each of the shop branch circuits.

   • Plug in a portable ground-fault interrupter when individual power tools are used.

   • Convert your two-conductor circuits to grounding-type circuits.

   • Replace your present tools with new ones equipped with double insulation.

   Get the advice of a competent electrician to help you decide which alternative is best or most economical for your farm.

2. Purchase tools designed to prevent shock: Look for tools that carry the approval label of a recognized inspection and approval agency. The label "UL Listed" indicates that the tool has been safety-approved by the Underwriters Laboratory. A UL label on a cord means only the cord has been tested. The PTI "Safety Seal" indicates approval by the Power Tool Institute. Approved tools are equipped with three-wire, grounding-type cords and plugs or with double insulation (Fig. 38). Buy either type if your shop has grounding-type circuits. If you have two-conductor circuits, with or without GFI protection, or for safety when you use it on a neighbor's unprotected circuits, buy double-insulated tools.

JDPX7774

Fig. 38 — Look for These Symbols when You Buy Power Tools

NOTE: DO NOT assume that a 3-prong plug is grounded. It may not be.

3. Avoid the use of grounding adapters (Fig. 39). You can buy adapters for plugging grounding-type plugs into two-conductor circuits. These are not recommended, and their use is prohibited in Canada by the Canadian Electrical Code. They are dangerous because two-conductor circuits don't have a grounding conductor to connect to the "pigtail" of the adapter. If you must use an adapter, have a competent electrician install a separate grounding conductor to the outlet to adequately ground the adapter.

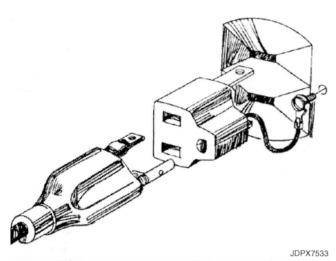

JDPX7533

Fig. 39 — Adapters Are Not Recommended. If Adapters Must Be Used, Install a Separate Grounding Conductor.

4. Inspect extension cords regularly. They have the same shock hazard as power tools. Keep them away from sharp objects, heat, oil, and solvents that can damage insulation. Do not patch a damaged cord — shorten it or get a new one. And use extension cords of adequate capacity. Undersized cords cause a loss of voltage (electrical pressure). Loss of voltage within the cord also makes the extension cord heat, possibly causing a fire. To determine the capacity needed for each power tool, check the nameplate for its ampere rating, and then refer to Fig. 40 to determine the recommended conductor size. Use extension cords for temporary connections only. Extension cords deteriorate, and are unsafe in permanent installations.

EXTENSION CORD SIZES

Ampere Rating (On Nameplate)	0 to 2.0	21.0 to 3.4	3.5 to 5.0	5.10 to 7.0	7.10 to 12.0	12.1 to 16.0
Extension Cable Length	Wire Size (American Wire Gauge)					
25 ft (7.6 m)	18	18	18	18	16	14
50 ft (15 m)	18	18	18	16	14	12
75 ft (23 m)	18	18	16	14	12	10
100 ft (30 m)	18	16	14	12	10	—
150 ft (46 m)	16	14	12	12	—	—
200 ft (60 m)	16	14	12	10	—	—

Fig. 40 — Use Extension Cords of Adequate Capacity to Prevent Overheating and a Possible Fire

5. Don't abuse tools: This may destroy the insulation on the conductors inside the tool. Don't drop power tools, throw them around, or pick them up by pulling on the power cord. Avoid overheating. And when tools become hot from continuous use or from temporary overloads, stop and let them cool.

Drill Presses

Three types of accidents are common with drill presses:

- A drill breaks and metal fragments are thrown into the operator's eyes

- Clothing is caught by a revolving drill or chuck

- A workpiece is caught and spun around with the drill, tearing skin from the operator's hands

Avoid injuries by following these suggestions:

1. Wear eye protection: Drilled-out chips or fragments from a broken drill could be flung through the air and injure your eyes (Fig. 41).

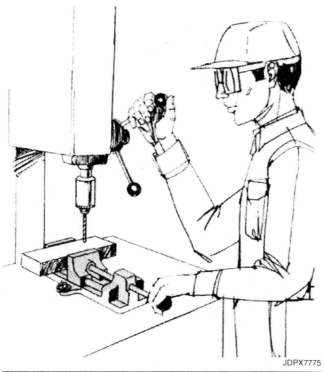

Fig. 41 — Wear Eye Protection and Hold Small Workpieces in a Drill Press Vise

2. Prevent drill breakage: Use sharp, straight drills; discard bent drills; and sharpen those that are dull or chipped. Mark the locations of holes to be drilled with a center punch to keep the drill from wandering when starting to drill (Fig. 42). Support the workpiece so it can't move or tip after you start drilling. Don't force the drill. If it doesn't cut, it is dull. Or you may need to drill a pilot hole before using a large drill. Finally, relieve the pressure when the drill starts to cut through. At this time, the drill may catch on the workpiece, breaking the drill, or spinning the workpiece if it's not held securely.

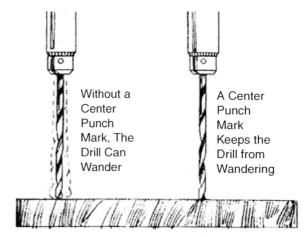

Without a Center Punch Mark, The Drill Can Wander

A Center Punch Mark Keeps the Drill from Wandering

Fig. 42 — Mark the Location of Holes with a Center Punch to Keep the Drill from Wandering

3. Remove the key from the chuck before switching on the motor: Wear close-fitting clothes and keep your sleeves buttoned. Loose clothing can get caught by a revolving chuck or drill. Tie or otherwise secure long hair to avoid entanglement, or wear it short.

4. Avoid injuries from workpieces: Clamp small pieces to the table or hold them with a vise, wrench, or pliers. Turn the motor off immediately if the workpiece starts to spin. Don't try to catch it — it could gash your fingers or hands. Remove metal chips and spirals with a wooden stick or small brush. Don't let spirals spin around with a revolving drill (Fig. 43).

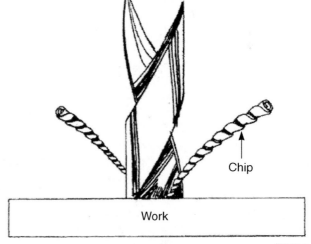

Chip

Work

Fig. 43 — Don't Let Spirals Spin Around a Revolving Drill; Remove Them with a Wooden Stick or Small Brush

Portable Drills

In addition to the precautions for drill presses, keep these in mind when using portable drills:

1. Keep a firm grip on the drill with both hands: Drills of 1/2-inch (12.7 mm) capacity or larger could throw you off balance into a fall.

2. Pull the plug when changing drills: If the switch is accidentally turned on, the chuck, key, or drill could tear skin from your fingers or hands (Fig. 44).

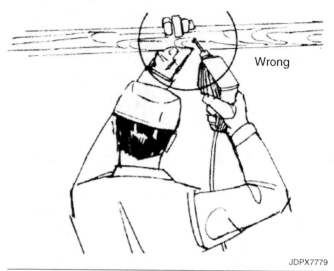

Fig. 45 — Never Drill Toward Your Hand or Any Other Part of Your Body

6. Wear eye protection so flying debris can't enter your eyes.

Fig. 44 — Pull the Plug before Changing Any Bit or Cutter in a Power Tool

3. Know where the drill will be going to be sure it doesn't penetrate hydraulic or electric lines or other hazardous areas.

4. Don't lock the switch in the on position: Use the lock only when the portable drill is mounted in a stand.

5. Never hold small work pieces in your hand: The drill is very likely to catch and spin them around. There is also the danger of drilling through a workpiece into your hand. Never put your hand behind the work in line with the drill (Fig. 45). The drill could go through the work and severely puncture your hand.

Stationary Grinders

There are several hazards associated with using grinders:

- If a grinding wheel explodes at high speeds, shattered pieces of it could fly into your face.

- If your hands touch the wheel, you will lose skin and flesh.

- If the workpiece gets very hot, your fingers could be burned.

- Flying particles can damage your eyes if you don't wear safety glasses.

- The tool can jam between the tool rest and the wheel if the tool rest is not against the wheel properly.

Protect yourself from these dangers:

1. Always wear eye protection (Fig. 46). Protect your eyes even if your grinder is equipped with a shatterproof eye shield.

2. Keep shields in place. The eye shield and the wheel shield are both needed to protect you from wheel fragments if the wheel breaks or shatters at high speed.

3. Check for a defective wheel before installing a new one. Tap the grinding wheel gently with a light metal object. A clear ring indicates a sound wheel. No ring indicates a defective wheel, and it should not be used.

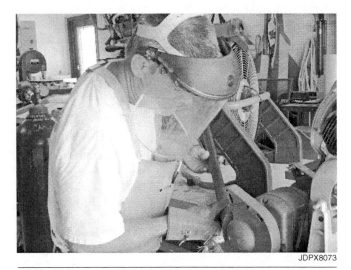

JDPX8073

Fig. 46 — Wear Eye Protection when Grinding

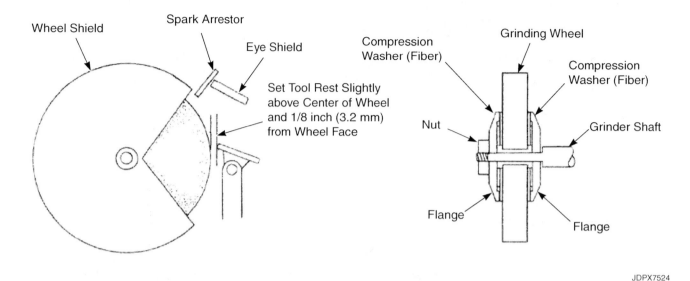

Fig. 47 — Install Grinding Wheels and Adjust Tool Rest Properly. Keep All Shields in Place.

4. Use compression washers and flanges on each side of the wheel (Fig. 47). Make sure the size of the arbor hole in the wheel matches the diameter of the grinder shaft, If not obtain and install bushings of proper size.

5. Make sure that the speed of your grinder doesn't exceed the recommended speed for the wheel. Grinder speed can be determined from the motor nameplate. The maximum recommended speed for the wheel is indicated on the label glued to the side of the wheel.

6. Set the tool rests slightly above center and 1/8 inch (3.2 mm) from the face of the grinding wheel. This position will help prevent thin work pieces and your fingers from getting wedged between the tool rest and the grinding wheel.

7. When starting the grinder, stand to one side of the wheel, turn on the switch, and let it run for a full minute before doing any grinding. Then grind with a light pressure, gradually, until the wheel warms up. Cold wheels may shatter.

8. Grind only on the face of the wheel. Side pressure may break the wheel if it's not specifically designed for side-pressure grinding.

9. Protect your fingers and hands. Never adjust the tool rests when the wheel is turning. Hold small pieces to be ground with pliers or a locking wrench. Position work pieces on the tool rest to prevent them from getting wedged between the tool rest and wheel.

10. Grind with moderate pressure. Forcing the work against the wheel heats the workpiece quickly, wears the grinding wheel out of round, and increases the chance that your fingers may slip onto the wheel. Grind with only moderate pressure, and dip the workpiece in water frequently to keep it cool. To eliminate the need for applying the workpiece against the wheel with more than moderate pressure, keep the wheel sharp and true by dressing it when needed.

Portable Grinders

Portable grinders are difficult to handle because of their size and weight. Extra care is needed to avoid injury and to protect the grinding wheel from damage. When using portable grinders, observe these precautions in addition to those listed for stationary grinders:

1. Hold the grinder firmly with both hands. And hold the grinder or position yourself to keep the stream of sparks and dust directed away from your body and away from persons who may be nearby (Fig. 48).

Fig. 48 — Hold the Portable Grinder Firmly with Both Hands

2. Before starting to grind, make sure that everyone within range, including you, is wearing eye protection.

3. Let the grinder come to a complete stop before laying it down.

4. Guard against blows to the wheel, either from dropping the grinder, from other shop tools, or by engaging the wheel too quickly or abruptly to the work.

Power Brushes

Hold Workpiece Angled in
Direction of Brush Rotation

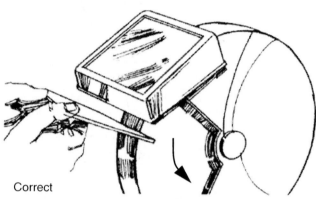 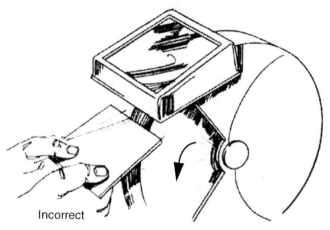

Correct   Incorrect

JDPX7780

Fig. 49 — Keep Workpiece Angled in the Direction of Wire Brush Flotation

To prevent accidents when using wire brush wheels, use these safe practices:

1. Follow grinding safety rules. The most important are: providing eye protection, using flanges to mount the brush, and setting the tool rest properly if one is used.

2. Hold the workpiece at the propel angle. Hold it with both hands at or below the horizontal center of the brush, and angled as shown in Fig. 49. Don't push the edge of the workpiece upward against the direction of wheel rotation. If you do, the wheel could jerk the piece out of your hands and cause an injury.

3. Let the brush tips do the work. Forcing the workpiece against the brush increases wire breakage and the chances of snagging the work. Force doesn't make the wheel clean faster — it merely bends the wires.

4. Hold small pieces with pliers or a locking wrench. This will save your skin if the workpiece catches or your hands slip against the brush.

Welding, Cutting, and Heating

Whether you are welding, heating, or cutting with an electric arc or oxyacetylene welder, take these special precautions:

1. Read the welding equipment operator's manual. Give special attention to the safety instructions.

2. Remove paint, paint stripper, or solvent before welding, cutting, or heating. Hazardous fumes can be generated when such substances are heated by welding, cutting, soldering, or heating. Do the work outside or in a well-ventilated area. If you use solvent or paint stripper, remove it with soap before heating the materials by welding, cutting, or heating.

3. Remove solvent or stripper containers or other flammable materials from the area and allow fumes to disperse at least 15 minutes before welding, cutting, or heating.

4. Avoid heating near any pressurized fluid lines such as hydraulic systems. The heat from welding or soldering operations can generate flammable spray which could severely burn the person doing the welding or persons nearby. Also. pressurized fluid lines can be accidentally cut when heat goes beyond the immediate flame or arc area.

Arc Welders

Always read your welding equipment operation manual. and follow directions.

The hazards of arc welding include intense heat, the brilliance of the arc, fumes, and working with a powerful electric welding current. Protect yourself from these hazards in the following ways:

1. Wear a helmet. Your eyes, face, and neck need protection from the burning rays of the arc and from the splatter of molten metal and slag. To protect your vision, make sure your helmet has a colored lens with at least a No. 10 shade. Lighter shades, indicated by lower numbers, will not protect your eyes from the harmful rays. Never strike an arc before your helmet is in place. And never look at the arc from any distance with naked eyes while another person is welding, not even for an instant! You could be permanently blinded even though you may feel no pain.

NOTE: *A No. 10 shade will protect your eyes adequately when welding with 200 amperes or less. Use darker shades when using higher currents. Refer to your welding instruction manual.*

2. Protect yourself from burns (Fig. 50). Always wear leather or heat-resistant gloves. Wear high-topped shoes to prevent leg and ankle burns. Button your shirt and collar, and turn down cuffs. Wear a skull cap under your helmet to protect your scalp and hair from welding splatter and hot fragments of flux when chipping the weld. Be careful where you stand. Recently welded or cut metal can melt your shoes.

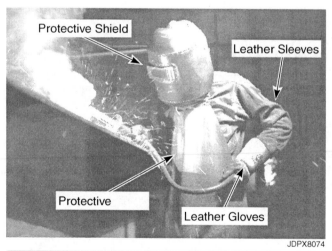
Protective Shield
Leather Sleeves
Protective
Leather Gloves
JDPX8074

Fig. 50 — Wear Protective Clothing To Prevent Burn When Arc Welding

3. Protect others. Warn others nearby when you are ready to strike an arc. Be sure their helmets are in place if they intend to watch (Fig. 51). When finished, don't leave pieces of hot metal or exposed and hot finished welds where they can be accidentally touched by others. Dispose of hot electrode stubs in a metal container — never drop them on the floor. They can burn if someone steps on them, and they can puncture tires.

JDPX7653

Fig. 51 — Never Expose Naked Eyes to a Welding Arc At Any Distance

4. Prevent fires and explosions. Before welding, clear away all combustible materials. Never weld on barrels, tanks, or other containers that once held flammable materials such as fertilizers or fuels. Vapors remaining in the container may explode. Let a professional welder tackle these jobs.

5. Provide ventilation. Clear away welding smoke by opening doors and windows or by switching on an exhaust fan. Avoid breathing welding fumes, especially from metals that produce toxic fumes (like zinc used to produce galvanized steel). Avoid breathing welding fumes from containers that have been cleaned with chlorinated hydrocarbons.

6. Protect your eyes and face at all times when chipping slag. Never chip slag when your eyes or those of others nearby are not protected by goggles, an eye-shield, or the clear lens of a welding helmet (Fig. 52). Fragments of hot slag burn. If they hit the eye, medical attention will be required to remove slag, and blindness may result. Remember that the risk of permanent eye injury is so great that you should never chip slag from a weld without protecting your eyes.

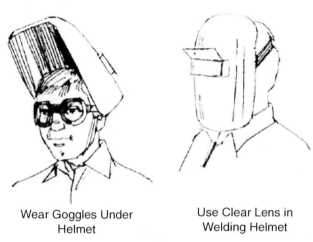

Wear Goggles Under Helmet   Use Clear Lens in Welding Helmet

JDPX7781

Fig. 52 — Protect Your Eyes while Chipping Slag

7. Guard against severe shock or electrocution. Weld in a dry location. Don't change electrodes with bare, sweaty hands. Wear dry gloves. Do not weld in a damp location. Remember that water is a good conductor of electricity, and that it increases the conductivity of your skin. To avoid a shock when welding, stay dry.

8. Keep your equipment in good condition. Protect welding cables from damage and keep all connections tight. Inspect them frequently for damaged insulation, frayed conductors, loose connections, and broken electrode holders or grounding clamps. Keep them in good repair.

9. Treat eye and skin burns promptly. Don't let any arc flash your eyes receive go unattended. If you receive a flash, get medical help right away. If you are fortunate, no lasting harm may have been done, but medical help can relieve the pain that usually appears several hours after the flash has been received.

10. Weld outside (Fig. 53). If possible, provide an outlet for your welder near the large service door of your shop. The cost will be repaid in safety and convenience. Fire hazards are minimized, there is better ventilation, and long welding cables are not needed to reach large pieces of equipment that can't be brought inside.

Fig. 53 — Provide an Outlet for Your Welder Near the Large Service Door so You Can Weld Outside

Oxyacetylene Welders Basic Safety Information

Follow operating recommendations carefully. Emergency situations resulting from misuse can be painful, costly, and disastrous.

1. Safeguard the fuel supply. Keep cylinders chained in an upright position (Fig. 54). Don't let them drop or fall. When stored, keep the caps in place to protect the valves from damage. If the valve assembly of a tank full of oxygen is broken off, for example, by tipping the tank over accidentally, the oxygen released could result in a fire or explosion.

Fig. 54 — Keep Cylinders Chained in an Upright Position

If a valve leaks. move the cylinder outside, keep all sources of fire away, and notify your supplier. Don't use cylinders as rollers, supports, or for any other purpose besides supplying oxygen or acetylene to your welding outfit. Never use the oxygen tank as a source of "compressed air." An oxygen-enriched atmosphere is extremely flammable.

2. Prevent equipment fires.

   a. Never use oil to lubricate any part of your welding equipment. Never weld or handle and adjust equipment with greasy hands. Oxygen and oil or grease is a perfect combination for a spontaneous fire. Never use compressed air to blow talc or clean out a new hose. Compressed air is treated with a lubricant.

   b. Test for leaks. Use a bucket of water to test for leaks in the hose. and a brush and soapy (non-petroleum base soap) water for leaky connections (Fig. 55). Keep hoses protected from hot and molten metal and your welding flame. Repair hose leaks by cutting the hose and inserting a splice. Don't try to repair a leaky hose with tape.

   c. Select gas pressures according to the size of the tip you're using. The operator's manual will give you this information. Remember that 15 psi (100 kPa) (1 bar) is the maximum safe pressure for acetylene. Normally, the pressure should never be adjusted this high. If there is a steady buildup of regulator pressure

   when the torch valves are closed, close tho cylinder valve and remove the regulator for repair.

   d. Open the oxygen cylinder valve to prevent leaks at the valve stem. The withdrawal rate from a cylinder of acetylene must not exceed one-seventh of the acetylene cylinder capacity. For example, a cylinder with 140 cubic feet (4 cubic meters) of capacity should not release more than 20 cubic feet (0.5 cubic meter) of acetylene per hour. Open the acetylene valve completely.

   e. In case of fire in the equipment, turn off the air cylinder valve first, and then the acetylene valve. Do not relight until the welding equipment has been inspected for damage and the cause of fire determined.

3. Wear goggles (Fig. 56). Use the specified goggles listed in your welding operation manual. When in doubt, use the darkest lens possible that allows you to clearly see the outline of your work.

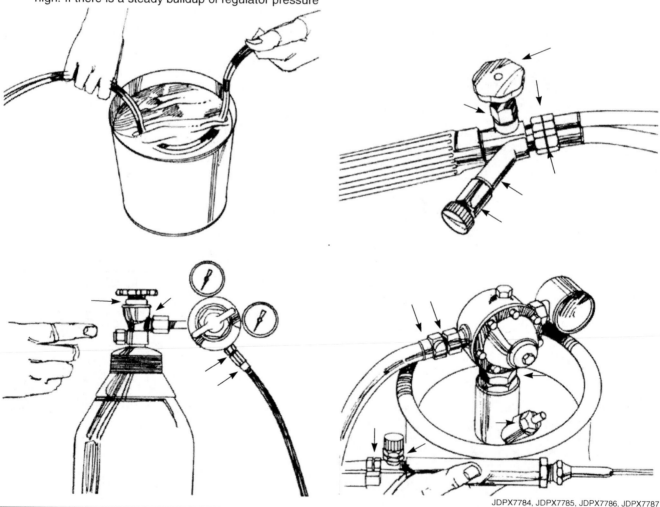

JDPX7784, JDPX7785, JDPX7786, JDPX7787

Fig. 55 — Use Soapy Water and Brush to Test for Leaks on Fittings, and a Pail of Water to Test Hose Leaks

Stand to One Side

Fig. 56 — Stand to One Side when Adjusting Regulators. Note Goggles Ready For Use.

4. Be sure the regulators are closed before opening the cylinder valves. If the regulators are not closed, and if the cylinder valves or the regulator diaphragms are damaged, gas pressure may shatter the glass on the regulator when you open the cylinder valves. Protect yourself from the possibility of being injured by shattered glass by following this procedure:

    a. Be sure the regulators are closed by turning each pressure adjusting screw counterclockwise until it fits loosely in the regulator.

    b. Stand to one side, away from the face of the regulator.

    c. Open each cylinder valve slowly. See step 2.d.

5. Prevent burns. Wear heat-resistant gloves to protect you from hot metal and the heat of the flame. Wear tight fitting, dark-colored, fire-resistant clothes that are free of oil and grease. Wear a long-sleeved shirt, and high-top boots or shoes with hard soles. Light the torch with the tip facing downward, away from your face and legs. Open the acetylene valve no more than a quarter turn to light the torch. Always use a friction-type lighter (Fig. 57). Lighting from a hot surface could result in a large flare-up, and the use of matches could burn your hands.

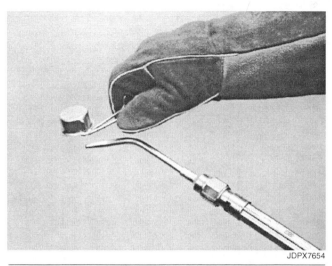

Fig. 57 — Light the Torch Safely with Tip Facing Downward

6. Prevent fires. Weld in an area clear of all combustibles. When cutting, remember that the shower of hot, molten metal can ignite materials some distance away.

7. Shut down the equipment properly. Close the cylinder valves and release the gas pressure from the hoses by opening and closing each torch valve one at a time. Then loosen the tension of each regulator pressure adjusting screw by turning it counterclockwise until it fits loosely in the regulator. If necessary, move your equipment to another location for safe storage.

8. Make sure your acetylene and oxygen supply lines are equipped with valves to help prevent flashbacks. See your welder supplier for more information

9. Never return a completely empty cylinder to your acetylene supplier. Internal cylinder characteristics make them difficult to refill when totally emptied. Consult your supplier for details.

10. Use ear protection.

NOTE: These are only basic safety procedures. Completely read and follow your operation manuals before you begin.

Jacks

Serious crushing accidents can result from the improper use of jacks. Each of the following recommendations is important. Be sure to observe them:

1. Don't overload the jack. Make sure any jack you use is strong enough to carry the load. If there is any doubt, check the rated capacity of the jack against the weight to be raised (Fig. 58). Jacks will lift some overload, but there is always the danger of immediate failure. Do not overload them.

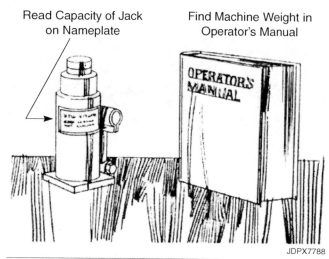

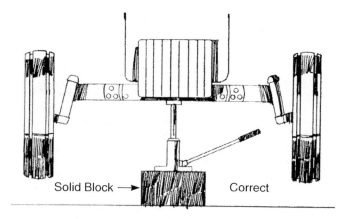

Read Capacity of Jack on Nameplate   Find Machine Weight in Operator's Manual

JDPX7788

Fig. 58 — Check Jack Capacity and Weight of Machine before Lifting

2. Lubricate jacks frequently, but only at points specified in the operating instructions. Some points may be designed to run dry.

3. Keep reservoirs of hydraulic jacks full of the recommended hydraulic jack oil. Don't use other fluids or dirty crankcase oil, as those may damage seals and valves within the jack. Take leaky jacks out of service and get them repaired or replaced.

4. Handle jacks carefully, Dropping or throwing them around may distort or crack the metal, and the jack may fail under load.

5. Position the jack properly. Select a point on the machine strong enough to carry the lifted weight. The lift point should be flat and level with the floor or ground supporting the base of the jack. Position the jack so the lift will be straight up and down.

6. If working on the ground, place a heavy block under the base of the jack. Select one long enough and wide enough to keep it from sinking, shifting, or tipping over when weight is applied. If the jack will not lift high enough, place additional blocking under the jack (Fig. 59). Don't put extenders between the jack saddle and the load.

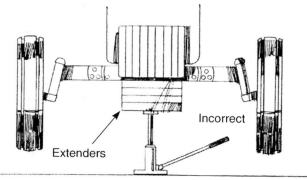

Solid Block →   Correct

Extenders   Incorrect

JDPX7789, JDPX7790

Fig. 59 — Place Heavy Blocking under the Jack

7. Stabilize the equipment. If the machine is self-propelled, place the transmission in gear or in park position, and set the brakes. Block at least one wheel that will remain on the ground. When lifting pull-type equipment, hitch it to a tractor drawbar to keep it in place.

8. Recheck the position of the jack after it has started to lift. If it starts to lean because the equipment rolls or shifts, lower the jack and reset it, and block the equipment more securely. If you do not, and continue to lift, the jack will push the equipment forward or backward, forcing it to roll off the jack and fall (Fig. 60).

Fig. 61 — *Keep Equipment as Level as Possible or the Jack May Slip*

Fig. 60 — *Avoid This Happening! Reset the Jack if it Starts to Lean, and Block the Equipment More Securely.*

9. Keep equipment level. Don't lift one side of a machine so high that there's danger of the jack saddle sliding on the equipment frame, permitting the equipment to fall (Fig. 61). Prevent this type of accident by observing these precautions:

   • Lift no higher than necessary.

   • Place a wooden shim between the equipment frame and the jack saddle to eliminate the metal-to-metal contact.

   • Keep equipment level by jacking or blocking each side, alternately or at the same time.

10. Beware of jack handles. The handles of some mechanically operated jacks can pop up and kick when the load is lifted or lowered. Stand to one side when jacking to avoid being struck by the handle (Fig. 62). Hold the handle firmly to prevent kicking, and do not release the handle from your hands when it is in a position where it can kick. Never straddle a jack handle with your legs. And remove the handle when it's not being used.

Fig. 62 — *Some Jack Handles May Kick. Stay Clear. Remove the Handle when Not Jacking.*

11. Support the load with blocks or stands. Never allow raised equipment to remain supported by jacks alone. Jacks can fail and tip permitting the equipment to fall unexpectedly. Place solid blocks or stands under the equipment immediately. Do not use concrete or cinder blocks. They will shatter under load.

   Two types of stands frequently used in high-lift applications are shown in Fig. 63. An adjustable stand is used with a hydraulic jack in Fig. 63. The support pipe serves as an extender for the jack.

When the equipment has been lifted to the desired height, a locking pin inserted through the support pipe converts the adjustable stand to a rigid one. Use stand rated for the specific job.

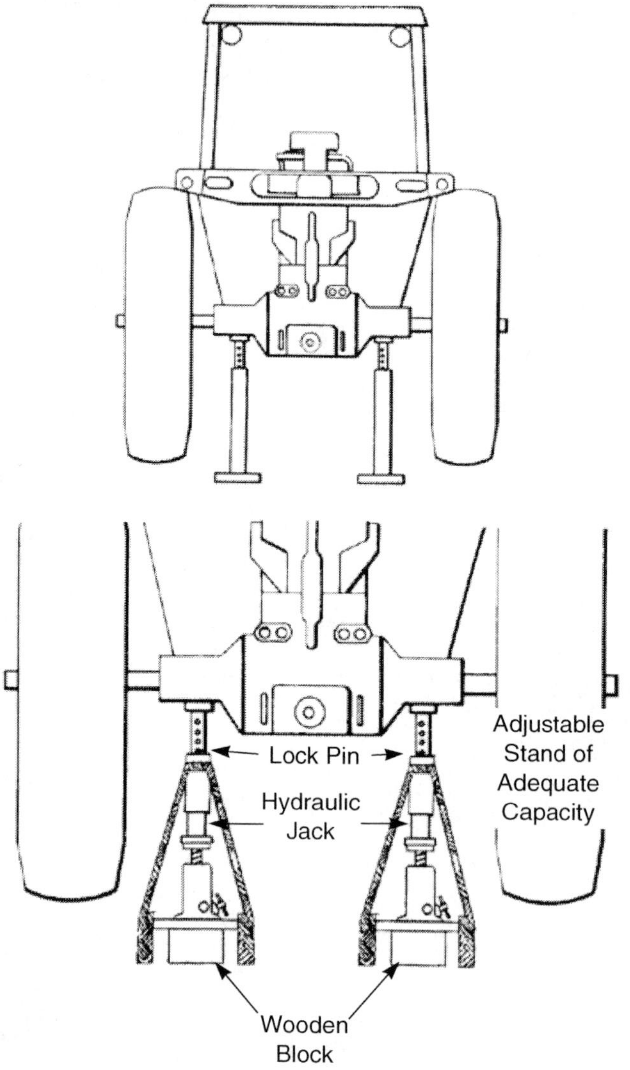

Fig. 63 — Support Equipment Immediately with Stands or Blocks After Jacking to the Desired Height

Lock Pin
Hydraulic Jack
Adjustable Stand of Adequate Capacity
Wooden Block
JDPX7794, JDPX7795

Hoists

Overloading the hoist and failing to rig lifting chains so they will not slip are the two most common causes of hoist-related accidents and injuries. When using hoisting equipment:

1. Use a chain hoist (Fig. 64). Chain is durable and has great strength for its diameter and weight. Don't use block and tackle and natural or synthetic fiber rope for lifting implements and machines. The fiber rope usually found on farms is not strong enough to carry the weight of heavy equipment, and you can never be sure how strong a piece of rope is. It weakens with age and abuse. Knots reduce its strength by one-half, and sharp bends made by looping the rope around some equipment frames break the fibers internally, making the rope unsafe for any use.

JDPX7620

Fig. 64 — Use a Chain Hoist. Do Not Use Rope.

2. Inspect chains often. Look for bent links, cracks, gouges, and extreme wear (Fig. 65). Don't repair chain by tying knots or by fastening with bolts and nuts, Obtain and use approved repair links available from hardware stores and farm equipment dealers. Ideally, damaged chain should be returned to the manufacturer for repair.

JDPX7796

Fig. 65 — This Chain Shows Extreme Wear and Should Be Replaced

3. Check the condition of the hooks. Replace those that are bent, sprung, or cracked. If in doubt about the condition of a hook, compare all dimensions of the questionable hook with a new one. If you find any differences in shape or size, replace the hook.

4. Know the capacity of your hoist and don't exceed it. Remember also that the beam or A-frame that carries the hoist must be strong enough to support the load. If in doubt about its strength, have a qualified person check it.

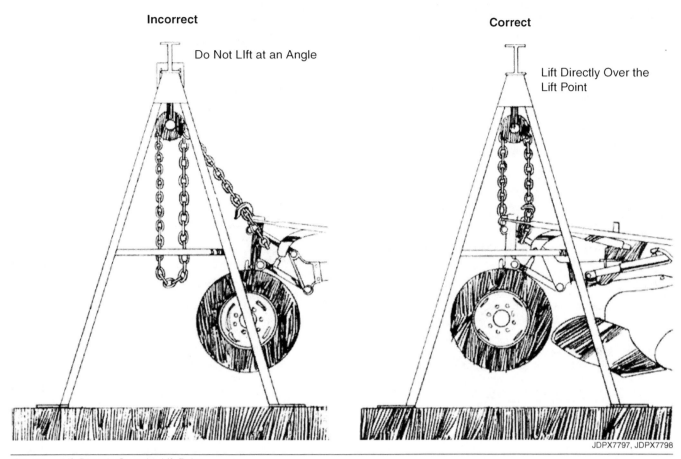

Incorrect

Do Not LIft at an Angle

Correct

Lift Directly Over the Lift Point

JDPX7797, JDPX7798

Fig. 66 — Lift Directly Over the Lift Point

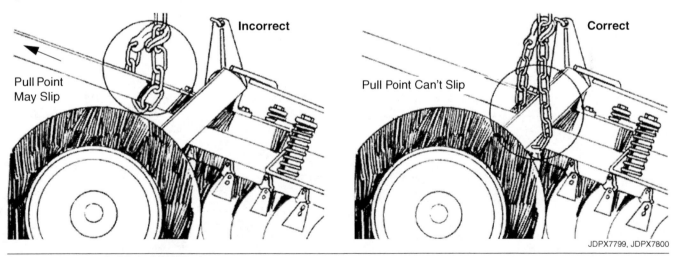

Incorrect

Pull Point May Slip

Correct

Pull Point Can't Slip

JDPX7799, JDPX7800

Fig. 67 — Keep the Pull Point from Slipping

5. Select a suitable lift point. Hook to a heavy section of the frame to prevent breaking parts or bending the frame out of alignment. Attach the hook directly over the point of lift on the implement or machine as some hoists, particularly A-frames, are not designed to withstand a sideways pull (Fig. 66). Attach the chain (or rig the lifting chains) to prevent the pull point from slipping (Fig. 67). if lifting an entire machine or implement, lift from a point where it will balance and not tip.

6. Protect yourself from injury while the equipment is being raised. Keep your hands away from pinch points when the lift chain tightens. Don't stand on or within the frame of any implement you are lifting. Make sure the implement doesn't swing and strike you as it leaves the ground. Have stands or blocks ready to use. Never get under equipment carried by a hoist unless it is securely supported on blocks or stands.

Chemicals and Cleaning Equipment

In service operations, cleaning is needed in order to:

- Keep dirt from entering the machine when parts are disassembled.

- Inspect parts for wear and damage.

- Install and adjust parts properly during reassembly.

We have already discussed the safe use of power brushes. Let's turn now to the safe use of solvents, steam cleaners, and high-pressure washers.

Solvents

Most solvents are toxic, caustic, and flammable. Be careful, therefore, to keep them from being taken internally, from burning the skin and eyes, and from catching on fire. Read the manufacturer's instructions and precautions before using any commercial cleaner or solvent.

Whenever possible, use a commercially available solvent (Fig. 68). Many types are available for general-purpose cleaning and for specific cleaning jobs. Always read and follow the manufacturer's instructions to get the best results and to be able to handle each product safely.

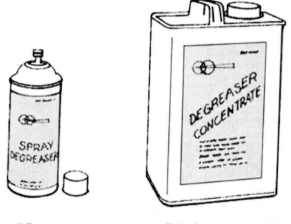

Small Pressure–Spray Can

Bulk Concentrate (Dilute before Application)

JDPX7801

Fig. 68 — Use Commercial Cleaning Solvents, Not Gasoline

Never clean with gasoline. It vaporizes at a rate sufficient to form a flammable mixture with air at temperatures as low as -50°F (-46°C). It can burn and it can explode (Fig. 69). Also, studies have linked cancer to contact with gasoline and other petroleum products and their vapors. If you can't use a commercial solvent, use diesel fuel or kerosene. These will burn, but not as easily or as explosively as gasoline.

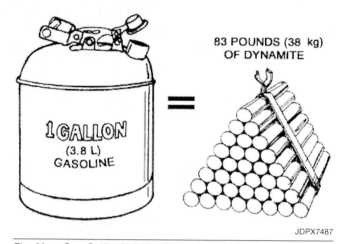

83 POUNDS (38 kg) OF DYNAMITE

JDPX7487

Fig. 69 — One Gallon (3.8 L) of Gasoline Mixed with the Right Amount o Air Can Have the Explosive Force of 83 Pounds (38 Kg) of Dynamite

Wear hydrocarbon-resistant gloves to avoid skin contact with petroleum products. If you do get gasoline or other petroleum products on your skin, wash immediately with soap and water or waterless hand cleaner. Also, promptly wash clothes, gloves, or shoes that become soaked with petroleum products to avoid prolonged skin contact with those products. Chemical cleaner safety practices include these:

1. Follow the manufacture's instructions. Different cleaners intended for the same purpose may be quite different in chemical composition. Read the label on each container you buy and follow instructions carefully.

2. Work in a well-ventilated area. Do the cleaning outdoors if possible. If you can't, provide ventilation by opening doors and windows.

3. Keep solvents away from sparks and flame. Don't let anyone smoke in the immediate area. Don't use solvents near heaters, sparks or open flames. Some solvent tanks, and tubs should be "grounded." Contact your solvent supplier for instructions on how to ground your solvent tank or tub.

4. Never heat solvents unless instructed to do so. And don't mix them, because one might vaporize more readily and act as a fuse to ignite the other.

5. Wipe up spills promptly. Keep soaked rags in closed metal containers and dispose of them promptly.

6. Store solvents in their original containers or in sealed metal cans properly labeled. In a sealed container, the solvent will be kept clean from further contamination, toxic fumes will be controlled, the fire hazard will be reduced, and there will be less likelihood of spillage. Never use open pans. Store solvents as flammable materials in closed storage away from workstations (Fig. 70).

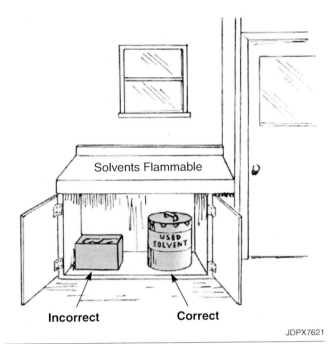

Incorrect   Correct

JDPX7621

Fig. 70 — Store Solvents in Sealed and Labeled Cans, Never in Open Pans; Use Closed Storage

7. If a commercially made parts washer is used, close the lid when you're finished cleaning. Never destroy the fusible link that closes the lid automatically in case of fire.

8. Protect your skin and eyes. Wear a face shield and rubber gloves when working with strong, concentrated cleaning solutions (Fig. 71). Some are caustic, and they can destroy the natural oils of the skin and burn it severely. Always read and follow the manufacturer's instructions carefully. Even when gloves are not called for, avoid long exposures to solvents and wash your hands when the job is done. Wear an apron if it's needed to keep your clothing dry.

NOTE: *If a solvent is splashed into your eyes, flush them thoroughly with water, keeping your eyelids open. Then get medical attention immediately. Some solvents may cause skin irritation or dizziness. If that happens, stop. Wash thoroughly with mild soap and water, and get some fresh air. If possible. use sol- vents in a location where a clean water source is available and ready for immediate use if needed.*

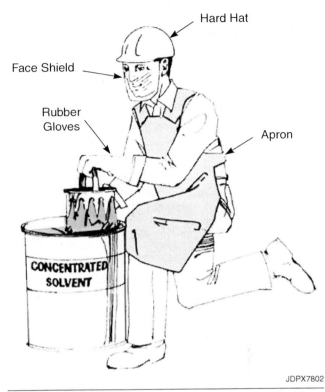

Hard Hat

Face Shield

Rubber Gloves

Apron

CONCENTRATED SOLVENT

JDPX7802

Fig. 71 — Wear a Face Shield and Rubber Gloves when Cleaning with Cleaning Solutions

9. Avoid accidental poisoning. Wash your hands and arms before eating or smoking. Keep all solutions in labeled containers. Never use empty containers, no matter how thoroughly cleaned, for carrying food or beverages. Keep poison containers sealed even when they are empty. Keep them properly stored out of the reach of children.

10. Be prepared for emergencies. Keep fire-fighting equipment near your cleaning area. Save the original containers for all solvents until they are completely used. In case of accidental poisoning, follow the instructions provided by the manufacturer immediately, and when going for medical aid, take the container with you so the chemical can be quickly identified.

High-Pressure Washers

High-pressure washers are useful for cleaning machines and buildings, removing paint, disinfecting, and even sand blasting. They involve several potential hazards discussed in Chapter 3, "Recognizing Common Machine Hazards." including electrical energy, hot liquid under pressure, chemicals, flying debris, and second-party hazards.

Before you install or use a high-pressure washer, read the operator's manual and the safety signs on the machines (Fig. 72). Follow them carefully. Some of the specific safety precautions are presented below:

Fig. 73 — Connect High-Pressure Washers Only to Properly Grounded Electrical Circuits

Fig. 72 — Study Safety Instructions in the Operator's Manual and Safety Signs on Your High-Pressure Washer

1. Avoid electric shock and electrocution. You will be working with electrically powered equipment around water. Don't risk electrocution. Connect the washer only to a properly grounded electrical circuit (Fig. 73), Always plug it into 3-wire grounded electrical outlets. It should also be protected by a ground-fault interrupter (GFI). A licensed electrician should check for grounding. Check local codes and ordinances and the electrical requirements in your operator's manual. It is best to have GFIs on all circuits for outdoor and shop use, even on a 3-wire grounded circuit. Do not spray on or near electrical power lines. outlets, switches, or bulbs.

2. Use the proper extension cords. Use of extension cords for high-pressure washers is NOT RECOMMENDED. If absolutely necessary, use only three-wire extension cords intended and labeled for outdoor use. Wires must be large enough to prevent excessive voltage drop. The receptacle should be protected by a GFI. Keep extension cords in good condition, and be sure all connections are kept dry. Refer to the operator's manual for electrical specifications.

3. Wear personal protective goggles, respirator, and clothing. Do the cleaning outside or with adequate ventilation. The high-pressure stream may get into your eyes or may drive particles, sludge, or chemicals into them. Wear safety goggles. If you use chemical solvents or remove chemical substances, wear a respirator. Gloves and protective clothing are also recommended. Wear all-rubber footwear to provide some electrical insulation.

4. Protect against injury from spray nozzle and high-pressure leaks. Spray from high-pressure spray nozzles can penetrate skin and flesh, causing serious injury similar to pinhole hydraulic leaks (Fig. 74). Don't put your hand over the nozzle or touch pinhole leaks. Any spray injected into the skin must be surgically removed immediately by a doctor familiar with this type of injury, or gangrene may result.

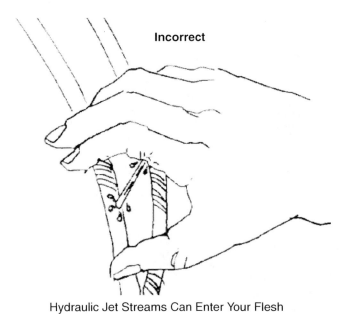

Incorrect

Hydraulic Jet Streams Can Enter Your Flesh

JDPX7554

Fig. 74 — A Jet Stream or Mist from Pinhole Hydraulic Leaks Can Penetrate Skin and Flesh. Don't Touch It.

5. Always direct spray away from your body and others. Warn others to stay away and watch for unexpected onlookers, Hold the wand securely so it won't whip violently, causing serious injury.

6. Use detergents and chemicals carefully. Read the labels and follow instructions. Some chemicals or detergents can give off harmful fumes or start an unpredictable chemical reaction when heated or mixed with another chemical on a sprayed surface.

    Never spray flammable liquids such as gasoline, fuel oil, alcohol, or naptha. Use only approved compounds. Wear safety goggles to protect eyes.

7. Sandblasting solution under pressure can penetrate skin and flesh. Always wear goggles, respirator, and protective clothes to guard against blasted sand.

8. Protect yourself and others from splatter and throwback. Hold the gun or wand at an angle to the surface so the spray will reflect away from you and others. Stand to one side when spraying into corners to avoid sludge and spray being thrown back.

Steam Cleaners

Safety practices for steam cleaners are similar to those for high-pressure washers, Read the operator's manual and safety signs on the machine. Following are some of the safety precautions to be observed:

1. Work outside or in a ventilated area. Steam must be ventilated away for good vision and to prevent shop tools and equipment from rusting.

2. Protect yourself and others from burns. Steam is extremely hot, Wear gloves if you light the burner by hand. Keep others away when cleaning. When shutting down, turn the burner off and keep the water running until all steam disappears from the nozzle in order to cool the equipment.

3. Keep the cleaner in good operating condition. Remember that tape does not make a safe repair for a steam gun or hose. Keep fuel connections tight. If the cleaner doesn't ignite readily, burn cleanly, or hold steam pressure within safe limits, have a qualified service person repair it.

Machinery Service Work

There are three requirements for safe service and maintenance work. We've outlined the first two: maintaining a hazard-free shop and using service tools and equipment safely. There is one more: using safe work procedures when servicing the equipment itself.

Let's look first at general recommendations that apply to all implements and machines. Then we will discuss precautions to observe when working on specific components and systems. Because we can't cover all components and systems, be sure to have operator's and service manuals on hand for all jobs you intend to do, and read them before tackling any job.

General Recommendations

1. Disengage the power and stop the machine before servicing (Fig. 75). This is the most basic rule of service and maintenance safety. Don't try to save time by keeping the machine running. You may be tempted to do this, thinking that, with care, you can do the job safely.

    A slip, a caught piece of clothing, or one false move could cut, chew, twist, or pull you into a machine before you realize what has happened. Do not clean, unplug, lubricate, adjust, or repair any machine while it's running, unless it's specifically recommended in the service or operator's manual. Lock out the ignition and put a warning sign over the ignition that tells everyone that you are working on the machine.

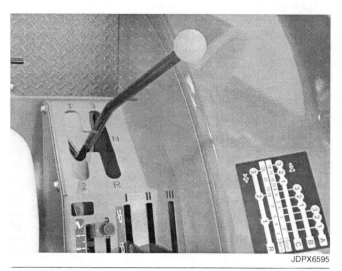

Fig. 75 — Disengage Power, Turn Off the Engine, and Remove the Key before Servicing Any Machine

2. Support equipment on blocks or stands. Don't rely on jacks, hoists, or hydraulic cylinders. They are designed for lifting only, and they can fail without warning. Never take the chance of being crushed by a machine or implement which could fall unexpectedly. Instead, protect yourself by following one or more of these procedures:

• Lower all equipment to the ground.

• Use support stands if provided for the equipment,

• Engage safety locks if the hydraulic cylinders are equipped with them.

• Support the equipment on metal stands or solid blocks (Fig. 76).

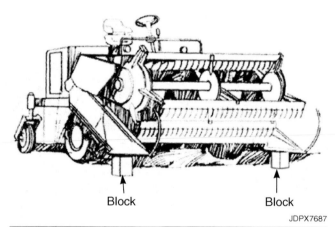

Fig. 76 — Support Equipment On Adequate Blocks Or Stands

3. Follow all recommendations outlined in operator or instruction manuals and sheets. Keep these materials in your shop or with your equipment for future use.

4. Use containers to store removed parts. Cans, small boxes, or metal pans work well. Having a place to put these parts will help you work in an orderly manner in an organized work area. Keep larger parts out of the way so you won't be stumbling over them.

5. Provide ventilation. Engine exhaust fumes contain carbon monoxide, a colorless, odorless, and deadly gas. Never run engines in buildings unless there's ample ventilation to carry exhaust fumes away (Fig. 77). If doors and windows don't provide sufficient ventilation, use also flexible metal tubing to help carry fumes outside. If you plan to run engines in your shop frequently, install an exhaust system equipped with a suction fan to pull the fumes through the tubing and safely outdoors.

Fig. 77 — Never Leave Engines Running Indoors without Adequate Ventilation

6. Protect yourself from sharp edges and protruding parts (Fig. 78). Knives, disk and coulter blades, sheet metal, and other sharp-edged parts can skin knuckles and cut severely. Wear gloves. Cover sharp cages with tape, rags, or wooden guards. Grasp all sharp-edged parts away from the cutting edge and hold firmly enough to keep them from slipping in your hand. Never push or pull toward a sharp edge without protecting yourself in some way.

Fig. 78 — Protect Yourself from Sharp Objects and Protruding Parts

7. Clean up spilled oil, grease, or fuel. In addition to removing a fire hazard, you won't be tracking it all over the shop where it could cause a slip or fall.

8. Make sure coolant level is okay in an engine before plugging in an engine block heater. A fire could result if coolant is low.

9. Use jacks and hoists to handle heavy components. Struggling with awkward and heavy parts may injure your back, or you might drop them, crushing your hands or feet. Before removing heavy parts, estimate their weight and be prepared to handle them safely.

10. Protect your helpers. Warn helpers of your intention to do anything that might affect their safety. Make sure they also warn you. Turning a shaft, for example, might crush fingers on the other side of the machine. Never start an engine, engage power, or raise or lower a machine without warning your helpers in advance. Don't proceed until you know where they are and that they know where you are.

11. Store attachments and large items carefully. Wheels, tires, and loaders and other attachments can fall on someone. Store them so they won't fall. Keep children and bystanders away from the area.

12. Make sure the job is done right. Your safety and the reliability of your equipment depend upon how well you do your job. Follow the service procedures outlined in your operator's and service manuals. Observe all precautions. As you work, check each step as you complete it to make sure it's right. Finally, turn all moving parts over several times by hand if possible before engaging power. Then gradually engage power to the machine.

Cooling System

Engines are equipped with pressure radiator caps. These caps maintain pressure of 6 to 7 psi (41 to 48 kPa) in the cooling system, which raises the boiling point of the coolant to about 230°F (110°C). If you remove the radiator cap when the coolant is hot, steam and violently boiling coolant may gush out on your hands and body.

When checking the coolant level:

1. Let the engine and radiator cool (Fig. 79).

2. Protect yourself if you must remove the cap from a hot radiator. Fold several thicknesses of a large rag and place it over the cap. Stand to one side and turn your face away. Then, before removing the cap, loosen it slightly and wait for all pressure to escape.

When handling antifreeze, keep these points in mind. Both types commonly used, alcohol-based antifreeze and ethylene glycol (permanent type), are poisonous. Ethylene glycol antifreeze is flammable in its concentrated form. And alcohol-based antifreeze will burn. For those reasons:

a. Don't spill ethylene glycol antifreeze on hot surfaces such as engine manifolds.

b. Keep alcohol-based antifreeze away from sparks and flame.

c. Keep both types of antifreeze away from children.

d. Never put foods or beverages in empty antifreeze containers.

Components and Systems

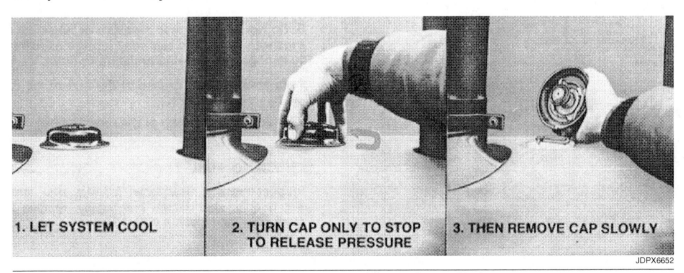

JDPX6652

Fig. 79 — If Possible, Let the System Cool before Removing the Radiator Cap. Then Remove Slowly.

Electrical System

Avoid these hazards when servicing an electrical system:

* Battery explosions

* Being burned by the battery electrolyte

* Electrical shock

* Bypass start hazard

NOTE: When servicing the electrical system, always follow the steps outlined in your operator's or service manual to avoid damaging the electrical system. Lock out or disconnect the electrical power source from the electrical system before you begin your service procedure.

Battery Explosions

When charging and discharging, a lead-acid storage battery generates hydrogen and oxygen gas. Hydrogen will burn, and is very explosive in the presence of oxygen. A spark or flame near the battery could ignite these gases, rupturing the battery case and throwing acid all over.

To prevent battery explosions:

1. Maintain the electrolyte at the recommended level. Check this level frequently. When properly maintained, less space will be available in the battery for gases to accumulate (Fig. 80). Refer to your operator's manual. Put only water or battery electrolyte in the battery.

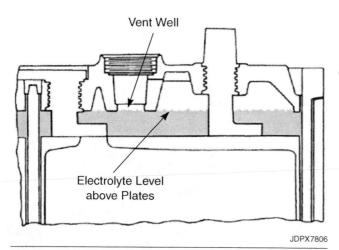

Vent Well

Electrolyte Level above Plates

JDPX7806

Fig. 80 — Keep Battery Electrolyte at the Proper Level to Reduce the Volume of Accumulated Hydrogen Gas

2. Use a flashlight to check the electrolyte level, Never use a match or lighter, because these could set off an explosion.

3. Do not short across the battery terminals. If the battery isn't completely dead, the resulting spark may set off an explosion if hydrogen gas is present.

4. Remove and replace battery clamps in the right order (Fig. 81). This is very important. If your wrench touches the ungrounded battery post and the tractor chassis at the same time, the heavy flow of current will produce a dangerous spark. To prevent this, follow these rules:

   a. Battery removal: Disconnect the grounded battery clamp first. Note: Some systems may be positive ground. Make sure you know which post is grounded.

   b. Battery installation: Connect the grounded battery clamp last.

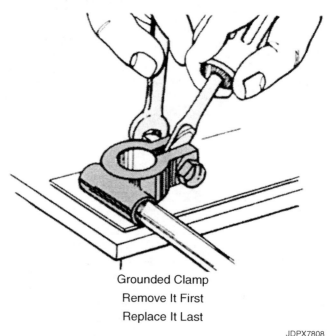

Grounded Clamp

Remove It First

Replace It Last

JDPX7808

Fig. 81 — Remove and Replace Battery Clamps in the Correct Order

5. Prevent sparks from battery charger leads. Turn the battery charger off or pull the power cord before connecting or disconnecting charger leads to battery posts (Fig. 82). If you don't, the current that is flowing to the leads will spark at the battery posts. These sparks could ignite the explosive hydrogen gas that is always present when a battery is being charged.

JDPX7807

Fig. 82 — Turn Charger Off before Disconnecting Clamps from Battery Posts

Connecting a Booster Battery

Improper jump-starting of a dead battery can be dangerous. Follow these procedures, or those in your operator's manual, when jump-starting from a booster battery.

1. Remove cell caps (if so equipped).

2. Check for a frozen battery. Never attempt to jump-start a battery with ice in the cells.

3. Be sure that booster battery and dead battery are of the same voltage.

4. Turn off accessories and ignition in both vehicles.

5. Place gearshift of both vehicles in neutral or park position and set parking brake. Make sure vehicles do not touch each other.

6. Add electrolyte if low. Cover the vent holes with a damp cloth, or if caps are safety vent type, replace the caps before attaching jumper cables.

7. Attach one end of one jumper cable to the booster battery positive terminal. Attach other end of the same cable to the positive terminal of the dead battery. Make sure of good, metal-to-metal contact between cable ends and terminals.

8. Attach one end of the other cable to the booster battery negative terminal. Make sure of good, metal-to-metal contact between the cable end and the battery terminal.

CAUTION: Never allow ends of the two cables to touch while attached to batteries.

9. Connect other end of second cable to engine block or vehicle metal frame below dead battery and as far away

from dead battery as possible. That way, if a spark should occur at this connection: it would not ignite hydrogen gas that may be present above dead battery.

10. Try to start the disabled vehicle. Do not engage the starter for more than 30 seconds or starter may overheat and booster battery will he drained of power. If vehicle with dead battery will not start, start the other vehicle and let it run for a few minutes with cables attached. Try to start second vehicle again.

11. Remove cables in exactly the reverse order from installation. Replace vent caps.

Acid Burns

Battery electrolyte is approximately 36 percent full-strength sulfuric acid and 64 percent water. Even though it's diluted, it is strong enough to burn skin, eat holes in clothing, and cause blindness if splashed into eyes. Fill new batteries with electrolyte in a well-ventilated area, wear eye protection and rubber gloves, and avoid breathing any fumes from the battery when the electrolyte is added. Avoid spilling or dripping electrolyte when using a hydrometer to check specific gravity readings (Fig. 83).

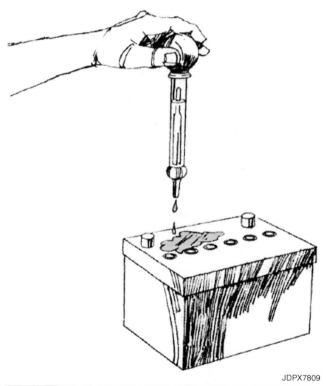

JDPX7809

Fig. 83 — Avoid Dripping Electrolyte when Reading Specific Gravity

If you spill acid on yourself, flush your skin immediately with lots of water for several minutes. Apply baking soda or lime to help neutralize the acid. If acid gets in your eyes, force the lids open and flood the eyes with running water for 10 to 15 minutes. See a doctor at once, and don't use medication or eye drops unless prescribed by the doctor.

Electrical Shock

The voltage in the secondary circuit of an ignition system may exceed 25,000 volts. For this reason, don't touch spark plug terminals, spark plug cables, or the coil-to-distributor high-tension cable when the ignition switch is turned on or the engine is running (Fig. 84). The cable insulation should protect you, but it could be defective.

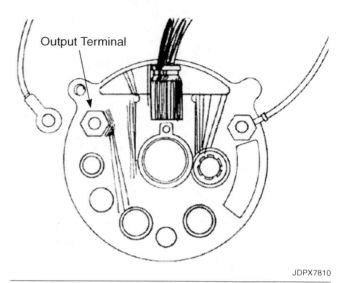

Output Terminal

JDPX7810

Fig. 85 — Don't Run the Engine when the Wire Has Been Disconnected from the Alternator or Generator Output Terminal

Hydraulic System

Damaged or improperly maintained hydraulic systems are hazardous in four ways:

- High-pressure oil could escape and hit you in the eyes or penetrate your skin and flesh.

- High-temperature oil could burn you badly.

- A raised implement could fall and crush you.

- A loosened component could be flung off as a projectile.

To prevent these hazards, protect the hydraulic system from damage and keep it in good repair.

1. Follow all maintenance procedures outlined in your operator's manual. Change oil and filters on schedule. Keep dirt, moisture, and air out of the system by following the maintenance recommendations in your operator's manual. Maintain the recommended level of oil in the reservoir by adding clean oil when necessary. Keep the fins of oil coolers clean.

2. Relieve system pressure before beginning maintenance work, (Fig. 86). Lower equipment to the ground, turn off the engine, and relieve the pressure as instructed in the operator's manual. This will help protect you from being crushed by falling equipment, and from being struck by high-pressure oil when disconnecting lines, hoses, and fittings. When disconnecting hydraulic fittings, loosen them slowly to relieve residual pressure. Never service the hydraulic system while the engine is running unless the service instructions indicate that you should.

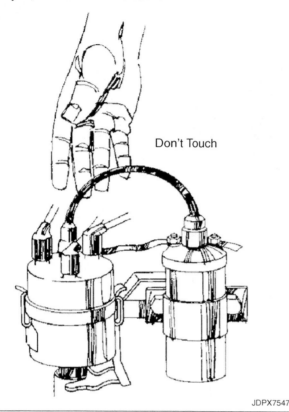

Don't Touch

JDPX7547

Fig. 84 — Never Touch Wires in the Secondary Circuit or Spark Plug Terminals while the Ignition Switch Is Turned On

Never run an engine when the wire connected to the output terminal of an alternator or generator is disconnected (Fig. 85). If you do, and if you touch the terminal, you could receive a severe shock. When the battery wire is disconnected, the voltage can go dangerously high, and it may also damage the generator, alternator, regulator, or wiring harness.

JDPX7811

Fig. 86 — Relieve All Hydraulic System Pressure before Starting Service Work

NOTE: *Equipment that has a hydraulic accumulator installed for pressure control requires special precautions because there is a pressure reserve which will need to be relieved to be able to perform service work on the system. Refer to the operator's manual for proper procedures for relieving hydraulic pressure.*

3. Avoid damage to lines and hoses. Keep sharp edges, sharp points, and heavy objects from striking or rubbing against lines and hoses. Rig hoses in clamps, avoiding twists and sharp bends. Allow slack for the changing distances between moving parts when installing new hoses. Replace damaged lines that could rupture under load (Fig. 87). Inspect hoses and lines frequently.

JDPX7622

Fig. 87 — Replace Damaged Hoses and Fittings

4. Maintain fittings. Keep fittings tight. Never reassemble fittings that are dirty, chipped, or distorted. Fittings must be clean and in perfect condition. Keep O-rings and other seals in place. They, too should be in perfect condition. Keep an assortment of O-rings in your shop so they are available when you need them. When tightening

fittings, use two wrenches to avoid twisting the lines, but do not over-tighten. Tighten the fittings just enough to prevent leaks (Fig. 88).

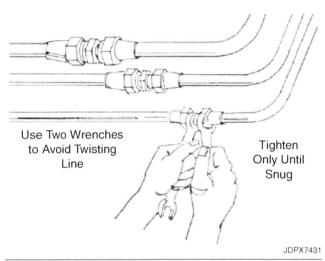

Use Two Wrenches to Avoid Twisting Line

Tighten Only Until Snug

JDPX7431

Fig. 88 — Tighten Fitting with Two Wrenches, but Only Enough to Prevent Leaks

5. Prevent burns. Let the hydraulic system cool before changing hydraulic filters or removing fittings or lines. Hot oil can burn you severely.

6. Replace parts with identical or approved replacements. Use replacements that have the same strength and capacity. Parts may look alike, but they may not be the same.

7. Avoid the hazards of pinhole leaks. Hydraulic fluid under pressure can penetrate skin and flesh and cause severe injury, including infection and gangrene (Fig. 89). If you see a jet stream or misty spray coming from a fitting or tubing, or just dripping fluid, don't touch it with your hand, not even with gloves on, To check for pinhole hydraulic leaks, wear eye protection, and use a piece of cardboard, paper, or wood to locate the leak or its source. The jet stream from a diesel fuel injector is just as dangerous as a pinhole hydraulic leak.

Relieve pressure before disconnecting hydraulic or diesel fuel lines, and tighten all connections securely before turning on the pressure again.

If ANY fluid is injected into the skin, it must be surgically removed within a few hours by a doctor familiar with this type of injury or gangrene may result. Gangrene often leads to amputation.

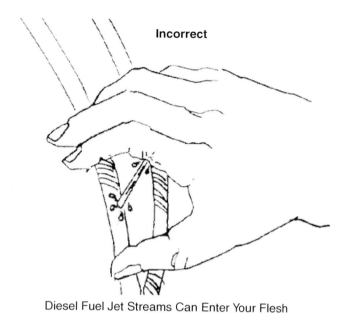

Incorrect

Diesel Fuel Jet Streams Can Enter Your Flesh

JDPX7554

Fig. 89 — A Jet Stream or Mist from Pinhole Diesel Fuel Leaks Can Penetrate Skin and Flesh. Don't Touch It.

Fuel System

Have all major fuel system service work done by your farm equipment dealer (Fig. 90).

JDPX7812

Fig. 90 — Take All Major Fuel System Work to Your Dealer

Unless you're qualified and have special service tools, never attempt to service diesel injection pumps or nozzles. In nearly all cases, when this type of service work is attempted by individuals without specialized training and service equipment, serious damage to the fuel system results. In addition, inexperienced personnel run the risk of injury from high-pressure oil.

Gasoline fuel systems may be serviced and adjusted by operators who have repair experience. However, service work is best left to trained service personnel.

When performing routine maintenance jobs, remember these general safety precautions:

1. Fire hazard is greatest with LP-gas and gasoline (Fig. 91). Diesel fuel and kerosene are less hazardous, unless spilled on a hot engine manifold. When fueling tractors, read and follow the precautions in Chapter 7, "Tractors and Implements."

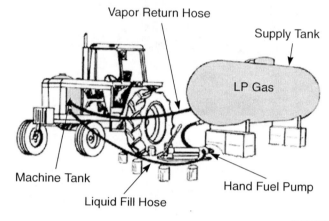

Vapor Return Hose

Supply Tank

LP Gas

Machine Tank

Liquid Fill Hose

Hand Fuel Pump

JDPX7813

Fig. 91 — Fueling with LP-Gas

2. Use an approved cleaner for dry element air filters. Get a package of filter element cleaner from your dealer and follow the instructions that come with the package. Don't use gasoline, fuel oil, or solvents. Gasoline is always too dangerous to use for cleaning. Fuel oil and solvents may destroy the fibers of the filter or leave an oily film.

3. Install the air filter before starting the engine (Fig. 92). The air filter is needed to protect the engine from dust and dirt drawn in with the air. If the engine backfires, the filter is needed to contain flame in the engine.

Fig. 92 — *Always Install the Air Filter before Starting the Engine*

4. Clean up spilled fuels. Turn off the fuel supply at the fuel tank when changing fuel filters and when cleaning sediment bowls. Promptly clean up fuel to reduce the fire hazard

5. Never adjust a carburetor when the tractor is in motion. Apply a load to the PTO instead. If you can't, hitch up and operate the tractor with a drawbar load between trial settings. Repeat your trial adjustments until the correct setting is obtained.

Working with LP-Gas Fuel

Read your operator's manual. Learn the procedures needed to work with LP equipment safely. If certain procedures are not explained in detail in the operator's manual, the manufacturer expects a dealer to do that service.

Restrict your maintenance operations to cleaning the fuel strainer and replacing valves and gauges. Never service the fuel tank except to install new valves or gauges. Don't repair valves or gauges or attempt to repair the converter.

If you use LP-gas equipment, follow these precautions carefully:

1. Refuel safely. Observe all precautions outlined in your operator's manual and in Chapter 7, "Tractors and Implements."

2. Do all service work outdoors. LP-gas is heavier than air. LP-gas will displace air and suffocate you. Moving air is needed to carry the gas away.

3. Don't place your hands in the path of the vapor stream when releasing pressure. The gas is so cold it will freeze flesh.

4. Empty the fuel tank before removing any line, valve, or gauge. If you can't run the engine to empty the tank, consult your fuel supplier for an approved method.

5. Don't store a tractor with a leaking LP-gas system in a building. The accumulated gas could be ignited easily and there is the suffocation danger (see above).

6. Watch for leaks. The appearance of frost at any point in the fuel system when the engine is not running indicates a leak. To test for leaks, use soapy water (non-petroleum-base soap) and a brush. Never use an open flame!

7. Don't use equipment that isn't approved for use with LP-gas. All equipment, including hose lines, should meet the safety standards of a recognized authority, such as the American Gas Association or the Underwriters Laboratory, Inc.

Working with Diesel Fuel

Since injection pumps and nozzles require special tools and equipment, restrict your service work to the low pressure side of the system: the fuel filters, sediment bowl, and fuel tank.

Air is left in the system when fuel lines and filters are drained. For this reason, you'll need to bleed the air out of the high-pressure lines. Normally. you can bleed half of the injection lines, and the others will bleed themselves after the engine has started. When bleeding the lines, loosen the injection line nuts only one turn to avoid excessive spray (Fig. 93). Use two wrenches to avoid bending or twisting the lines. After cranking the engine with the starter to remove the air, tighten the connections until they are snug and free of leaks.

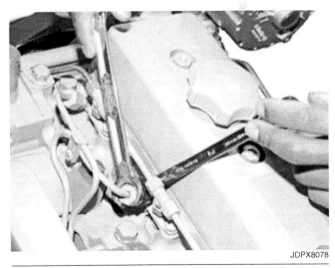

Fig. 93 — *When Bleeding Injection Lines at the Injectors, Loosen One Turn Only*

Before running the engine, be sure all fuel connections are tight. Diesel fuel escaping under high pressure can penetrate the skin (Fig. 94).

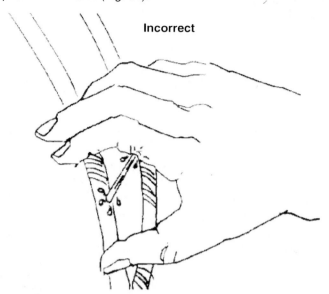

Incorrect

Diesel Fuel Jet Streams Can Enter Your Flesh

JDPX7554

Fig. 94 — A Jet Stream or Mist from Pinhole Diesel Fuel Leaks Can Penetrate Skin and Flesh. Don't Touch It.

Avoid the hazards of pinhole leaks. Diesel fuel under pressure can penetrate skin and flesh and cause severe injury, including infection and gangrene (Fig. 89). If you see a jet stream or misty spray coming from a fitting or tubing, or just dripping fluid, don't touch it with your hand, not even with gloves on, To check for pinhole hydraulic leaks, wear eye protection, and use a piece of cardboard, paper, or wood to locate the leak or its source. The jet stream from a diesel fuel injector is just as dangerous as a pinhole fuel leak.

Relieve pressure before disconnecting hydraulic or diesel fuel lines, and tighten all connections securely before turning on the pressure again.

If ANY fluid is injected into the skin, it must be surgically removed within a few hours by a doctor familiar with this type of injury or gangrene may result. Gangrene often leads to amputation.

When using an ether starting aid in cold weather, remember ether is extremely flammable. Keep sparks and flames away. Store ether in a cool, protected place. Prevent accidental discharge. Keep the safety cap on the pressurized container. Do not burn empty cans or attempt to crush them. They will blow up!

Power Transmission System

Repair and adjustment of transmissions and other major drive line components should be performed by your dealer. These jobs require specialized training and service equipment. Put your efforts into maintaining and servicing the drive mechanisms that don't require major tear downs, specialized training, or special tools.

To get these jobs done safely:

1. Observe general safety recommendations for doing service work. These are discussed earlier in this chapter. All of them apply when servicing drive line and power transmission components. The first precaution is especially important: Never clean, unplug, lubricate, adjust, or repair any machine while it is running. Always disengage power, turn off the engine, and remove the key before servicing.

2. Make your work accessible. Remove shields, guards, and other components to get access to the parts you want to service. Use good judgment. Your work will proceed more easily and safely if you have room to use your tools. Always replace guards and shields that were removed.

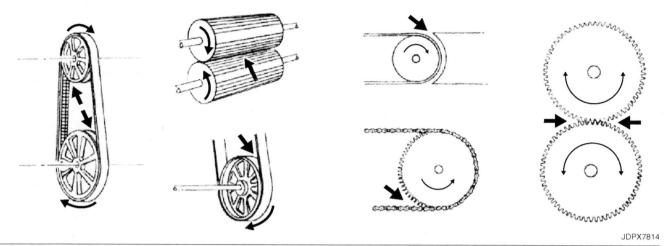

Fig. 95 — Don't Turn the Machine Over by Hand if Your Fingers Are at One of These Pinch Points

3. Avoid pinch points. Turning a shaft can pinch your fingers if they are caught between meshing gears, a belt and pulley, or a sprocket and chain (Fig. 95). A common error is to grasp and pull on a V-belt. As the belt moves and rotates the pulley, the mechanic's fingers are pinched between them.

4. Pull bearings cautiously. Removing bearings can be dangerous. Hardened steel bearings can shatter under pressure. Pieces fly outward with deadly force. Always wear eye protection when pulling bearings or striking bearings (Fig. 96). Shield the bearing if you can and keep others away. Always apply force to the tight bearing ring. Exerting force on the free ring may crack the bearing, causing pieces to fly with dangerous force.

force could cause a ball or roller to fly outward with enough force to cause an injury.

6. Clean with a commercial solvent, fuel oil, or kerosene. Remember that these are flammable liquids, and keep sparks and flame away. Don't clean with gasoline. It's too hazardous. Refer to "Chemicals and Cleaning Equipment," earlier in this chapter, for additional precautions, when using solvents.

7. Keep rotating edges smooth. File or grind sharp edges smooth. When installing sprockets, pulleys, and gears, drive the keys all the way in past the end of the shaft. Clip and bend the ends of cotter keys. These procedures will help prevent cuts from sharp edges, and keep clothing and gloves from being caught in revolving parts (Fig. 97). Always replace guards and shields that were removed.

Fig. 96 — Wear Goggles when Pulling Bearings or Striking Bearing Parts

5. Don't spin bearings with compressed air. The rollers or balls in a bearing are held to the inner ring with a bearing cage. If the cage is spun with compressed air, centrifugal

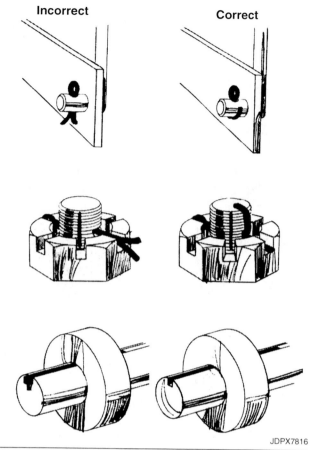

Incorrect   **Correct**

JDPX7816

Fig. 97 — Install Keys Properly to Keep from Tearing Skin or Clothing

Tires

Unless you have the experience and the proper tools for mounting and removing tires, it's better to have your tire work done by a farm equipment dealer or a tire service company.

A professional is essential for tires mounted on split or multipiece rims. Changing split rims is hazardous, especially if proper equipment is not available. If a tire explodes, a lock ring blows off, or a hydraulic bead breaker slips, you could be struck with explosive force. Leave split rims to professionals.

If you mount and remove tires on farm equipment, follow the step-by-step instructions given in a tire repair manual, such as John Deere's Fundamentals of Service: "Tires and Tracks." Following recommendations in these types of books will make your job much easier and safer.

If you mount and remove tires, observe these basic safety practices:

1. Support the machine safely. Before jacking the equipment off the ground, take steps to keep it from

rolling and falling. If the machine is self-propelled, put the transmission in gear or in park position and set the parking brake. When changing tires on pulled equipment, block the wheels. It is usually safest to hitch pulled and semi-mounted equipment to a tractor hitch to keep the raised equipment in place.

2. Use a rubber lubricant. After tire beads have been broken from the rim, apply a recommended rubber lubricant to the bead you need to remove (Fig. 98). Lubricant will make tire irons easier and safer to use because less force is needed to slip the bead over the rim. To remove a tire, apply lubrication to both beads. Likewise, when mounting tires, use a lubricant to help slip the beads over the rim, and seat the beads against the flanges of the rim.

JDPX7571

Fig. 98 — Apply Rubber Lubricant to Help Slip Tire Bead over Rim

3. Handle tire irons safely. When prying with tire irons, keep these points in mind:

   • Never let go of an iron when prying. The iron may spring back with terrific force.

   • Don't try to remove large sections of the tire from the rim with each bite. Instead, take smaller bites to avoid hard prying (Fig. 99).

   • Keep your balance. Tire irons may slip, forcing you to fall.

Fig. 99 — Take Small Bites with Irons to Avoid Hard Prying

4. Use chains and a hoist, or loader, and get help to handle large tires. Large tires, rims, and wheels are heavy and awkward to handle. They can fall, injuring or killing you or others. Get another person to help you, stand on each side of the tire as shown (Fig. 100) when moving large tires, and clear others from the area. Also, store the tires so they cannot fall on someone. It is best to have a professional tire service work with large tires.

Fig. 100 — Get Help with Large Tires and Wheels

5. Inspect tires and rims before mounting. If necessary, clean rust, corrosion, and old rubber off the rim with a chisel or wire brush. Also, inspect the rim for cracks and other damage. Check the tire casing and tube to make sure they're in serviceable condition, and clean and dry. Moisture trapped inside a tire can deteriorate the cord fabric and cause a premature failure.

CAUTION: Don't try to fit a tire to a rim that does not match the rim diameter molded into the sidewall of the tire. Example: A 24.5-32 tire must be mounted on a 32-inch rim, and a metric size 710/70R38 tire must be mounted only on a 38-inch rim. (Rim sizes are given in inches even for metric-sized tires.)

6. Watch for pinch points: Whenever you're working with a partially mounted tire, there is the danger of the tire slipping and crushing your hands or fingers between the tire and the rim. Block tires securely to avoid being pinched or crushed, especially when installing or removing the tube. Never place your fingers between the tire bead and the rim when inflating a tire. The beads usually "pop out" against the flange with crushing force.

7. Stand to the side when inflating tires. Use an extension hose with a clip-on chuck (Fig. 101). Then you will not need to hold the chuck on the valve stem when you are filling the tire with air. If the beads do not seat at the highest recommended pressure given by the tire dealer, deflate the tire, reposition the tire slightly on the rim, lubricate the beads again, and reinflate. Never exceed the highest recommended pressure given by the tire dealer when seating beads. Higher pressure could break the bead or even the rim with explosive force. Never fill tires with flammable gas (Fig. 102).

When Inflating Tires, Use a Clip-On Chuck and Extension Hose. Do Not Stand Over or in Front of the Tire

Fig. 101 — Stand Clear when Inflating Tires

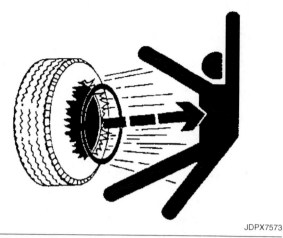

JDPX7573

Fig. 102 — Never Exceed Recommended Inflation Pressures. Exploding Tires Can Injure or Kill.

8. Store tires so they can't fall over and hurt someone. Tie them to the wall or lay them down flat.

9. Always follow tire and rim manufacturer handling procedures when working on tires and rims. Multi-piece rims and liquid filled tires can present special problems and hazards if they are not handled properly.

Additional information can be obtained from the following:

Rubber Manufacturer's Association
1400 K Street, N.W.
Washington, D.C. 20005

The Commercial Vehicle Solutions Network (CVSN)
3943-2 Baymeadows Road
Jacksonville, FL 32217

U.S. Department of Transportation
National Highway Traffic Safety Administration
1200 New Jersey Avenue, SE
West Building
Washington, D.C. 20590

Avoid Harmful Asbestos Dust

When asbestos fibers are inhaled they may cause lung cancer. Some machine parts or components contain asbestos fiber, including these:

• Brake pads

• Brake band and lining assemblies

• Clutch plate linings

• Gaskets

The use of asbestos in machine parts is being discontinued. However, since it may be encountered in servicing and repairing machines, use caution in working with the types of parts mentioned above.

Normal handling of parts containing asbestos is not hazardous as long as airborne dust containing asbestos is not generated.

To guard against the hazard when you work with parts or components that may contain asbestos, avoid creating dust. Never use compressed air for cleaning. Avoid brushing or grinding material containing asbestos fibers. When servicing, wear a respirator approved for asbestos. A special vacuum is also recommended, but if one is not available, apply a mist of oil or water on the material to avoid creating dust. Keep bystanders away from the area.

Check The Safety Features

Whenever you service or repair any farm equipment, be sure the safety features are in place and in proper working condition.

CHECK SAFETY FEATURES	
❑ Lights, reflectors, turn signals	❑ Shields and guards
❑ SMV emblem	❑ Safety signs
❑ Brakes	❑ Safety switches
❑ Steering	❑ Seat Belts with ROPS
	❑ Others

Manufacturers provide safety features for your protection. Unfortunately, many men, women, and children have been the tragic victims of farm machinery accidents that could have been prevented merely by keeping the original safety features in proper working order.

If a safety feature has been removed for any reason, re-attach it. If one has become damaged, repair it or replace it with a new one. If someone else has disconnected a safety feature, such as a seat switch interlock, reattach it and tell the person who disconnected it how important it is. Here are SOME of the safety features you should check:

• Lights, reflectors, turn signals. SMV emblem

• Brakes and steering systems

• Shields and guards including those on rotating shafts, gears, chain and belt drives, auger intakes

• Safety signs

• Neutral start safety switches and operator presence systems, interlocks, and emergency stop systems

• Seat belts and rollover protective structures (ROPS)

To complete your machinery service or repair job, check to be sure the safety features are in place and in good working condition. Don't wait. It could cost a life.

Test Yourself

Questions

1. The efficiency, performance, and life of a machine can best be improved by:

   a. repainting every year

   b. following manufacturers' maintenance recommendations

   c. trading in equipment for new machinery every two to three years

   d. allowing only federally licensed machine operators to run the equipment

2. Which of your senses might alert you to a change in machine performance?

   a. sight

   b. smell

   c. touch

   d. all of the above

3. When is the ideal time to perform routine scheduled maintenance, if possible?

   a. during planting season

   b. during harvest season

   c. during slack times

   d. when the machine has a malfunction

4. (T/F) Many times, a farmer can actually save time and money by taking equipment to a dealer for service or repair.

5. (T/F) The best way to have a hazard-free farm shop is to carefully plan and recognize potential hazards in advance, before construction.

6. Class _____ fires will often spread when water is used in an attempt to extinguish them.

   a. A

   b. B

   c. C

   d. D

7. An ABC fire extinguisher be used on fires that include:

   a. burning wood paneling

   b. flames surrounding a 200-amp fuse box

   c. box of greasy rags

   d. all of the above

8. Which option would be the safest if no wrench was available to remove a particular size bolt?

   a. vise-grips

   b. pliers

   c. drill the bolt out with a portable drill

   d. go buy the appropriate wrench

9. (T/F) Eye protection is recommended when using nearly all types of tools.

10. (T/F) Cold chisels and punches may be used safely as pry bars.

11. When arc welding, it is a good idea to wear safety glasses under a welding helmet because:

    a. glasses will protect your eyes from fumes

    b. the glasses will already be on when you need to chip slag

    c. molten metal might get under the welding mask

    d. none of the above

12. To turn off oxygen and acetylene regulators, adjusting handles should be turned:

    a. clockwise until tight

    b. clockwise until loose

    c. counterclockwise until tight

    d. counterclockwise until loose

13. When checking for defects in a grinding wheel by tapping with a metal rod, which sound indicates a defective wheel?

    a. clear ringing

    b. loud bong

    c. no ringing

    d. none of the above

14. (T/F) Gasoline does not vaporize enough in cold winter months to form an explosive mixture with air.

15. Which of the following provides the best protection from electric shock?

    a. ground-fault interrupters

    b. black electrical tape

    c. rubber-soled shoes

    d. 3-prong adapters

16. Grounding conductors are indicated by what color of insulation?

    a. white

    b. red

    c. green

    d. black

17. What action should be taken if chemical solvents are splashed into the eyes?

    a. cover the eyes with gauze and get medical attention

    b. cover the eyes with gauze and seek medical attention if pain persists

    c. use pain relieving eye drops

    d. thoroughly flush the eyes with water, then seek medical attention

18. (T/F) A pinhole hydraulic leak can cause fluid to actually be injected into a person's flesh under high pressure.

19. Which of the following methods is recommended to check for LP-gas leaks?

    a. use a match to see if leaking vapors ignite

    b. use soapy water and a brush

    c. feel for escaping vapors with your fingers

    d. sniff for leaking gas around connections

20. (T/F) Install cotter keys by bending the ends so they fit closely against the shaft or nut.

References

1. Fundamentals of Machine Operation (FMO) – Tractors. Deere & Company.

2. Fundamentals Of Service (FOS) – Tires and Tracks. Deere & Company.

Waste Recycling
and Disposal

6

JDPX8060

Introduction

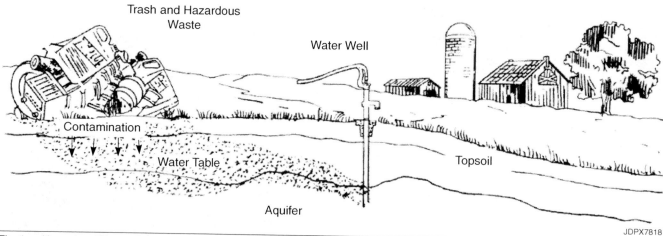

Fig. 1 — *Hazardous Wastes Can Leach Into Soil and Endanger People and Animals*

In operating, servicing, and repairing farm equipment, some materials and products need to be discarded. It is increasingly important to properly dispose of all waste materials.

Many materials can become health hazards when dumped on the land (Fig. 1). They leach down into the soil and groundwater. Ultimately, they contaminate the drinking water and endanger the health of people and animals.

According to EPA (Environmental Protection Agency) reports, 216 million tons (196 million metric tons) of waste are generated each year throughout the United States. For example, over one billion gallons (3.8 billion liters) of used oil are generated each year; and much of it is dumped improperly to the detriment of the environment, including farms and ranches.

Agricultural Pollution Is Costly

Farmers and ranchers contribute to the national problem. However, their greatest concern should be the contamination of their own land. It endangers the health and lives of their own families, friends, and workers. It can also cost them a lot of money.

When farmland is sold, buyers and money lenders want to be sure the soil and underground water are not contaminated by hazardous substances. They often require landowners to pay for professional assessment and analysis of the soil and water (Fig. 2). In 1992, environmental assessment costs were nearly $2,000 for a typical Midwestern farm.

Fig. 2 — *When Land Is Sold, Owners Must Pay for Soil and Water Sampling, Analysis, and Decontamination*

If there is apparent contamination such as by fuel, oil, lead, or pesticides, the laboratory costs for analyzing contamination can be considerable. If contaminated materials must be removed, the additional cleanup costs can be staggering. A Midwestern farmer had to pay $30,000 for professional removal and aeration of subsoil contaminated by fuel that had leaked from an underground tank for several years.

Hazardous Waste

Federal, state, and local regulations control the disposal of certain waste materials to help protect the environment and public health. Although regulations may not yet cover the smaller quantities of waste materials generated by farmers and ranchers, common sense tells you waste should be properly disposed of voluntarily, especially the hazardous wastes.

What Is a Hazardous Waste?

According to a U.S. Environmental Protection Agency (EPA) publication, Understanding the Small Quantity Generator Hazardous Waste Rules, a waste is any solid, liquid, or contained gaseous material that you no longer use and either recycle, throw away, or store until you have enough to treat or dispose of properly. Wastes are considered hazardous if they can:

- Cause injury or death

- Damage or pollute land, air, or water

RCRA (Resource Conservation and Recovery Act of 1976)-EPA regulations list more than 400 hazardous wastes.

Generally, a waste is classified hazardous if it is ignitable, corrosive, reactive (violent chemical reaction with water or other materials), or EP-toxic. (Extraction-procedure testing reveals dangerous concentrations of pesticides or heavy metals such as lead or mercury.)

It is illegal to dump "regulated hazardous" wastes. They must be taken to a facility licensed to process the material. To find out if a specific substance is considered "regulated hazardous," call the EPA-RCRA HOTLINE number: 1-800-424-9346.

EPA INFORMATION HOTLINE
1-800-424-9346
FOR INFORMATION ON DISPOSAL OF SOLID WASTE, HAZARDOUS WASTE, UNDERGROUND STORAGE TANKS, AND SOIL CONTAMINATION PROBLEMS.

Wastes From Equipment Service and Maintenance

Waste materials and products resulting from servicing and maintaining farm equipment should be recycled or disposed of so they won't endanger personal or animal health or pollute the environment (Fig. 3). Here is information on some of them:

Dispose of Waste Properly

JDPX7601

Fig. 3 — Recycle or Safely Dispose of Waste From Farm Equipment Service and Maintenance

- **Liquids:** Liquids should not be disposed of in landfills or on the ground.

- **Used oil:** It should be recycled. Some states consider waste oil to be hazardous and illegal to dump. Some service stations and oil change stations will accept oil for recycling, although they may charge if they don't do an oil change. Some licensed waste haulers will collect used oil from farms if quantities are large enough. Used oil is sometimes sold to industry for fuel or to oil companies who process it for blending stock. It is illegal to apply oil to roadways to suppress dust.

- **Oil filters:** State regulations vary on oil filter disposal. However, they should be recycled if a recycling facility is available. Filters can be taken to a landfill only if they have been "hot drained" at engine temperature.

NOTE: Terne (an alloy of tin and lead) plated filters must be disposed of as hazardous waste. Most oil filters are not terneplated. Check the filter label.

- **Used batteries:** They must be recycled. Battery retailers are required to take one old battery for each one you buy. Manufacturers recycle batteries to produce new ones.

- **Used tires:** Tire dealers are required to take your used tires when you buy new ones. Some states have collection programs to clean up existing tire dumps and are researching the use of old tires for fuel and for mixing in road surfacing materials.

- **Pesticides and pesticide containers:** See Chapter 9, "Chemical Safety" in this book about handling and disposing of pesticides and their containers. Always read the label on the container for instructions on use and disposal.

  You should triple-rinse empty pesticide containers with water, and return them to your chemical supplier or to a local recycling program. Store and use the residue rinse so it will not have significant impact on the environment. Some state and county governments, fertilizer and chemical associations, and manufacturers have developed pesticide container recycling programs. Recycled containers are often ground up to make new ones.

- **Antifreeze:** Ethylene glycol mixtures are considered special (sometimes hazardous) wastes in some states, requiring disposal at special or hazardous waste or disposal sites. It should not be dumped. There are commercial antifreeze recyclers.

- **Refrigerant from air conditioners:** By law, chlorofluorocarbon (CFC) refrigerants can no longer be released into the environment. They must be recovered by licensed air conditioner services. They will either return refrigerant to your air conditioner during repair, or send it to a recycling operation. New air conditioning systems will use a special refrigerant (HFC R134a) that will not harm the ozone layer of the atmosphere.

- **Paint:** Liquid paints containing lead and chrome are hazardous, and even if they are lead- and chrome-free, they may be hazardous if they are flammable. Dispose of liquid paints through licensed disposal firms or local approved programs.

- **Solvents:** Most solvents are hazardous wastes by federal law and must be disposed of through licensed disposal facilities. Mineral spirits with flash points below 140°F (60°C) are also considered hazardous.

- **Shop rags:** If rags are soaked with chlorinated or low-flashpoint solvents, or with other flammable materials, they are probably hazardous and should be disposed of accordingly.

Local Recycling and Disposal Sites

State departments of agriculture, Cooperative Extension Service offices, county governments, farm organizations, and others are beginning to organize collection of waste materials and products. They often designate special dates and places for you to take specific waste materials for recycling or safe and legal disposal.

Fairgrounds, county highway garages, and county landfills are sometimes designated for special collections. Commercial recycling is expanding as the use of recycled materials becomes more profitable, or is required.

Guidelines for Recycling and Disposal

By properly recycling or disposing of waste, you can help protect the health of your family, friends, and fellow workers, and future generations. You can do that by following these recommendations:

Be sure to check and follow federal, state, and local requirements. Regulations, facilities, and guidelines for waste recycling and disposal are being revised, expanded, and publicized more each year. Learn about them and use them (Fig. 4).

JDPX8080

Fig. 4 — Learn About Regulations, Facilities, and Guidelines for Waste Disposal in Your Area

Also, follow common-sense guidelines such as these:

1. Don't dump liquids or other waste materials or products on the land.

2. Collect waste in safe containers and in locations where they will not harm people or the environment. Label materials so they will not be misused and for proper disposal.

3. Check the labels on containers for chemicals, cleaners, solvents, and similar materials for special disposal instructions.

4. Recycle everything you can. Find out when and where collection or recycling sites will be open in your area.

5. Return waste material to the place you bought it, if possible. For example: Take used oil, tires, and batteries to your suppliers. By law, some retailers and distributors must take specific waste materials and products. If they can't take your waste materials, ask for their advice. They may know of recycling or safe disposal facilities.

6. Dispose of paper, steel, and other scrap metal through local recycling facilities or scrap dealers.

7. Dispose of hazardous waste only through approved or licensed disposal facilities.

8. Sensibly dispose of waste for which you cannot locate a collection or recycling facility. Use approved landfills.

9. Find out what wastes are considered "special" or "hazardous" in your state and county, and follow the applicable regulations and guidelines.

JDPX7619

Fig. 5 — Check Your Telephone Directory for Organizations to Advise You on Waste Disposal

Information Sources

There are federal EPA (Environmental Protection Agency) laws, regulations, and guidelines for waste recycling and disposal. However, state and county governments also have special regulations and programs. Be sure you know what they are.

Here are some of the organizations that can provide information and advice. Look for them in your telephone directory's government section (Fig. 5).

• Federal, regional, and state EPA (Environmental Protection Agency).

• EPA Information Hotline: 1-800-424-9346. For information on disposal of solid waste, hazardous waste, underground storage tanks, and soil contamination problems

• State and county Cooperative Extension Service

• State departments of agriculture and natural resources

• State and county health departments

National, state, and county farm organizations, such as American Farm Bureau Federation, can also provide helpful information about waste programs.

For additional information on the Environmental Protection Agency and regulations, and for addresses of EPA offices, see the Appendix near the end of this book.

Test Yourself

Questions

1. If you need to know whether a substance is a "regulated hazardous" waste, what agency should you contact?

   a. OSHA

   b. EPA

   c. OPEC

   d. FIFRA

2. When farmland is sold, who is likely to be required to pay for an environmental assessment?

   a. Federal government

   b. State government

   c. Township assessor

   d. The farmer selling the land

3. (T/F) Retailers are required to take your used batteries and tires when you buy new ones.

4. (T/F) Small amounts of hazardous wastes may be dumped on private property.

5. Wastes are considered hazardous if they can:

   a. Cause injury or death

   b. Damage or pollute land or water

   c. Pollute the air

   d. All of the above

6. (T/F) Chlorofluorocarbon refrigerants in air conditioners should be released into the air and replaced with HFC R134a refrigerant.

7. Laws or regulations concerning waste recycling or disposal can be developed by:

   a. EPA

   b. State governments

   c. County governments

   d. All of the above

8. After using a pesticide, where would be the first place to check for information about disposal of the container?

   a. Cooperative Extension Service

   b. Pesticide Handlers' Association

   c. The pesticide label

   d. Farm magazines

9. (T/F) Federal, state, and local requirements and regulations concerning waste disposal are likely to change in the future.

10. (T/F) An excellent way to recycle used motor oil is to spray it on gravel roads for dust suppression.

References

Understanding the Small Quantity Generator Hazardous Waste Rules: A Handbook for Small Business. 1986. U.S. Environmental Protection Agency, Office of Solid Waste and Emergency Response, Washington, D.C. 20460

Tractors and
Implements

JDPX8061

7

Introduction

Based on surveys by National Safety Council and others, it is estimated that nearly 500 people were killed each year in farm tractor accidents in the 1980s and early 1990s. Many others are seriously injured or killed each year in accidents involving towed and self-propelled implements.

How can you avoid accidents involving tractors and implements? First, decide you won't let an accident ruin your chances of living a happy and productive life (Fig. 1). Commit yourself to avoiding accidents. Acquire the knowledge and skills you need to perform safely. Then, always use your knowledge and skills to protect yourself and others.

JDPX7820

Fig. 1 — Practice Safe Machinery Operation for a Happy, Productive Life

This chapter discusses general tractor and implement safety. Other chapters discuss various specific safe practices. The most important safety information is in the operator's manual of each tractor and implement you operate. Please read it.

NOTE: *For operating procedures, refer to Fundamentals of Machine Operation — Tractors, Combines, and other manuals in this John Deere series. Also, refer to the operator's manuals provided with your machines.*

Tractors

Safe operators assume seven basic responsibilities. They meet these responsibilities by using the knowledge and skill acquired. The safe tractor operator:

- Performs proper maintenance according to the operator's manual

- Uses tractors only for the jobs they were designed to perform

- Makes pre-operation checks

- Removes risk of fire or explosion when refueling

- Follows recommended procedures when starting and stopping

- Takes special care to avoid accidents during operation

- Observes special precautions when towing implements and machines

Using Tractors for Their Intended Functions

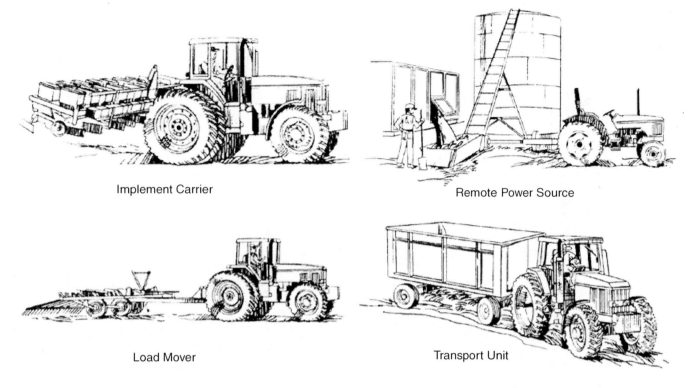

Implement Carrier

Remote Power Source

Load Mover

Transport Unit

JDPX7821, JDPX7822, JDPX7823, JDPX7824

Fig. 2 — Use Tractors for Their Intended Functions Only

Tractors perform a wide variety of jobs in farming operations (Fig. 2). They can function as:

- Load movers that pull heavy implements for tillage operations

- Transport units that move equipment from place to place

- Implement carriers that carry mounted equipment

- Remote power sources that power other machines through their power takeoff shaft or hydraulic system

To handle such a wide variety of jobs, tractors have many features that call for special attention. These include:

- Adjustable wheel tread spacing

- Quick steering response

- High-power engines

- Multi-speed transmissions

- High chassis clearance

- Unequal weight distribution between front and rear wheels

- Individual brakes for each rear wheel

- Adjustable drawbar hitches

- Hydraulically controlled hitches

- Power-assisted controls

- Provisions for adding weights and dual wheels

- Hydraulic circuits to power or lift implements and machines

- PTO shafts to operate trailed and mounted equipment

- Differential locks

- Provisions for mounting equipment on the tractor chassis

Features like these let a machine perform a wide variety of jobs in farming operations and increase the tractor's efficiency and ease of operation. They also make tractor operation safer. The better a machine is matched to its job, the fewer operating hazards there will be.

Features designed into tractors can become operating hazards if the operator misuses or ignores recommended operating procedures. Turning quickly and braking only one rear wheel, for example, can overturn a tractor. Engaging a gear quickly when starting a heavy load can tip a tractor over backwards. And failing to disengage the power and shut off the engine before cleaning, servicing, or adjusting PTO-driven machines frequently results in terrible injuries. To benefit from the features your tractor provides, and to prevent any of these features from contributing to an accident, always:

1. Use your tractor only for the specific jobs it was designed to perform. It is not a recreational vehicle (Fig. 3).

Danger: Not a Recreational Vehicle

JDPX7672

Fig. 3 — A Tractor Is Not a Recreational Vehicle

2. Follow all operating precautions and recommendations outlined in your operator's manual (Fig. 4).

JDPX7825

Fig. 4 — The Most Important Safety Instructions — The Operator's Manual and Safety Signs

Making Pre-Operation Checks

Before checking your tractor, examine yourself. Loose, frayed clothing can catch on moving parts and jerk you into machinery. Check your safety boots. Are they in good condition? Do they have slip-resistant soles? Are laces tied tight?

Check your tractors, and also other machines, like a pilot checks a plane (Fig. 5). Use a checklist.

Check these:

- Tires
- Shields
- Platform and steps
- Fuel and hydraulic lines
- Visibility from within operator enclosures
- Brakes
- Steering
- Lighting equipment, reflectors, SMV emblem
- Seat adjustment
- Safety interlock systems/seat switches
- Neutral-start safety switches

JDPX8083

Fig. 5 — Safety Starts With a Pre-Operation Check

Tires

Look for cuts in the tread and sidewalls. Damaged tires can blow out and throw your tractor out of control.

Check tire pressure. Underinflated tires develop internal damage. Overinflated tires make the front wheels bounce on rough ground and damage the tire internally. They may also cause loss of steering control.

Shields

Make sure all shields are in place. They protect you from moving parts and pinch points. Always make sure the tractor PTO shaft is shielded, and that all shields provided for PTO-operated equipment are in place (Fig. 6).

JDPX8084

Fig. 6 — Make Sure All Safety Shields Are in Place

Refer to "Power Takeoff Accidents" later in this chapter for more information on shielding the tractor PTO shaft. Chapter 3, "Recognizing Common Machine Hazards", gives details on shielding moving machine parts.

Platform and Steps

Clean off slippery mud, grease, and any crop residue that may have accumulated on the steps and operator's platform of your tractor (Fig. 7). Remember that chains and tools carried on the platform may interfere with pedal operation or cause a slip or fall from the tractor. Remove them.

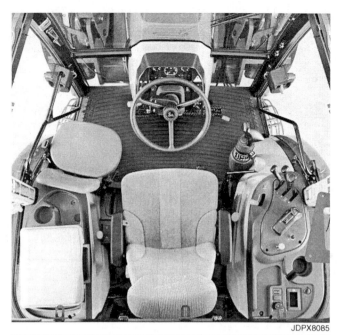

JDPX8085

Fig. 7 — A Safe Platform Is Clean and Free of Tools and Debris

Fuel and Hydraulic Lines

Leaky fuel lines and connections are fire hazards and they waste fuel. Loss of pressure or fluid from hydraulic lines can cause loss of power steering, power brakes, and control of the three-point hitch. Keep hydraulic lines in good repair.

Hydraulic oil and diesel fuel released under high pressure are especially hazardous. Many hydraulic and diesel fuel injection systems develop pressures over 2500 psi (17 200 kPa) (172 bar). That is over three times the pressure needed for a fine jet stream of oil from a pinhole leak to penetrate skin and flesh (Fig. 8). See "Pinhole Leaks" in Chapter 3, "Recognizing Common Machine Hazards".

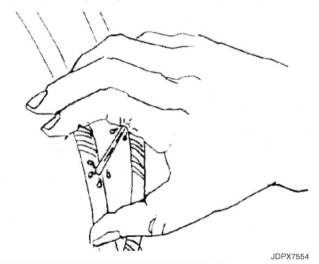

JDPX7554

Fig. 8 — A Jet Stream From a Hydraulic Pinhole Leak Can Penetrate Skin and Flesh. Do Not Touch It!

Do not touch any hydraulic pinhole leaks. They are often invisible but very dangerous. To locate the source of hydraulic leaks, use a piece of cardboard, paper, or wood, but never your hand. If ANY fluid is injected into the skin, it must be surgically removed within a few hours by a doctor familiar with this type of injury, or gangrene may result.

To prevent injury from escaping high pressure oil and to avoid loss of power steering, power brakes, and control of the three-point hitch, replace defective lines, fittings, and seals. Keep all connections tight. When making replacements, make sure all pressure is released from the system before loosening any connection or fitting. Diesel fuel system pressure is released when the engine stops. To release hydraulic system pressure, lower equipment to the ground, turn off the engine, and move all hydraulic controls back and forth in each direction a few times. Refer to Chapter 5, "Equipment Service and Maintenance", for additional service procedures and precautions.

NOTE: On some tractors, you must adjust the load to a position of low hydraulic pressure before you stop the engine. Refer to your operator's manual for the proper procedure for releasing pressure.

Visibility

Safe tractor operation requires good visibility in all directions. Check for clean windshield or wiper operation and mirror adjustment on tractors equipped with operator enclosures. If the windows need cleaning, take time to complete this chore (Fig. 9).

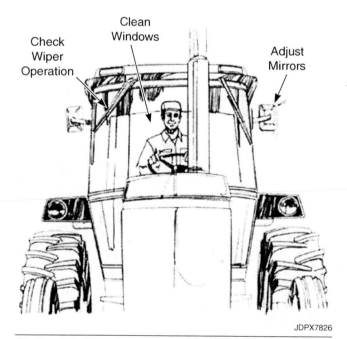

Fig. 9 — With Windows Cleaned, Mirrors Adjusted, and Wipers Functioning, This Tractor Is Safer to Operate

Brakes

Test your tractor's brakes before starting tractor movement (Fig. 10), and again at slow travel speed. If you have power brakes, start the engine, press each pedal, and observe its operation. Brakes should have a solid feel, and the distance of pedal travel should not exceed that specified in your operator's manual. Then lock the brake pedals together, drive ahead slowly, declutch, and press on the brakes. If brakes aren't equalized, the tractor will swerve. If the brakes do not pass these tests, check your operator's manual for adjustment procedures, or have a service shop adjust the brakes for you.

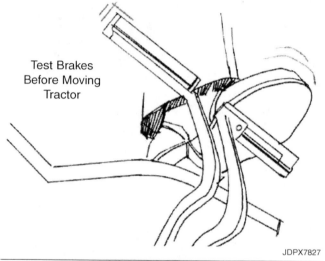

Fig. 10 — Brakes Should Have a Solid Feel

Steering

Turn the steering wheel in each direction and note the amount of rotation before the wheels start to turn. Steering should be quick and responsive without excessive play in the steering linkage. Units equipped with power steering should respond when only finger pressure is applied to the steering wheel.

Lighting Equipment

Lights are required during daytime periods of low visibility, after sunset, and when traveling on public roads (Fig. 11). If you anticipate the need for lights, check their operation before you start. Check them periodically for proper adjustment by following the procedures outlined in your operator's manual.

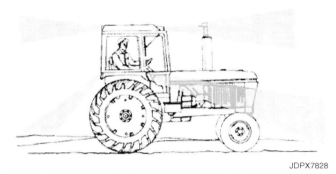

JDPX7828

Fig. 11 — Lights for Seeing and Being Seen Are Essential for Safe
Operation

Seat Adjustment

Check the position of the seat if another person has been
operating your machine. Your arms and forearms should
form a 90-degree angle when your hands are on the steering
wheel (Fig. 12). Your legs should remain slightly angled after
depressing the foot pedal.

JDPX7829

Fig. 12 — Proper Seat Adjustment Keeps the Operator Comfortable
and Within Easy Reach of Controls

Safety Interlock Systems/Seat Switches

Some tractors and self-propelled implements have a seat
safety switch system. The system may provide a signal,
such as a light, if the PTO or some other function is
engaged. It may prevent starting the engine if a specific
component is engaged; or it may shut off a machine function
if the operator leaves the seat. Such systems vary by
machine, but are intended to prevent mistakes. If your
machines have such safety interlocks, be sure they work
according to the operator's manual. If they do not, get them
repaired immediately.

JDPX8086

Fig. 13 — Check Controls to Be Sure Neutral-Start and Seat Safety
Switches Work Properly. Never Bypass Them

Neutral-Start Safety Switches

Tractor neutral-start safety switches keep the starter from
cranking the engine if the transmission or clutch is engaged.
If they do not work right, a tractor in gear can suddenly move
as the engine starts. It can run over and kill a person.

Neutral-start safety switches may be designed to prevent
starting the engine unless:

- The clutch or inching pedal is depressed

- The shift lever or direction-speed control is in neutral or
  park position

- Any combination of the above

Here's how to check neutral-start switch operation:

1. Drive to an open area with no one nearby. Then,
   determine from your operator's manual which controls
   are connected to the starting switches.

2. From the operator's seat, try to start the engine with
   controls in each position that should prevent starting
   (Fig. 13). Follow the operator's manual.

3. If the starter cranks in any of the positions, the neutral-
   start system is not working properly. Have your dealer
   check and repair it immediately.

NEVER bypass a neutral-start safety switch by shorting
across starter terminals. It can be fatal. Start a tractor engine
only from the operator's station.

Refueling

Prevent fires and explosions when refueling. The greatest danger occurs when handling gasoline or LP-gas. These fuels vaporize at temperatures as low as –50°F (–46°C), and mix with surrounding air to form explosive mixtures at relatively low (6%) fuel-vapor concentrations.

Diesel fuel is safer to handle. It vaporizes at approximately 110°F (43°C). But it's good practice to observe the same precautions when handling diesel fuel that you do for gasoline.

Safety begins with the proper installation and use of fuel storage equipment. Review the recommendations presented in Chapter 5, "Equipment Service and Maintenance".

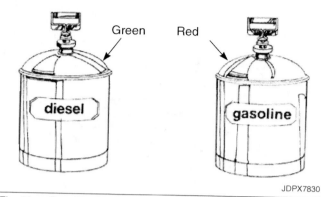

JDPX7830

Fig. 14 — Carry Fuel in Cans Labeled and Painted to Identify the Type of Fuel

Gasoline and Diesel Fuel

1. Identify fuel containers. Do not run the risk of someone confusing gasoline with diesel fuel. Gasoline has almost no lubricating qualities, and its use in a diesel tractor would quickly seize and damage the injection pump. Tractor fuels are flammable and explosive, so people should be able to identify them quickly and positively. Label your containers, and paint gasoline containers red and diesel fuel containers green.

2. Carry gasoline and diesel fuel in approved safety cans (Fig. 14). These cans are made from heavy-gauge metal or tough polyethylene. They are equipped with caps that seal completely and with pressure-relief valves that open when vapor pressure within a can exceeds 3 to 5 psi (21 to 35 kPa) (0.21 to 0.35 bar). In addition, safety cans are frequently equipped with flash-arresting screens in the filler opening and pouring spout. These screens reduce the possibility of fire or explosion. Identify approved safety cans by reading the label. Those approved by the Underwriters Laboratories Inc. or Factory Mutual, or those meeting United States government standards are acceptable safety cans.

Never carry fuel in glass or plastic containers designed for household or farm chemicals. These containers are not strong enough for fuel and do not identify by their shape or color the fact that they may contain flammable, explosive fuel. Thin-gauge metal economy gasoline cans often found in discount stores are not recommended.

3. Let hot engines cool before refueling. Wait at least 5 minutes before refueling a hot engine, and longer if possible. If you can't wait for a hot engine to cool at the end of a day's work, refuel it the next morning or after supper. Refueling after supper is preferred. Moisture may condense in the fuel tank during the night if you wait until morning to refuel. If you spill fuel, wait for it to evaporate before starting the engine.

4. Stay away from flame. Welding or smoking in the area of your fueling operation could touch off a flash fire or explosion. Turn off the engine and all electrical switches. Do not refuel your tractor near fires or sparks. Gasoline fumes are heavier than air. They sink down. A hot engine or spark can ignite the fumes.

LP-Gas Fuel

1. Refuel outdoors. Never refuel in or near buildings or over pits. There is too much risk of heavy LP-gas collecting and increasing the possibility of an explosion.

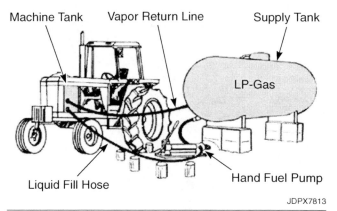

Machine Tank   Vapor Return Line   Supply Tank

LP-Gas

Liquid Fill Hose   Hand Fuel Pump

JDPX7813

Fig. 15 — Refuel With LP-Gas Outdoors and With Vapor-Return Line Attached

2. Always connect the vapor-return line. The vapor-return line equalizes the pressure between the tractor and supply tanks and permits the tractor tank to fill without loss of fuel vapor to open air.

3. Never fill an LP-gas tractor tank more than 3/4 full. Overfilling can cause excessive pressure buildup in the tractor tank, and this will open the relief valve. Gas will then escape from the tank, wasting fuel and creating hazardous explosive conditions. LP-gas tanks are equipped with either automatic fuel tank gauges or manually operated fixed tube-type gauges. To operate a fixed tube-type gauge, open it momentarily while the tank is being filled and then close it. When the tank is filled, the mist that comes from the outlet changes to a spray of liquid fuel. Do not overfill. Never permit the gauge to remain open. Doing so is extremely hazardous.

4. Do not refuel an LP-gas tractor when it's hot, when the engine is running, or when the ignition switch is on.

5. Do not smoke during the refueling operation. Wait until all of the colorless LP-gas vapors have ventilated away.

Starting

After completing your pre-operation safety checks, take the following precautions when starting the engine:

1. Provide adequate ventilation if starting the tractor indoors (Fig. 16). Carbon monoxide, a colorless, odorless, and deadly gas, is present in the exhaust gases of all engines. Be sure you have plenty of good, fresh air when starting up indoors. Open doors and windows. Do not take a chance.

   If you use ether, follow the instructions printed on the can and those in your operator's manual. These instructions

are designed to prevent fire and damage to your engine. Ether is extremely flammable. Keep sparks and flame away when using it. To prevent accidental discharge when storing the pressurized can, keep the cap on the container, and store it in a cool, protected place.

Before Starting, Open All Doors

JDPX7831

Fig. 16 — Provide Ventilation Before Starting

2. Never bypass-start a tractor. If you start a tractor by shorting across the starter terminals with a screwdriver or other tool, you could bypass the neutral-start safety system (Fig. 17). Do not do it! If the tractor is in gear, it may suddenly move when the engine starts. A runaway tractor can run over and kill you or others. If the starting system does not work properly, have your dealer check and repair it. (See "Neutral-Start Safety Switches" earlier in this chapter.)

Chapter 5, "Equipment Service and Maintenance", in this book, discusses proper use of a booster battery and jumper cables, if that is necessary. Follow instructions in the operator's manual.

No! Never Bypass Start!

JDPX7832

Fig. 17 — Never Start the Engine While Standing on the Ground, or Short Across Starter Terminals

3. Start the tractor only from the operator's seat. Never start the engine while standing on the ground. Neutral-start switches occasionally fail, and some older tractors do not have them. If the transmission is in gear when the engine starts, the tractor can move more quickly than you. You could be run over and crushed. (See "Neutral-Start Safety Switches" earlier in this chapter.)

4. Make sure everyone is clear. Avoid having others injured when the tractors or implements start up. Be especially careful when starting articulated four-wheel-drive tractors to keep others away from the wheels. Do not move the steering wheel as you enter the cab or before starting the engine. When the engine starts, the tires, chassis, and mounted implements can swing quickly, striking a bystander. Also, the front and rear wheels can come together quickly, crushing anyone who stands between them.

5. Be sure the PTO is disengaged. The power takeoff shaft may have been shifted into gear after you last parked the tractor. If engaged and connected to a machine, the machine might start when the engine starts, creating a definite hazard.

6. Move remote hydraulic controls to neutral position. Make sure they are in neutral so attached implements won't move as the engine starts.

7. Check the position of forward-reverse controls. Make sure forward-reverse controls are in neutral or in the direction you intend the tractor to move.

8. Shift to neutral or park and disengage the clutch before starting the engine (Fig. 18). Make sure all other persons are clear. A tractor started in gear will lurch forward or backward, possibly causing serious personal injury or property damage. Most tractors and self-propelled machines are equipped with neutral-start switches and other systems that make it impossible to start the engine

with the key when the transmission is in gear. But make it a habit to disengage the clutch or inching pedal before starting, even if starting switches are provided. They can fail!

JDP7833

Fig. 18 — Before Starting Engine, Shift to Neutral or Park and Disengage the Clutch

9. Look ahead and to the rear before moving the tractor. Make sure no people or obstructions are ahead or behind. Sometimes machines are accidentally started in the wrong direction. Small children's actions are unpredictable. Sound the horn to warn people that your tractor is about to move.

10. Engage power to the drive wheels slowly. Engaging the clutch suddenly, or quickly shifting a hydraulic transmission to high speed could tip the tractor over backwards, especially when starting up a slope. Be especially careful when starting a heavy load and when the rear wheels are frozen to the ground. If they're frozen, break them loose by starting out slowly in reverse.

Stopping, Parking, and Towing

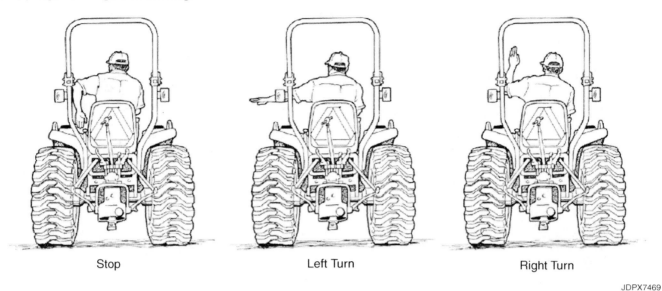

Stop　　　　　　Left Turn　　　　　　Right Turn

JDPX7469

Fig. 19 — Signals Prevent Accidents

Your ability to stop your tractor and park it safely is just as important as being able to get in underway safely.

Tractor upsets, collisions, runaway tractors, and people being crushed under machines and implements can happen when operators ignore safety recommendations.

To avoid these accidents, special precautions are necessary when stopping and parking tractors. Here are some important ones:

1. Signal before stopping, turning, or slowing down. If other drivers are not warned in advance of your intentions to change rate or direction of travel on the highway, they may run into your equipment or be forced to swerve into a dangerous situation themselves.

   Avoid this risk by giving the proper signal before you slow down, turn, or stop.

   Newer tractors are equipped with turn signal lights. Use those if your tractor has them. If not, use the standard hand signals for motor vehicle drivers on public roads (Fig. 19). Remember to signal well in advance and for the distance of travel required by the laws in your state. And remember that towed equipment may obstruct the vision of drivers behind you. If it does, they won't be able to see the signals you give.

2. On highways, pull over to the right-hand shoulder before stopping if possible. This is usually the safest procedure. Before doing so, however, make sure the shoulder is wide enough and sound enough to support your tractor and equipment.

3. Slow down before braking. Use engine braking to reduce speed before applying the brakes and disengaging the clutch. This will help keep your tractor under control. Latch brake pedals together for highway use.

4. Pump the brakes when stopping on slippery surfaces like ice, snow, mud, loose gravel, or manure. If brakes are applied and the tractor skids, release the brakes and apply them again with a slight pumping action — on and off — until the tractor stops (Fig. 20).

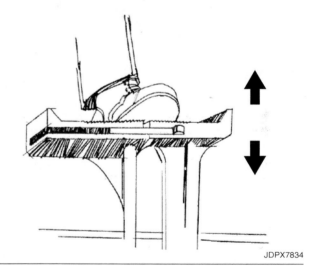

JDPX7834

Fig. 20 — Pump Both Brake Pedals to Stop in Slippery Conditions. Pedals Should Be Locked Together

5. Be careful when towing and stopping heavy loads. Remember that a combination of factors determines your stopping distance:

   - Your reaction time
   - Travel speed
   - Weight of the tractor and load being towed
   - Efficiency and condition of the brakes
   - Traction of the tires that have brakes
   - Surface conditions

The preceding factors become important when towing heavy loads. To maintain your stopping ability and to avoid jackknifing towed equipment, observe these precautions carefully:

- Check brakes on towed equipment, if present.

- Lock the tractor brake pedals together for equal braking in each wheel.

- Tow heavy loads at less than road-gear speeds. Use safety chains in case the load becomes unhitched.

- If heavy, rolling loads are not equipped with their own brakes, keep travel speed slow and avoid steep hills and slopes if possible (Fig. 21).

- Give yourself plenty of time and distance to stop.

- Slow down by reducing engine speed before applying the brakes.

- Travel more slowly in slippery conditions — on ice, mud, gravel, and manure.

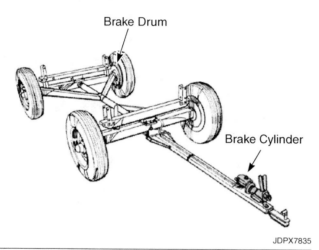

JDPX7835

Fig. 21 — Equipment Brakes Are Desirable When Pulling Heavy Rolling Loads

6. Shift to park or set the parking brake (Fig. 22). If your tractor has a parking brake, use it. This is most important on hillsides and grades. Do not depend on leaving the transmission in one of the driving gears to keep your tractor from rolling. Some tractors with hydraulic clutches can roll when parked in gear position.

Many tractors are not equipped with parking brakes. Instead, one of the transmission shift levers is shifted to the park position. It locks the transmission with positive action, keeping the tractor stationary.

Make it a habit to secure your tractor every time you leave it. Set the parking brake, or shift to park.

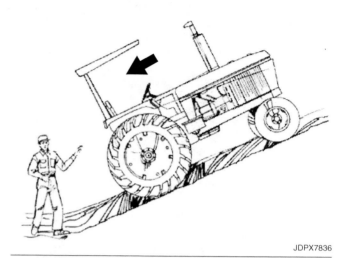

JDPX7836

Fig. 22 — To Avoid Rolling on Incline, Set the Parking Brake and Shift Into Park

7. Lower all equipment when leaving your tractor. Do not rely on the tractor's hydraulic system to support a raised implement or machine temporarily, or for service or storage. Hydraulic lines may rupture, and seals and valves may fail. Children or others could move the controls when someone is under the machine. To eliminate any chance of an implement or machine falling unexpectedly, follow one or more of these procedures:

a. Lower the equipment to the ground or onto sturdy blocks.

b. Use mechanical safety stops or transport links, if provided.

c. Use support stands if provided for the equipment.

8. Turn off all electrical switches and remove the key. Take the key with you to prevent unauthorized people from starting the tractor. See that all lights and electrical components are turned off. This will prevent a drain on the battery and starting problems the next time the tractor is used.

Tractor-Related Accidents

Unfortunately, many people are seriously or fatally injured in accidents involving tractors and implements each year. Here are some of the types of tractor-related accidents in which people are involved:

- Tractor upsets or overturns
- Falls from tractors
- Run over by tractors or implements
- Crushing accidents (other than run over)
- Collisions with motor vehicles on public roads
- Entangled in power takeoff shafts

Essentially all farm machinery-related accidents can be prevented if people working with and around machinery will do the following:

- Follow safe working practices.
- Recognize and understand potential safety hazards associated with farm machinery, and know how to avoid them.
- Select machinery with adequate safety features.
- Read and follow the safety instructions in machinery operator's manuals and safety signs.
- Keep original equipment safety features in place and in good working condition. Do not modify or bypass safety features designed for your protection.
- Stay alert and work deliberately.
- Do not take unnecessary risks.
- Look out for the safety of others in the area.
- Do not allow extra riders on farm equipment.

The following pages discuss the types of accidents listed above and how to avoid them.

For an understanding of basic types of hazards, also refer to other chapters in this book, including Chapter 2, "Human Factors and Ergonomics", and Chapter 3, "Recognizing Common Machine Hazards". The safe operator understands the causes of accidents and takes every precaution to avoid them. Let's turn our attention now to some of the most common types of tractor accidents and see how they can be prevented:

- Tractor upsets
- Falls from tractors
- Crushing and pinching accidents during hitching operations
- Collisions with motor vehicles
- Getting entangled in PTO shafts

Tractor Upsets (Overturns)

Many fatalities result from upsets or overturns. They are the number one type of fatal farm work accident in the United States. Most of those fatalities can be prevented with overturn protection.

The chances of surviving an upset without injury are not good, unless the tractor is equipped with rollover protection and the operator uses the seat belt. In a backwards tip, the tractor hood can hit the ground less than 1-1/2 seconds after the front wheels begin rising. After the wheels begin to rise, the operator has less than 3/4 second to realize what is happening and to take preventive action. Usually, the tractor is past the point of no return before the driver can do anything to keep from being crushed (Fig. 23). However, most tractor overturn accidents are upsets to the side. You can prevent tractor upsets by knowing the causes and how to avoid them.

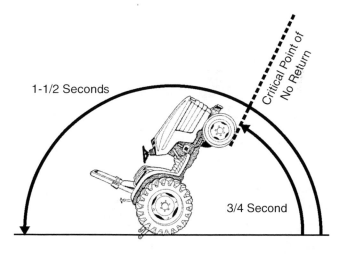

JDPX7403

Fig. 23 — Reaction Time in a Backwards Tip. You Can't Always Stop It in Time

Causes of Upsets

To understand the causes of upsets, learn how these forces act on a tractor:

- Gravity
- Centrifugal force
- Rear-axle torque
- Leverage of the drawbar or hitch

When a tractor is being operated, all these forces act at the same time. Each one is capable of upsetting the tractor by itself, but most upsets are caused by a combination of these forces.

Gravity

The earth exerts a pull. This pull is called the force of gravity. Gravity is measured in units of weight, such as ounces, pounds, kilograms, and tons.

A tractor's center of gravity is the point where all parts balance one another. The center of gravity of a block of wood is easy to visualize, being in the center. If suspended by a string attached to the center, the block will hang straight (Fig. 24). But if the string's hitch point is at another location, the block will move until the hitch point and the center of gravity form a straight line. The block will then hang in a tipped position (Fig. 24).

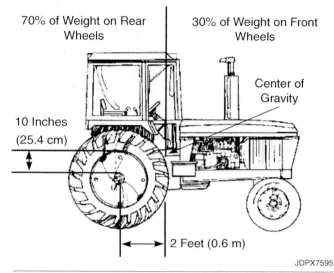

Fig. 25 — The Center of Gravity for a Two-Wheel-Drive Tractor

When rear-mounted weights or duals are added for increased traction, the center of gravity is moved even farther to the rear unless weight is also added to the front. On the other hand, addition of mechanical front-wheel drive will shift the center of gravity toward the front. The weight of four-wheel-drive articulated tractors is quite evenly distributed between front and rear.

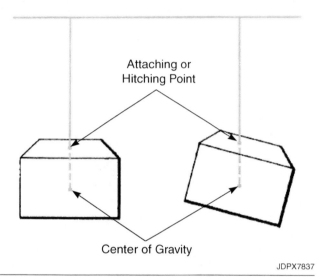

Fig. 24 — The Block Tips Until the Line of Suspension Passes Through the Center of Gravity and the Attaching Point in a Straight Line

A tractor's center of gravity depends on the location of its components. On conventional two-wheel-drive tractors, the rear wheels, differential, final drive, transmission, and cab weigh more than the rest of the tractor. The center of gravity is behind the midpoint of the tractor's length. And since most of a tractor's weight is carried above the axle height, its center of gravity is above this level.

For a typical row crop tractor with two-wheel drive, the center of gravity is about 2 feet (0.6 m) in front of, and 10 inches (25.4 cm) above, the rear axle. This means that about 70 percent of the tractor's weight is on the rear wheels, and 30 percent on the front wheels (Fig. 25).

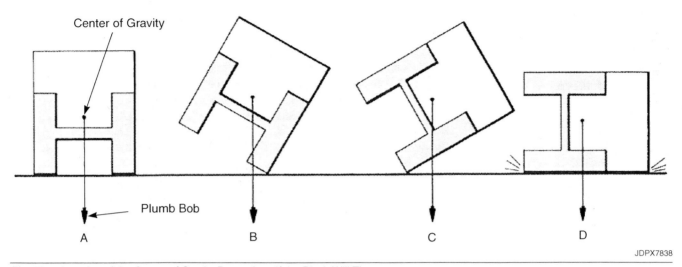

Center of Gravity

Plumb Bob

A   B   C   D

JDPX7838

Fig. 26 — Location of the Center of Gravity Determines If the Block Will Tip

To help understand the role that the center of gravity plays in causing upsets, let's try to tip the block of wood we mentioned earlier. Before we do, let's attach a string and plumb bob at its center of gravity and place the block on a table edge with the plumb bob hanging below. Notice in Fig. 26A that the plumb bob hangs directly below the center of gravity, and that the entire bottom surface of the block is supported by the table.

If we tip the block a little, the plumb bob swings toward the edge supporting the block (Fig. 26B). If the block of wood is released, it will fall back to its original position. But if we tip the block far enough, the block will tip over on its side (Fig. 26C and D). Tipping occurs when the block's center of gravity moves outside the limits of its base of stability.

Tractor upsets happen the same way. If the tractor's center of gravity moves outside its base of stability, the tractor will overturn.

A tractor's base of stability is determined by the location (width) of its rear wheels and the type of chassis support on the front wheels. Because the chassis of tractors with wide-front axles are supported on pivot points on the front axle, the base of stability is triangular, as it is for tricycle-type tractors, but the pivot point is higher on wide-front axles and thus provides more stability (Fig. 27).

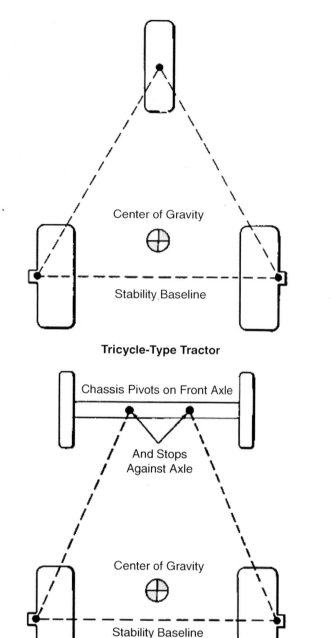

Tricycle-Type Tractor

Center of Gravity

Stability Baseline

Chassis Pivots on Front Axle

And Stops Against Axle

Center of Gravity

Stability Baseline

Wide-Front Tractor

JDPX7596

Fig. 27 — Base of Stability for Tricycle and Wide-Front Tractors

To prevent upsets caused by the force of gravity, avoid all situations that might put your tractor in an unstable position. These situations are described in the section on preventing tractor upsets.

Centrifugal Force

An object in motion will travel in a straight line unless there is some force exerted upon it. The force that resists change in direction is called centrifugal force.

There are many illustrations of centrifugal force in everyday life. For example, whirl a weight around on the end of a string. The string is kept tight because it pulls in on the weight, keeping it in a circular path. But if the string is cut, the weight will travel outward in a straight line. In this example, centrifugal force straightens the weight's circular direction of travel into a straight line.

Let's look at the reverse situation and change the direction of travel of an object moving in a straight line to move it around a curve. To change its direction of travel, we must overcome the object's centrifugal force by placing a lateral force on it.

Consider a locomotive on a curve. Its forward motion makes the flanges on the wheels press out against the inside edge of the rails. The rail, in turn, pushes in against the flanges, forcing the locomotive to follow the curve and stay on the track.

Tractors are made to change direction in the same way. When the operator turns the steering wheel on a conventional two-wheel-drive tractor, its forward motion places an outward lateral force between the front tire treads and the ground's surface. If the tires do not skid, the ground resists this outward force, forcing the tractor to turn. If the front tires do skid, of course, the tractor continues its travel.

Centrifugal force is a major factor in tractor upsets. If strong enough, centrifugal force can tip a tractor sideways using the outside wheels as pivot points (Fig. 28).

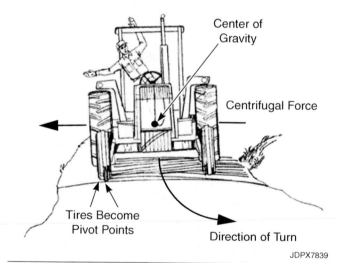

Center of Gravity

Centrifugal Force

Tires Become Pivot Points

Direction of Turn

JDPX7839

Fig. 28 — Centrifugal Force Tries to Pivot the Tractor on Its Outside Wheels

Centrifugal force varies in proportion to the square of the tractor's speed. This means, for example, that doubling tractor speed increases the strength of centrifugal force four times (2^2 = 2 x 2 = 4). Tripling speed, say from 5 to 15 mph (8 to 24 km/h), increases the strength of centrifugal force nine times (3^2 = 3 x 3 = 9).

Centrifugal force also varies in reverse proportion to the length of the turning radius. For example, if you make a short turn using half the turning radius of a wider turn, centrifugal force doubles.

Centrifugal force becomes more dangerous in tractor operation as the tractor's center of gravity is moved higher. This happens, for example, when a front-end loader is raised, or when spray tanks are mounted high on tractor chassis (Fig. 29). The center of gravity moves forward as well as up when the scoop is raised.

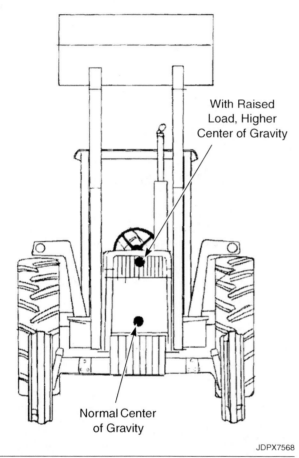

With Raised Load, Higher Center of Gravity

Normal Center of Gravity

JDPX7568

Fig. 29 — Using Some Mounted Equipment Raises the Center of Gravity

Remember these two points: short, quick, high-speed turns upset tractors, and tractor-mounted attachments that raise the tractor's center of gravity increase the risk of tractor upsets caused by centrifugal force.

Rear Axle Torque

Engaging the clutch applies torque (twisting force) to the rear axle of the tractor. Normally, the axle rotates and the tractor moves ahead. But if the axle rotation is restrained in some way, for example, when starting a heavy load or when the drive wheels are frozen to the ground, the twisting force of the axle tries to lift the front wheels off the ground and rotate the tractor backward around the rear axle. If it is easier for engine power to lift the front of the tractor than it is to move the tractor ahead or spin the wheels, the tractor will tip over backwards (Fig. 30A).

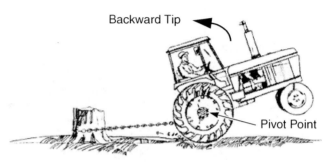

Backward Tip

Pivot Point

A. Rear Axle Torque Will Upset the Tractor If Rear Wheels Can't Spin or Move Forward.

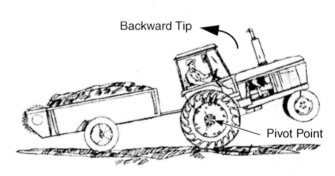

Backward Tip

Pivot Point

B. Hitching Above Normal Drawbar Height May Tip a Tractor Backward.

JDPX7842, JDPX7840

Fig. 30 — Hitching Mistakes That Can Cause Tractor Upset

Because rear-axle torque reacts against the tractor chassis, the following practices increase the chances of tipping a tractor over backward when driving ahead:

- Starling off in low gear with high engine speed

- Engaging the engine clutch quickly

- Accelerating quickly, especially when traveling uphill or pulling a heavy load

- Engaging power to drive wheels when unable to move forward or spin wheels

Leverage of the Drawbar or Hitch

When pulling a heavy load, the rear tires push against the ground with considerable force. At the same time, the load attached to the tractor hitch point pulls back against the forward movement of the tractor. When this happens, the point of contact between the rear tire and the ground serves as a pivot point, and the pulled load attempts to pivot the tractor, raising the front wheels off the ground (Fig. 30B).

The draft (amount of pull) and the height of the hitch point determine the force of the pivoting action. Increasing the pull and raising the hitch point increases the danger of upsetting the tractor.

An excessive load attached to a drawbar set at the recommended height and with the recommended ballast will usually slip the rear wheels or stall the engine before overturning occurs, but even a light load attached too high can quickly tip a tractor backward. Drawbar heights are selected by tractor engineers to provide efficient and safe hitching points for towed loads. If the drawbar height on your tractor is adjustable, check your operator's manual for proper adjustment and for proper ballast.

Preventing Tractor Upsets

Put your knowledge of the causes of upsets to work to keep yourself alive and prevent tractor damage. Follow operating practices that utilize the force of gravity to keep your tractor upright and minimize the forces that work to tip it over.

About 75% of the tractor overturn fatalities result from overturns to the side. Let's review the recommendations for preventing these, and then go on to those for preventing rearward upsets.

Rollovers to the Side

1. Set the wheel tread at the widest setting suitable for the job you must do. This provides the widest possible base of stability for your tractor (Fig. 31).

JDPX7841

Fig. 31 — Wide Wheel Spacing Increases Lateral Stability

2. Lock brake pedals together before driving at transport speeds (Fig. 32). Applying uneven brake pressure to the rear wheels during high-speed operation can cause a tractor to roll over. When only one wheel is braked, the other is driven at a correspondingly higher speed by the differential. Braking one pedal severely can force the tractor to swerve abruptly to the right or left into a rollover situation (Fig. 33).

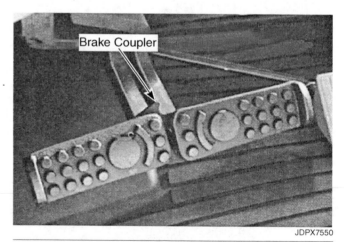

Brake Coupler

JDPX7550

Fig. 32 — Typical Coupler for Locking Brake Pedals Together in Highway Travel

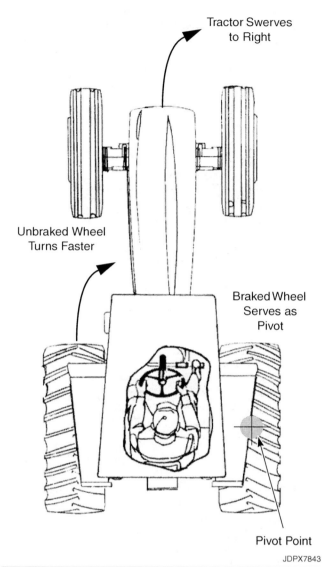

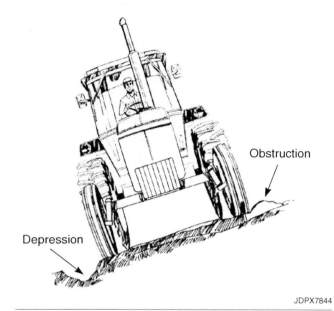

JDPX7844

Fig. 34 — Watch for Both Obstructions and Depressions That Might Upset the Tractor

4. Do not let your tractor bounce. This causes loss of steering control, and you do not have full control of the tractor. Slow down to avoid bouncing.

5. Drive slowly in slippery conditions. If your tractor starts to skid sideways to your direction of travel, you could easily tip over in a ditch or crash into an obstruction. Beware of this hazard: If your tractor slides sideways and the tires hit an obstruction or a surface that brakes the sliding of the tires, your tractor may roll over.

JDPX7843

Fig. 33 — The Unbraked Wheel Can Cause a Tractor to Swerve Out of Control

3. Restrict your speed according to operating conditions. To avoid an upset, you need complete control over your tractor, and a travel speed slow enough for you to see and to react to hazards in your path. Always watch ahead for bumps, rocks, stumps, and other obstructions that could tip one side of your tractor up past its tipping point (Fig. 34). Also, watch for holes, ruts, and depressions that could drop one side down, putting your tractor in an unstable position. That has caused many overturn accidents in transport.

6. Pull heavy loads and equipment at safe speeds. Reduce travel speed if towed equipment fishtails (weaves back and forth behind the tractor). The whipping action of the equipment is transferred to the front wheels of the tractor. If it becomes severe, or if you hit a slippery spot, the tractor could be thrown out of control into a sideways tip or into a ditch. You can get the same results when you try to stop a heavy load rolling behind a tractor. When the brakes are applied, the load pushes forward, forcing the tractor to skid. If it skids sideways, the tractor could overturn (Fig. 35).

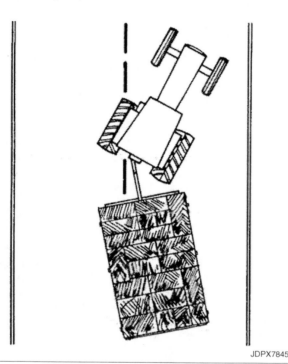

JDPX7845

Fig. 35 — Fishtailing or Severe Braking at High Speed Can Cause Jackknifing and Rollover

7. Turn safely. Slow down before turning. Remember that centrifugal force affects the stability of your tractor. The chances of tipping are greatest when crossing steep slopes while turning uphill or near ditches, but tractors can also tip on level ground, traveling at moderate speeds. Do not apply individual brakes for turning when driving faster than normal field-working speeds. Remember that quick, short, brake-assisted turns at high speeds can overturn tractors. Here's the safest procedure: Slow down by reducing engine speed before turning. Apply both brakes if braking action is required. Then turn as widely as you can with engine power pulling the load.

8. Use engine braking when going downhill. Runaway tractors often tip over. Shift to a lower gear and reduce speed for downhill travel (Fig. 36). If in doubt as to what gear to use, select the lowest-speed gear. Shift before you start downhill. Do not disengage the clutch and try to shift after you have started. The brakes may not hold, and you may not be able to shift back into gear. If the engine brakes too much, increase speed slightly with the

throttle. If it doesn't brake enough, apply the tractor brakes. If your tractor is equipped with front-wheel drive, engage it to gain the advantage of four-wheel braking.

JDPX7833

Fig. 36 — Shift to a Lower Gear on Level Ground, Before Going Downhill

9. Some tractors "freewheel" and provide no engine braking in certain speed ranges. If your tractor is one of these, travel downhill using only those shift positions that provide engine braking action. Know what these shift positions are before you first drive the tractor. Check your operator's manual.

10. Avoid crossing steep slopes, if possible. Tractor stability is reduced on steep slopes. If you must cross a steep slope, take every precaution to avoid rolling over. Drive slowly, avoid quick uphill turns, and watch for holes and depressions on the downhill side and for bumps on the uphill side. Keep side-mounted equipment on uphill side of tractor, if possible (Fig. 37). Turn downhill if stability becomes uncertain. Space the rear wheels as far apart as you can to increase stability.

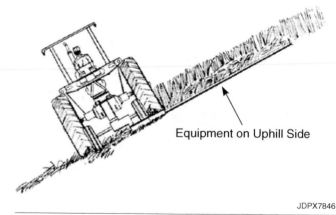

Equipment on Uphill Side

JDPX7846

Fig. 37 — Keep Side-Mounted Equipment on Uphill Side, if Possible

11. Stay away from ditches and riverbanks. Stay at least as far away from the bank as the ditch is deep. The weight of the tractor could cave the bank in if you approach or cross the shear line (Fig. 38). As you approach ditches and banks, look ahead rather than at your equipment. Give yourself plenty of room for turning and watch for holes, gullies, and washouts that could place your tractor in an unstable position.

Correct: Bucket Low While Transporting

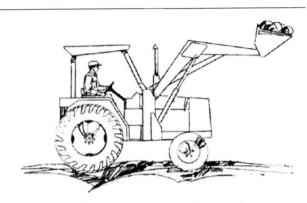

Incorrect: Bucket High While Transporting

JDPX7848, JDPX7849

Fig. 39 — Always Lower the Bucket When Transporting

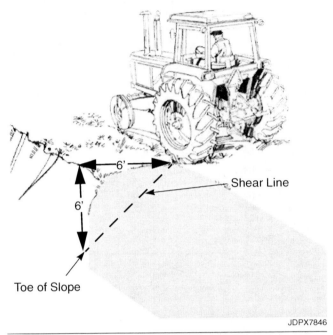

Shear Line

Toe of Slope

JDPX7846

Fig. 38 — When Operating Near Ditches and Banks, Always Keep Your Tractor Behind the Shear Line

12. Operate front-end loaders cautiously. Mounting a front-end loader on a tractor raises its center of gravity. In addition, loader-tractors often operate in confined areas that make short turns unavoidable. Both of these factors make loader-equipped tractors susceptible to rollovers caused by centrifugal force. To avoid an upset:

a. Keep the bucket as low as possible when turning and transporting (Fig. 39).

b. Watch carefully for obstructions and depressions.

c. Handle the rig smoothly, avoiding quick starts, stops, and turns.

NOTE: Mounted loaders should be removed before the tractor is used for field or transport work. If this is not practical for a short job, keep the bucket positioned low, watch for obstructions, and drive slowly.

Rollovers to the Rear

1. Hitch towed loads only to the drawbar. The regular drawbar or the drawbar attachment available for the three-point hitch are the proper hitching points on a tractor for pulling towed loads. Never attach a towed load to the axle, to one of the lower links, or to the top link of the three-point hitch. Doing so can pivot the tractor into a rearward tip.

When using a drawbar attachment between the draft arms (lower links) of a three-point hitch, keep the hitch low, at the height recommended in your operator's manual. Install stay braces, if available, and lock the hitch control lever in position to keep the hitch from raising. Raising of a three-point hitch can be caused by hydraulics, or by the towed load pushing against the hitch when braking or traveling downhill.

NOTE: Holes are provided on the frames of some tractors for attaching mounted implements and machines. Do not use these holes to attach a clevis, chain, or other device to pull a towed load.

2. Limit the height of three-point hitch drawbars. Do not pull with three-point hitch drawbars positioned higher than about 13 to 17 inches (330 to 430 mm) above the ground (Refer to your operator's manual). To avoid accidentally raising or lowering the three-point hitch above or below the desired height, set the height limits on the rockshaft control. Some older systems require engaging stay braces to prevent inadvertent action (Fig. 40).

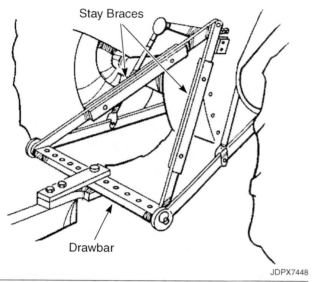

JDPX7448

Fig. 40 — Stay Braces Will Prevent Raising the Hitch Accidentally on Older Type Three-Point Hitches Such as This

3. Use weights to increase stability. Add front-end chassis weights to counterbalance rear-mounted implements and heavy vertical drawbar loads like two-wheeled trailers, manure spreaders, and rear-mounted loaders (Fig. 41). Do the same for pulling heavy loads uphill to help offset the loss of tractor stability. Add rear-wheel weights or tire ballast to counterbalance front-end attachments like spray tanks or front-end loaders. Refer to your operator's manual for recommendations.

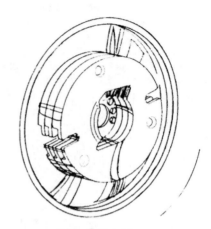

Wheel Weights

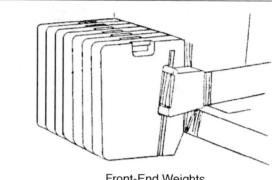

Front-End Weights

Tire Ballast

JDPX7850, JDPX7449, JDPX7851

Fig. 41 — Three Ways to Add Weight: As Wheel Weights, Front-End Weights, and Tire Ballast

4. Start forward motion slowly. Gunning the engine and jerking your foot off the clutch to start a heavy load is one of the surest ways to tip a tractor over backward. To start safely, use only enough speed to prevent engine lugging, and engage the clutch slowly. You can increase speed once you are underway.

5. Change speed gradually. Move the throttle lever slowly when changing engine speed. Accelerating a heavy load quickly, especially on uphill grades, can lift the front wheels off the ground, causing loss of steering control. And it could be the start of a rearward upset.

6. Brake cautiously when backing down a grade. Braking hard will turn the tractor over backwards. If you must go up a steep grade, back the tractor up in reverse, and come down by driving forward (Fig. 42). If you must back down a slope, do it slowly, keeping the tractor in a low gear, so braking will not be required. If brakes are applied backing downhill, the tractor could rotate around the rear axle, tipping it backward. The faster the speed and the steeper the grade, the easier it is for the tractor to tip.

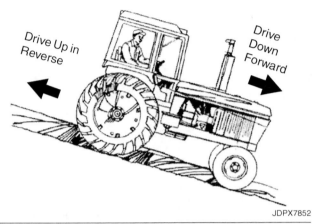

JDPX7852

Fig. 42 — Direction of Travel Up and Down Steep Slopes

NOTE: *If your tractor should ever roll backward down a steep grade with the clutch disengaged, think twice about engaging it. Engaging the clutch is nearly the same as applying the brakes, and could result in a backwards tip. If you find yourself in this situation, riding the tractor to the bottom of the slope without engaging the clutch or applying the brakes is usually the safest procedure.*

7. Plan for safe uphill pulls. Pulling heavy loads uphill calls for a number of precautions, since the slope and drawbar leverage both work to tip the tractor backward. The safest procedure is to add front-end weights, set the drawbar in its lowest and longest position, engage the clutch slowly, and accelerate gradually.

8. Drive around ditches. Do not cross them unless it's absolutely necessary. If you must cross, recognize the hazards involved. Situations vary so much that it's impossible to outline all recommendations, but keep these points in mind:

   a. Drive forward when going downhill, using low gear. If the slope is steep or slippery enough for the tractor to slide, do not go down.

   b. Back the tractor up steep grades. If you can't, shift to a low gear and drive forward very slowly. Maintain uniform engine speed and try to avoid moving the throttle. Rapid acceleration could throw your tractor over backward. Keep in mind that the governor may accelerate your tractor automatically even though you do not move the throttle lever.

9. Back your tractor out if it gets mired down in mud (Fig. 43). If necessary, do any or all of the following to enable you to back out:

   a. Dig mud from behind the rear wheels.

   b. Unhitch any towed implement or machine and pull it away with a chain to a dry location.

   c. Place boards behind the wheels to provide a solid base and try to back out slowly.

   d. When backing out is impossible, dig mud away from the front of all wheels, unhitch any towed load, if necessary, and drive slowly ahead.

   e. If you need a tow from another tractor, use a long chain (not a nylon rope) and caution your helper to engage the tractor clutch slowly and to use the tractor power cautiously to prevent the towing tractor from upsetting rearward.

Danger of Tipping

JDPX7853

Fig. 43 — Follow Recommendations to Get Out of This Situation Safely

 CAUTION: Never put boards or logs in front of the drive wheels and attempt to drive ahead. If the drive wheels catch on them and cannot turn, the tractor may tip over backward.

10. Back your tractor out if the drive wheels get lodged in a ditch. Tractor stability is reduced by the tipped angle of the tractor, and the drive wheels are not as easily turned by the axle as they would be on level ground. Trying to drive the tractor ahead may tip it over backward.

Rollover Protective Structures

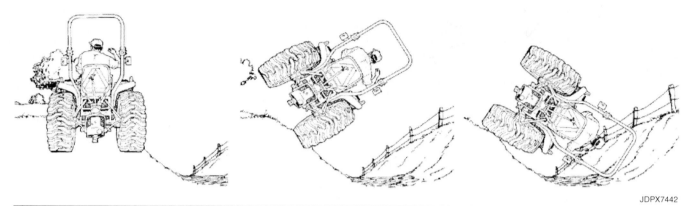

Fig. 44 — ROPS Are Designed to Limit Most Upsets to 90°, and to Protect the Operator in Upsets Beyond 90°

Rollover protective structures (ROPS) with seat belts were introduced for many tractors in the 1960s. They were intended to meet two major goals (Fig. 44):

- Limit many upsets to 90°

- Provide protection for the operator in upsets or overturns

ROPS, with seat belts, have proved to be highly effective in preventing fatalities in tractor overturns. New tractors are equipped with them. ROPS can also be fitted on many older tractors. They should be installed by a dealer. If your tractor has no ROPS, ask your dealer if one is available.

There are two basic types of ROPS for farm tractors: protective frames and protective enclosures.

Rollover Protective Frames

Protective frames are generally two-post or four-post structures attached to the tractor chassis (Fig. 45). Two-post structures may be rigid-frame types, or they may be foldable or adjustable, to accommodate low clearance needs such as doorways or tree branches. Some frames may be equipped with overhead canopies to protect against sun and rain. A seat belt is provided to keep the operator inside the protective zone to help prevent crushing injuries in an upset.

Four-Post Protective Frame

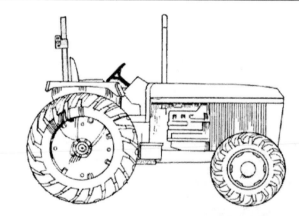

Two-Post Protective Frame With Folding Hinge

JDPX7854, JDPX7855

Fig. 45 — Four-Post and Two-Post Rollover Protective Frames. Some Frames Can Be Folded Down for Low Clearance

Rollover Protective Enclosures

Rollover protective enclosures are cabs or enclosures built around a protective frame. A strong metal ROPS frame is incorporated into their design (Fig. 46). In addition to protecting the operator during upsets, these ROPS enclosures provide other safety and health benefits. Since they're enclosed by windows and doors and have fans to bring in filtered air, enclosures are slightly pressurized to help keep dust from entering the compartment. Many enclosures reduce engine noise and vibration. They help prevent falls from the tractor. And they can be equipped with heaters, air conditioners, entertainment systems, radios, and cellular phones to provide additional comfort, convenience, and entertainment for the operator.

JDPX8087

Fig. 46 — The Heavy Metal Frame of This Enclosure Provides Rollover Protection

It is important to realize that some enclosures being used and sold are designed primarily to protect the operator from the weather. They are not designed to provide overturn protection. Consequently, these enclosures may be crushed if the tractor tips.

Enclosures and frames designed to provide overturn protection are built to meet the standards and regulations of various agencies. If an enclosure or frame has not been certified as meeting these standards, you have reason to question whether it will provide adequate protection in an upset. To check for overturn protection, check the manufacturer's literature, or look for a label on the frame or enclosure for evidence that it meets the standards of one or more of these organizations or governmental units:

- The American Society of Agricultural and Biological Engineers (ASABE)

- The Society of Automotive Engineers (SAE)

- The State of California (or other state certification)

- The federal government (such as established by the Occupational Safety and Health Act of 1970-OSHA)

- European Economic Community (EEC or EC)

- Organization for Economic Cooperation and Development (OECD international)

Replacement and Maintenance of ROPS

ROPS designed, manufactured, and tested to meet tough standards are made of steel specified to meet those standards. ROPS are designed to absorb a lot of energy in an overturn. To absorb energy they must bend somewhat, but must maintain a defined zone of operator protection in an overturn. After a ROPS is bent, it will have less capacity for providing protection and should be replaced. Observe the following:

- Do not attempt to cut, drill, weld, bend, or otherwise modify a ROPS. That will weaken it.

- Do not try to build your own frame or enclosure. You will not be able to test the design or strength of construction to be sure of adequate protection.

- After an upset, have your dealer inspect the entire ROPS. An upset places severe stress on the structures of ROPS frames or enclosures (Fig. 47).

JDPX7664

Fig. 47 — When a ROPS Is Damaged, It Must Be Replaced, Not Repaired. This One Did Its Job

NOTE: If any structural member is damaged, the entire ROPS should be replaced. Do not attempt to repair it. Consult your dealer and follow the manufacturer's instructions for repair or replacement.

- If a rollover protective structure (ROPS) is loosened or removed for any reason, make sure all the parts are installed correctly. The specified mounting bolts must be used and tightened to proper torque, as recommended in the operator's manual.

Using Foldable and Telescoping ROPS

Foldable ROPS are available for many tractors (Fig. 48). They can be folded down to accommodate low overhead-clearance situations such as low-clearance buildings. Telescoping-type ROPS, which can be telescoped to a lower position, are recommended for orchard and vineyard operations.

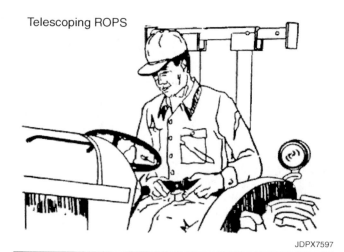

Telescoping ROPS

JDPX7597

Fig. 49 — Buckle Up When Your Tractor Is Equipped With ROPS Enclosure or Fully Upright ROPS Frame

JDPX8088

Fig. 48 — Foldable ROPS Hinges Down for Low Clearance; Pin It Fully Upright as Soon as Clearance Permits

If ever a tractor is operated with the ROPS folded or telescoped down, operate with extreme caution and DO NOT USE THE SEAT BELT. When in the lowered position, ROPS cannot provide overturn protection, so you should NOT be belted into the seat. Be sure to pin the ROPS securely in its fully extended or vertical position as soon as overhead clearance permits, and use the seat belt (Fig. 49).

Operating ROPS-Equipped Tractors

1. Use the seat belt. The belt is intended to hold you within the safety zone of the ROPS if an upset occurs. Without the seat belt, the risk of injury increases from being thrown out of a ROPS frame or tossed about in a ROPS enclosure. But, do NOT wear a seat belt if your tractor is not equipped with ROPS or if a foldable ROPS is not in its fully upright position. In an upset, you would be held in the seat without overturn protection.

2. Stay alert. An enclosure will muffle some of the warning sounds from machinery, other people, livestock, or motor traffic. Be aware that you are insulated from some warnings, and you need to stay alert at all times. Exercise all precautions for safe operation.

3. Beware of surface conditions outside enclosures. The comfort and shelter of an enclosure can make you forget how rough or slippery it really is outside. Exercise all precautions for safe operation.

Pesticides and Operator Enclosures

Avoid contact with pesticides. An operator enclosure typically does not protect against inhaling harmful pesticides, even if it has an air filter. When operating in an environment where harmful pesticides are present, follow the safety precautions provided by the pesticide manufacturers and in equipment operator's manuals. Here are some precautions:

• Outside the operator enclosure, wear protective clothing as required by pesticide use instructions.

• If the pesticide use instructions call for respiratory protection, wear a respirator inside the cab. Do not depend on cab filters to remove harmful particles. Change filters frequently to avoid pesticide buildup in the filter. Air passing through contaminated filters can carry harmful particles into the air you breathe.

• Avoid leaving the enclosure to enter a pesticide-treated area, if possible. If you must leave, for instance to work on contaminated equipment such as sprayer nozzles, or for mixing or loading, wear personal protective equipment as required by pesticide use instructions.

• Avoid bringing pesticide contamination into the enclosure. Before re-entering the cab from a pesticide environment, remove protective equipment and store it outside the cab in a sealable container, or inside the cab in a pesticide-resistant container such as a plastic bag. Also, clean your shoes or boots to remove soil or other contaminated particles.

For more details on chemical safety, see Chapter 9, "Chemical Safety".

Falls From Tractors and Implements

Falling from tractors and implements is a common cause of serious accidents.

Falls are needless and preventable accidents. Falls occur from moving and parked tractors and implements. Let's explore some methods for preventing these accidents.

Mounting and Dismounting

1. Keep the steps and platform clean and dry. Take time to clean off mud, ice, snow, grease, and other debris that accumulate on the platform and steps. Do not carry tools or log chains on the platform. Equip your tractor with a toolbox or use a trailer to carry tools for doing maintenance work.

2. Do not jump from the machine. There is always the danger of catching your clothing on pedals, levers, or other protruding parts. And you could land on an uneven surface and injure your ankles, legs, or back.

3. Use handrails, handholds, and steps to pull yourself up to the operator's platform (Fig. 50). Try to keep three points on the machine at all times — either two hands and one foot, or two feet and one hand. Wear heavy-treaded safety boots.

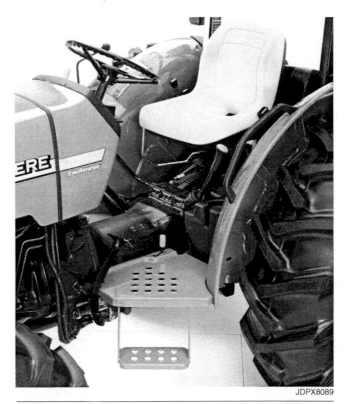

JDPX8089

Fig. 50 — Use Steps, Ladders, and Handholds to Get On and Off Farm Machinery Safely; Do Not Jump

4. Face the steps when you dismount to avoid falling.

Four-Wheel-Drive Tractors

When getting on and off, and operating articulated steering four-wheel-drive tractors, you should exercise the same precautions as for other tractors.

However, there are some special considerations.

The front and rear chassis sections can swing quickly to the right or left when the steering wheel is turned (Fig. 51). A mounted implement will also swing. Take these precautions:

1. Avoid moving the steering wheel as you mount the tractor. If you must enter or exit the cab when the engine is running, do not grasp or pull the steering wheel as you do so. The power steering responds quickly, and could cause the chassis and mounted implements to swing into someone.

2. Do not allow anyone to be in the immediate area before or during start-up. They could be struck by the swinging chassis or mounted implement.

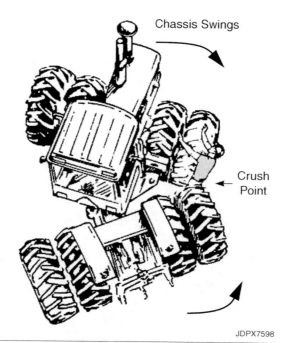

Fig. 51 — Do Not Allow Anyone Near the Tractor or Mounted Implements When Starting Four-Wheel-Drive Articulated Tractors

3. Never stand between the tires, and never let anyone else stand between the tires when the engine is running. If the steering wheel is accidentally moved a little, anyone standing between the tires could be crushed.

4. Lock the tractor hinge with proper pins or bars to prevent pivoting in these situations: before operating a stationary PTO implement, before performing service work near the hinge area, before towing the tractor, and before lifting the tractor or transporting it on another vehicle. Remove the lock pin or bars before resuming operation.

Operator Falls

Operators sometimes fall from tractors and are crushed under the wheels or mangled by trailing equipment. This won't happen if you observe the following safety practices:

1. Operate the tractor from the operator's platform only. Fasten the seat belt if your tractor is equipped with a protective frame or enclosure. If you get tired of sitting, stop and take a break.

   Never operate the tractor, while riding on the drawbar, sitting on the fender, standing on the steps, or sitting on the backrest of the operator's seat.

2. Maintain safe operating speeds. Never drive so fast that the front wheels of the tractor bounce. Watch ahead for obstructions and avoid them. Slow down before making turns.

3. Rest when you are tired. Stop the tractor and lie down for a short nap if you feel sleepy. Stop for 10 or 15 minutes every 2 to 2-1/2 hours. Do not drive when you feel like dozing. In this condition you are not alert enough to operate a tractor.

Carrying Passengers

Most tractors are designed to carry one person, the operator. The proper place to ride is sitting in the operator's seat.

Runover accidents are a major cause of farm work fatalities. Most of the victims are "extra riders," or passengers riding along with the operator. Some victims have fallen off tractors and were crushed under the tires. Others have been mangled by trailing implements such as rotary cutters. Children have been backed over as they approached a tractor for "another ride." Do not let it happen!

You will be tempted to transport a helper. Children often plead for rides, but do not give in to them (Fig. 52). The risk is too great. An unexpected bump or turn could toss a passenger off the operator station or through the cab doorway onto the ground where he or she can be run over. Some victims have accidentally unlatched a cab door, fallen out, and been run over.

Fig. 52 — Avoid Extra Passenger Accidents. Observe The "No Rider" Rule

Passengers, especially infants and children, may be unnecessarily exposed to chemicals, dust, noise, and vibration, which could have long-term, chronic effects.

Tractor ROPS, intended primarily for operator protection, will generally not provide the same degree of overturn protection for a passenger as for the operator.

In addition, passengers can interfere with operation of controls. They can accidentally move a control, and they generally distract the operator.

Plan your work to avoid carrying passengers unless the machine manufacturer has provided a passenger seat. It's safer to use a different form of transportation for helpers, such as another tractor or a pickup truck.

Power Takeoff Accidents

Severe injury or death can result from accidents involving the power takeoff. There are two types. The most common is getting loose clothing caught by the rotating shaft. And, on occasion, an operator is struck by a broken or disconnected shaft as it swings violently behind a tractor. Avoid these accidents by following proper installation, maintenance, and operating procedures.

Shielding

An unguarded PTO shaft is dangerous. It can catch on your clothing before you realize it. If you're lucky, your clothes will tear, freeing you without serious injury. If you're not so lucky, you could be strangled, mutilated, or killed by the high-speed shaft.

For the Tractor

Master PTO shields help prevent accidental contact with the tractor stub shaft and the front universal joint of the attached machine's driveline (Fig. 53). On some tractors, this shield also provides a point of attachment for older-style tunnel shields (Fig. 54).

PTO stub shaft guards are provided for many tractors to completely enclose the tractor stub shaft when the PTO is not being used.

Fig. 53 — Prevent Injuries; Keep Tractor PTO Shields in Place and in Good Condition

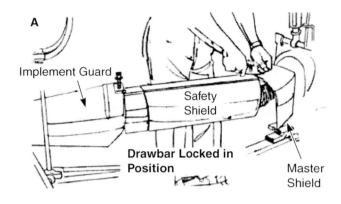

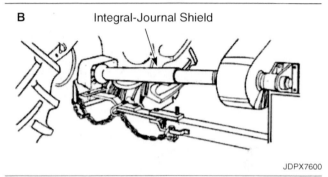

JDPX7600

Fig. 54 — A: Old-Style Inverted-Tunnel PTO Shielding. B: Integral-Journal PTO Shielding Stops Rotating When a Person Contacts It

For the PTO-Operated Machines

Tubular, integral-journaled shields completely enclose the PTO shaft (Fig. 54). These shields are metal or plastic tubes supported on bearings so the shields rotate independently of the shaft. When the PTO shaft is turning, the shield also rotates with it. But when contacted externally, shield rotation stops while the shaft continues to rotate inside the shield. Also, on some machines, the shield is chained or tethered to the implement so it cannot rotate, but the shaft rotates within the shield.

Keep both types of tubular shields in good condition, and they will protect you from the grabbing or wrapping action of revolving shafts and universal joints. Make sure the shaft and joints can rotate freely within the shields.

The fully-enclosed PTO shield system also uses integral-journaled rotating shields in combination with complete shielding of the tractor-attached universal joint and coupler (Fig. 55). When in place, the driven, rotating parts are not exposed to catch fingers, gloves, or clothing. If a person accidentally contacts the shield or coupler enclosure, shield rotation stops, giving protection while the driven parts rotate within the shield.

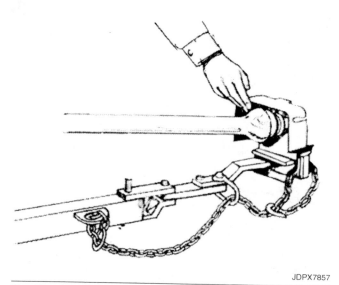

Fig. 55 — The Fully-Enclosed, Shielded Driveline Completely Covers the Tractor-Connected Universal Joint and Coupler

Entanglement With the PTO

To prevent injury from entanglement with the power takeoff shaft, follow these safety practices:

1. Always disengage the PTO, shut off the engine, and take the key before getting off the tractor (Fig. 56). This gives you three-way protection: from shaft rotation; from moving machine parts; and it prevents the unexpected engagement of power by another person when you are cleaning, lubricating, adjusting, or making repairs.

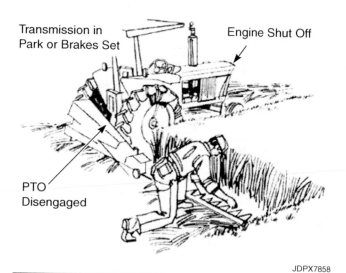

Transmission in Park or Brakes Set

Engine Shut Off

PTO Disengaged

Fig. 56 — Always Disengage the PTO, Shut Off the Engine, and Take the Key Before Dismounting the Tractor

2. Keep the tractor PTO master shield in place. The tractor master shield is to be removed only when required for special attachments that provide equivalent shielding. Some master shields are hinged so they can be moved upward out of the way while attaching an implement PTO. After the implement is attached, return the shield to its proper operating position. PTO stub shaft and coupler area should be shielded at all times.

3. Check to see that integral shields are in good condition. When the powershaft has stopped, you should be able to rotate the shield freely by hand. If the shield or bearings are damaged, make the necessary repairs immediately. Also check tethered shields to be sure the tether is secure and the shaft can turn freely inside the tubular shield.

4. Keep tunnel shields in proper alignment over the powershaft. Keep them in good condition. Make sure they're securely fastened in place all the way from the implement gearbox to the tractor master shield.

5. Never step across a rotating powershaft. When PTO-driven machines are running, always walk around the machine. Safety devices could malfunction.

6. Wear close-fitting clothing and short or restrained hair around PTO-operated machinery. Do not wear scarves, hooded jackets, pull strings and other dangling or loose-fitting clothing or long hair around farm machinery. They can be caught and wrapped by rotating PTOs and other rotating components. Many victims have been severely injured, handicapped, and disfigured. Some died. Dress for safety and avoid such tragedies.

Electronically Controlled PTO

Many tractors have interlocks that prevent the PTO from engaging when the engine is being started. Follow these safety precautions:

1. Do not attempt to modify or disconnect such safety interlocks or any other safety features.

2. Have PTO control systems repaired only with parts recommended by the manufacturer. Some electrical/electronic PTO systems have special switches, relays, wiring harnesses, and computers with built-in safety features. Do not substitute or improvise. Check the operator's manual, and see your dealer for repairs.

3. Be careful when checking electrical PTO circuits. If you must move a wire or wiring harness or otherwise check for a loose connection, do not attempt it with the PTO switch turned on, unless everyone is clear of the PTO or driven parts. The PTO could turn when electrical contact is made.

Broken or Separated Shafts

A PTO shaft may break or separate during operation if improperly used or adjusted. If it does, the tractor-driven end swings violently behind the tractor, often causing severe tractor damage or injury to the operator. These accidents are most likely to occur when powershaft assemblies are improperly installed, when the tractor drawbar is incorrectly positioned, or when the PTO shaft is abused during operation. To prevent these accidents:

1. Keep universal joints in phase. Universal joints are properly phased or aligned when the end yokes are positioned in the same plane. Fig. 57 shows correct assembly. Some shafts can be separated; others can't. Separable shafts must be reassembled in phase to avoid driving the machine at irregular speeds and placing a severe strain on the shaft. Many shafts are designed so they cannot be reassembled improperly.

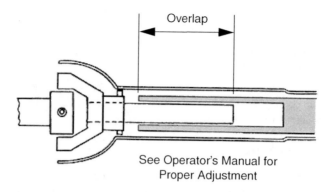

See Operator's Manual for
Proper Adjustment

JDPX7599

Fig. 58 — Shaft Ends Must Overlap Properly to Permit Telescoping Without Compression or Separation

Phasing

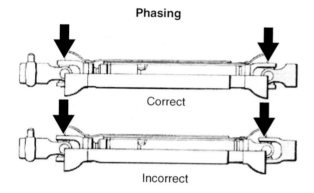

Correct

Incorrect

Note Position of Right-Hand Joints

JDPX7674

Fig. 57 — Correct and Incorrect Assembly of the PTO Shaft Assembly

2. Always use the driveline specified for the implement. Each implement requires a specific driveline. Drivelines must overlap properly (Fig. 58). Otherwise, they may separate when the tractor passes over a ridge (Fig. 59), or telescope together so much that they bottom out when the tractor goes through a depression or a sharp turn. If that happens, the shaft and bearing supports may be damaged. Drivelines are also made in different diameters, depending on power requirements. Use only the PTO driveline recommended for each of your machines, and keep PTO shafts properly positioned as instructed in the operator's manual.

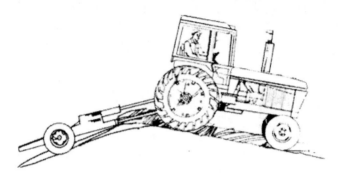

Drivelines Lengthen When Driving Over
the Crest of a Hill

Drivelines Shorten When Driving
Through Depressions

JDPX7859, JDPX7860

Fig. 59 — Powershafts Must Lengthen Without Separating and Shorten Without Bottoming Out

3. Position the drawbar correctly. You may have to position the tractor drawbar differently for each machine used in your farming operations, To do this:

   a. Find the following measurements in the tractor operator's manual (Fig. 60):

   • The recommended tractor drawbar height of the PTO-driven machine you are using

   • The recommended horizontal distance between the end of the tractor PTO shaft and the center of the drawbar hitch hole

   b. Using these measurements, position the drawbar parallel to and directly below the tractor shaft.

   c. Lock the drawbar in this position.

JDPX8084

Fig. 60 — Refer to Your Operator's Manual for Adjustment of Drawbar Length and Height

4. Make sure all connections are tight. Fasten universal joints securely to the tractor shaft. After connecting the driveline to the tractor, check by pulling firmly on the driveline assembly to make sure it is securely fastened (Fig. 61). Check all universal joints and connections on the shaft regularly.

JDPX7958

Fig. 61 — Be Sure Coupler Locking Mechanism Is Securely Engaged

5. Do not abuse the PTO shaft. Specifically:

   a. Avoid sharp turns when the driveline is rotating, Disconnect the driveline if the corner is too tight.

   b. Avoid driving through ditches or depressions if telescoping will be excessive.

   c. Do not overtighten overload protection clutches on PTO-driven machines.

   d. Apply power to the driven machine gradually.

   e. Do not jerk the PTO shaft by applying power suddenly to clear machines that are plugged.

Overspeeding PTO Implements

Tractors are equipped with either a 540 rpm or 1000 rpm PTO stub shaft, or both. The 540 rpm stub shaft has 6 splines. There are 20 or 21 splines on the 1000 rpm shaft. This is intended to prevent powering a 540 rpm implement from a 1000 rpm tractor stub shaft, and vice versa. Do not override that safety feature with a PTO adapter.

Overspeeding can damage implements, and people can be injured by failed parts flying from them. Operate implements only at PTO speeds for which they are designed, as instructed in the operator's manual.

Some tractors with interchangeable stub shafts have a shiftable, two-speed PTO system. Be sure the speed selector lever is in position for the correct implement PTO speed to avoid overspeeding.

Hitching Accidents

Farm equipment is big and heavy, so be careful when hitching, unhitching, mounting, or removing it. Three risks are involved: crushing your hands and feet, being crushed between the tractor and the equipment, and being crushed by an implement or machine that falls unexpectedly.

Hitching accidents take many lives every year. To avoid them, follow these safe hitching and mounting practices:

Hitching

1. Position your tractor for hitching from the operator's seat (Fig. 62). Never operate the tractor from the ground. The risk of it jerking and pinning you is too great.

Fig. 62 — Align the Hitch Holes From the Tractor Seat

2. Turn the engine off and shift to park or set the parking brake before dismounting to hook up. This precaution is especially important on sloping ground. Pulling or pushing the tractor or equipment to align it could bring it rolling toward you with crushing force.

   If you have problems aligning the hitch, you may find this procedure helpful: Back the tractor beyond the hitch point with the implement and tractor drawbars in line. Place the top of the implement clevis on top of the tractor drawbar, and insert the drawpin in the implement hitch hole. Then, inch forward. When the holes align, the pin can drop into place.

3. Shift to park and set the parking brake before permitting a helper to go behind the tractor to complete the hitch. If you can't align the hitch by yourself, back the tractor past the hitch alignment point, shift to a forward gear, and then complete the hitch by driving forward (Fig. 63). This eliminates the risk of crushing your helper behind the tractor if your foot slips off the clutch. Capable operators take pride in being able to hitch and unhitch by themselves.

4. Do not try to lift heavy equipment by hand. Put a jack under the hitch or equipment frame to position the hitch for attachment to the tractor. Many implements and machines are equipped with a jack or stand to keep the hitch at the proper height. Use these if they are provided. For some implements, the remote hydraulic cylinder can be used to lift the hitch into position.

When hitching, keep your fingers and hands away from pinch points. Never put your fingers in the holes in the tractor drawbar or in the holes in the equipment hitch when they are being aligned.

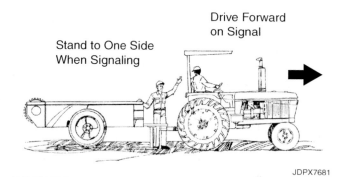

Fig. 63 — Drive Forward to Align the Hitch

5. If you are helping hitch, never stand behind a tractor when it is backing up. Stand to the side. Let the driver align the hitch and put the tractor in park or a forward gear. Then step in and make the hitch. The driver may inch forward to make the final alignment, but only on the assistant's signal!

6. Avoid being pinched or crushed by a load-sensing hitch. The three-point hitch draft arms of many tractors can be placed under three different modes of operation: floating, depth control, and draft or load control. When floating, the draft arms are raised and lowered, without interference of the hydraulic system, by the attached equipment as it follows the contour of the ground. When operating in depth control, the draft arms remain positioned at the height selected by the tractor operator. In draft or load control, the hitch is responsive to changes in the amount of draft (backward pull) of the implement. Gauge wheels are not necessary. When the draft decreases, as when plowing through light soil, the draft arms lower the implement. When draft increases, as when plowing through a tough spot, the arms raise the implement to reduce the draft of the implement on the tractor.

If your tractor is equipped with a draft- or load-sensing hitch, move the hydraulic control lever to a position that maintains a fixed draft arm height before attaching equipment to a three-point hitch. If you do not, the draft arms may rise unexpectedly when the equipment's weight is placed on the hitch, pinching or crushing your hands, arms, or legs.

7. Do not allow people or pets on the tractor operator station while you attach an implement to an electronically controlled hitch. If they move the switch accidentally, the three-point hitch may move quickly, crushing you against the implement.

8. Use caution when checking malfunction of electronically controlled three-point hitches, Do not stand in the hitch area while moving, wiggling, or adjusting a wire or wiring harness to check for a loose connection. The hitch may move quickly up or down when electrical contact is made.

9. Secure the hitch with a locking device that can't jump out and release the trailed equipment. Several types are available. Replace damaged pins (Fig. 64).

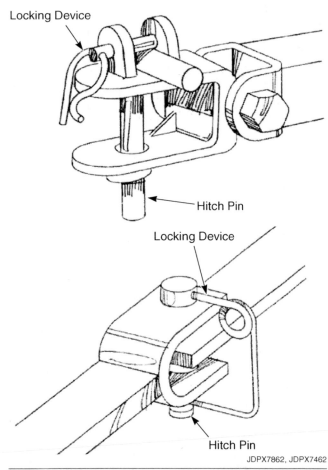

Fig. 64 — Use Hitch Pins That Have a Locking Device to Keep Them in Place

 CAUTION: Trash and ground contact can unhook hitch pins.

10. Attach a safety chain between the tractor and the equipment when transporting on public roads. The safety chain will help guarantee the security of the hitch. Safety chains are available from dealers. The chain and its attaching points should be strong enough to pull the gross weight of the towed load, and it should have a safety hook to prevent accidental unlatching.

Attaching Equipment

1. Use the proper hardware. Be sure you have the right brackets and the recommended sizes and types of bolts for attaching equipment to your tractor. Then you'll be able to complete the job quickly and in a safer manner, and you will be sure that the equipment is mounted securely. Do not use undersized bolts, bolts of the wrong grade of hardness, or improvised brackets. Use the hardware recommended for the equipment.

2. Use adequate jacks, hoists, or blocks to position the equipment (Fig. 69). This will eliminate the need to physically lift, pull, or push heavy equipment into alignment for mounting. It will also reduce the risk of getting your hands and fingers pinched, or having the equipment fall or tip on you. Many implements have their own stands and jacks. If these are provided, use them to make mounting safer and easier.

Fig. 65 — Use Stands or Blocks Provided to Support Equipment Adequately and for Ease of Mounting

3. Stay out from under equipment if it's not adequately supported. It should be fully attached to the tractor and resting on the ground, supported on solid blocks, or supported on the stands provided for the equipment. Do not rely on jacks. And do not rely on the tractor's hydraulic system to support raised equipment while you are under it. Hydraulic lines, valves, or seals may fail, or another person might accidentally lower the equipment by moving the hydraulic control lever. Remember that you may be pinched or crushed if the equipment tips, rolls, slides, or falls.

4. Protect your arms and legs. Watch for all possibilities of pinching or crushing your fingers, hands, arms, feet, and legs. Do not get them between the equipment and the tractor, between assemblies to be fastened together, or between linkages that move in a scissors-like manner. Be careful when prying with aligning punches and prybars. These tools are brittle and will snap if used with excessive force.

5. Make sure safety locks and catches are fastened. Many types of equipment and quick-attach frames are equipped with safety locks to secure the equipment in place. Be sure these are always in the locked position. If they aren't, the equipment may fall or tip when you get underway.

Unhitching Equipment

1. Select a good location. A firm, level, well-drained area away from livestock and traffic is best. In such an area, your equipment will not roll when uncoupled. It will be protected from damage by livestock and moving vehicles. Also, the stands and jacks, and the equipment itself, will not sink as they would into mud or soft soil. If you must uncouple wheel-carried equipment on sloping ground, block the wheels adequately to keep them from rolling. If you are in soft or muddy conditions, put boards under the support stands and the equipment, if necessary, to prevent them from sinking.

2. Lower support stands and jacks, or block the equipment. Many implements and machines are not balanced when uncoupled from the tractor. They may tip to the front, rear, or sideways. To prevent tipping, lower all stands provided by the manufacturer before you uncouple the equipment from the tractor. If stands or other devices are not provided, support the equipment on blocks. The direction of tip of load-carrying machines like grain drills, fertilizer spreaders, and sprayers depends on machine design, weight of the load and, frequently, as with manure spreaders, the location of the load within the machine. If you unhitch a machine that's not completely empty, be sure it is adequately supported in the direction it may tip. Your operator's manual may give you more information.

> ⚠ CAUTION: A shift in the center of gravity can make the implement tongue move up or down violently.

3. Use transport links or lower equipment to the ground (Fig. 66). Many implements and machines are equipped with transport links or safety locks to keep them supported in a raised position. If you store equipment in the raised position, engage these links and locks if they're provided, or use solid blocking, Do not rely on the pressure in the hydraulic system.

JDPX8090

Fig. 66 — Transport Links May Be Used to Keep Equipment in a Raised Position

Accidents on Public Roads

Many tractor fatalities occur on public roads and highways. Surprisingly, most of these accidents happen when operating conditions seem safest (Fig. 67). You can't afford to take chances, even when conditions seem ideal.

JDPX7863

Fig. 67 — Not as Safe as It Looks

Collisions with motor vehicles, driving off the road, and upsets are the most common types of machinery accidents on public roads. To prevent them, the operator must follow safety precautions when getting equipment ready for highway travel and when driving.

Highway Safety Equipment

To avoid accidents on the road or highway, you must be able to see clearly, and motorists must be able to see your machinery. They need to recognize that it is a slow-moving vehicle, it may be wide and long, and it may change directions. The safety equipment described here will help accomplish that.

SMV Emblem

Most state vehicle codes require that slow-moving-vehicle (SMV) emblems be displayed on tractors and farm equipment traveling on public roads. If farm employees are driving, federal law (the Occupational Safety and Health Act of 1970) requires display of the emblem.

Many new tractors and implements come factory-equipped with SMV emblems (Fig. 68). If you need to add or replace an emblem on your machinery, mount it at the rear, near the center, from 2 to 6 ft (0.6 to 1.8 m) above the ground, with the point at the top.

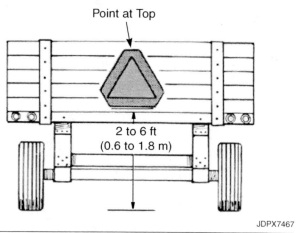

Fig. 68 — The SMV Emblem Correctly Displayed

Mount it perpendicular to the direction of travel. Proper mounting is important so the emblem can be seen by motorists behind you and so it will reflect back to them in their headlights.

The SMV emblem is NOT a substitute for warning lights and reflectors. They make your equipment visible, and the emblem tells them it moves slowly. Dirty, damaged, or improperly displayed emblems do little good. Keep them in good, clean condition. Replace them when they lose reflectiveness.

LIGHTS AND REFLECTORS FOR TRACTORS AND SELF-PROPELLED IMPLEMENTS

JDPX8091

Fig. 69 — Recommended Lighting and Marking for Tractors and Self-Propelled Machines (Check State and Local Regulations)

Several organizations have cooperatively developed standards for providing lights, reflectors, and SMV emblems on farm equipment: American National Standards Institute (ANSI); American Society of Agricultural and Biological Engineers (ASABE); and Society of Automotive Engineers (SAE). Most state laws generally conform to the standards, or allow them to be applied. Check your state requirements. ANSI/ASABE/SAE standards recommend the following highway safety equipment (Fig. 69).

- Two white head lamps

- One red taillight, visible from rear, on left extremity

- Two amber flashing warning lamps, visible from front and rear, at left and right extremities

- Two red reflectors, visible from rear, at extreme left and right extremities

- One SMV emblem, visible from rear

- Seven-terminal electrical connector on tractors to operate lights on towed implements

Lights, Reflectors, and SMV Emblem for Implements That Are Wide or Obscure Tractor Lighting and Markings

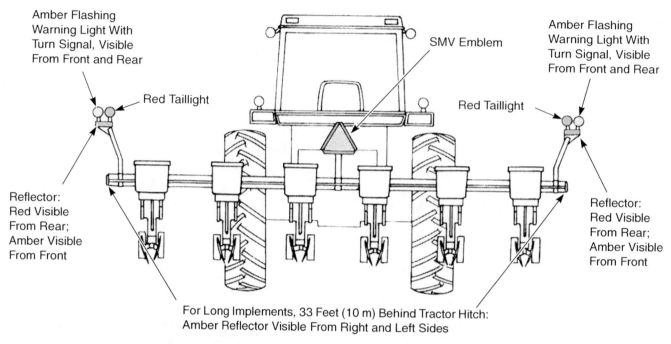

Amber Flashing Warning Light With Turn Signal, Visible From Front and Rear

Red Taillight

Reflector: Red Visible From Rear; Amber Visible From Front

SMV Emblem

Amber Flashing Warning Light With Turn Signal, Visible From Front and Rear

Red Taillight

Reflector: Red Visible From Rear; Amber Visible From Front

For Long Implements, 33 Feet (10 m) Behind Tractor Hitch: Amber Reflector Visible From Right and Left Sides

JDPX7864

Fig. 70 — Recommended Lighting and Marking for Towed Equipment (Check State and Local Regulations)

- One amber reflector, visible from front, at left extremity — For implements more than 4 feet (1.2 meters) left of tractor centerline

- Amber reflectors visible from left and right sides, 16.4 ft (5 m) apart on both sides — For implements more than 33 feet (10 meters) behind tractor hitch point

- Two red reflectors, visible from rear, at extreme left and right — For implements beyond 4 feet (1.2 meters) to the rear of tractor hitch point or more than 4 feet (1.2 meters) to right or left of tractor centerline.

- SMV emblem, visible from the rear — For implements that obscure SMV emblem on tractor

- Two amber flashing warning lamps, visible from front and rear at extreme left and right, flashing with tractor lights and turn signals — For implements that obscure any flashing warning lamp or extremity lamp on tractor; and for implements over 13 feet (4 meters) wide, or 79 inches (2 meters) left of center

- One red taillight, visible from rear, left extremity — If implement obscures tractor tail lamps

- A seven-terminal electrical connector to connect with tractor flashing lights and turn signals — For implements that obscure taillights on tractor; also implements over 13 feet (4 meters) wide, 79 inches (2 meters) left or right of tractor center, or beyond extremities

Lights, reflectors, and SMV emblems are available from many dealers. Equip your machines for safer travel on roads, and keep that equipment in good condition. Check your state regulations.

Headlights and Flashing Warning Lights

Keep headlights working properly and aimed low enough to avoid blinding approaching motorists.

For work at night, front and rear field lights are recommended, but they should never project rearward toward motorists. It would confuse them. Field lamps aimed downward to illuminate an area close to the machine can help motorists recognize it.

Some states require steady-burning amber warning lights, but most allow or require them to flash. Use your warning lights day and night on the road. Ask your dealer about state and local requirements.

Mirrors

Mirrors are effective safety devices (Fig. 71). They let you see all around you all the time.

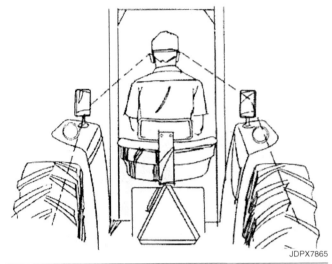

Fig. 71 — Mirrors Enable the Operator to See All Around

Mirrors designed specifically for use on tractors without operator enclosures are not always readily available. However, some mirrors designed for trucks or tractor cabs can be used by building a mounting bracket. Follow installation instructions to mount them firmly.

Tractor and Implement Condition

Keep your tractor and self-propelled implement in top condition for highway travel.

1. Brake pedals must apply equal braking force to each rear wheel without excessive foot pressure. Check pedal adjustment. Pedals get out of adjustment when used in the field where one brake is used more than the other.

2. Brake pedals should be locked together for highway travel.

3. Steering must be responsive, with no excess play in the steering linkage.

4. Tires must be in good condition, without cuts, cracks, or bulges that could cause a blowout.

5. All lights, reflectors, and SMV emblems must be in good condition: at least one red taillight on the left, two amber flashing lights visible from front and back, two red reflectors.

Driving Practices for Highways

1. Maintain full control of your tractor and equipment on the highway. The safety practices discussed in this chapter that deal with safe travel speed, turning, braking, hitching, and preventing upsets are very important for safe highway travel.

2. Stay alert. Keep your eyes moving and watch for hazards in the road and on the shoulders ahead (Fig. 72). Use the mirror or look back frequently to check for traffic coming from the rear, but avoid pulling the steering wheel in the direction you turn your head. Keep in mind that motorists often misjudge the speed of farm equipment and, to avoid hitting you from behind, they may be forced into panic braking or passing you on either side. Evaluate everything you see, forward and rearward, in terms of possible accident causes. Anytime the situation looks potentially hazardous, take the necessary precautions.

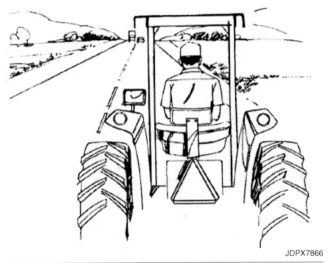

Fig. 72 — Keep Your Eyes Moving to Watch for Hazards on the Road and Shoulders Ahead

3. Wait for traffic to clear before entering a public road. Wait your turn (Fig. 73). Do not take a chance and pull out in front of moving traffic. A car traveling at 60 mph (97 km/h) on dry pavement needs approximately 600 feet (183 m) to stop, and approximately 700 feet (213 m) at 70 mph (112 km/h) — after the other driver realizes you're in his way. Remember that it's not easy to estimate the speed of approaching vehicles, and that the other driver may not see you. Misjudgment on your part or lack of attention by the other driver could result in a very serious accident.

Danger!
Wait for Road to Clear

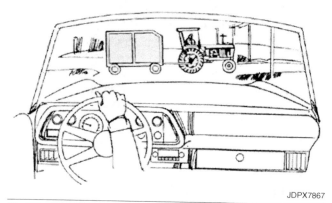

JDPX7867

Fig. 73 — This Tractor Operator Should Have Waited for the Road to Clear

4. Beware of blind intersections. Slow down or stop before entering blind intersections, where traffic from the right or left cannot be seen (Fig. 74). Intersections of country roads without stop signs are especially dangerous, since motor vehicles often speed through these intersections. Always be prepared to yield the right-of-way to other vehicles.

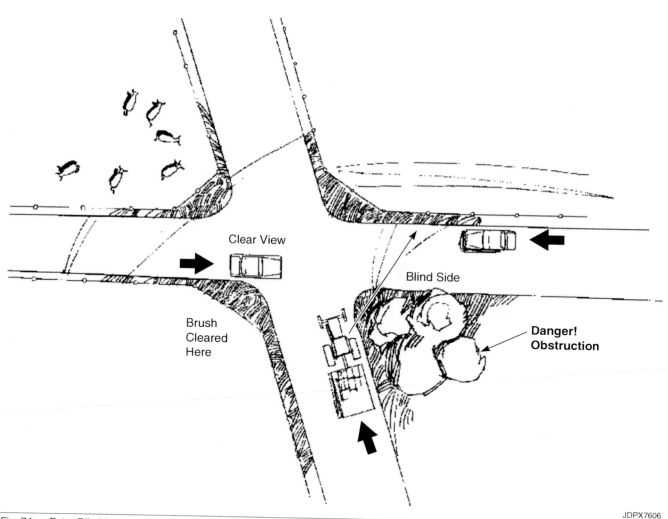

JDPX7606

Fig. 74 — Enter Blind Intersections Cautiously

5. Pull off the road and let traffic pass. (Fig. 75). However, be sure the shoulder is wide enough and smooth enough. Watch for road signs, bridge abutments, and also depressions, holes, and washouts that could cause you to upset in a ditch. If the shoulder is not wide or smooth enough for safe driving, do not drive partially on the road and partially on the shoulder. That encourages motorists to pass in unsafe situations. Occupy a full lane on a highway. Before turning into a traffic lane, make sure no vehicles are approaching.

JDPX7869

Fig. 77 — Do Not Signal Motorists to Pass

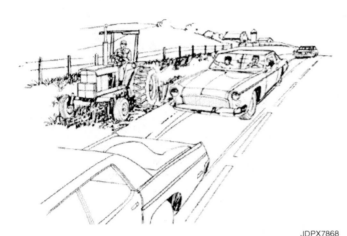

JDPX7868

Fig. 75 — Pull Over and Let Traffic Go By

6. Keep the approaching traffic lane clear. If your equipment is wider than your traffic lane, keep it over on the shoulder to enable motor vehicles to pass in either direction (Fig. 76).

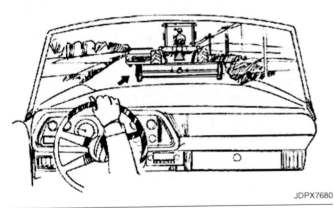

JDPX7680

Fig. 76 — Do Not Obstruct the Approaching Traffic Lane

7. Do not encourage motorists to take chances (Fig. 77). Signaling them to pass is risky. The traffic situation could change so quickly that they might not be able to pass safely.

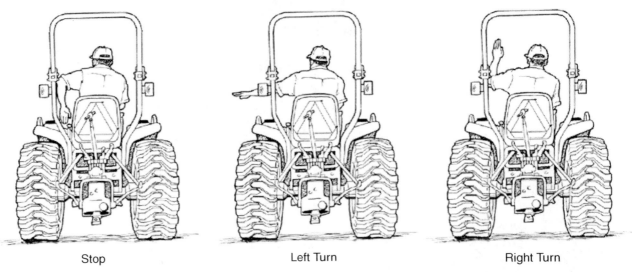

| Stop | Left Turn | Right Turn |

JDPX7469

Fig. 78 — Standard Hand Signals Warn Others of Your Intentions

8. Use hand signals or turn signal lights if available. Amber warning lamps are used as turn indicators. When signaling, the indicator light blinks, and the light on the opposite side burns steady. Keep your driving behavior predictable by giving advance warning of your intentions to turn or stop (Fig. 78). This lets other drivers govern their driving behavior accordingly. Except when making left-hand turns, do not stop in a traffic lane. Pull over to the shoulder if a sound one is available. If not, continue traveling in the traffic lane until you find a safe place to turn off.

9. Obey all traffic signs. They warn drivers of hazardous conditions like intersections, hills, curves, narrow bridges, and railway crossings (Fig. 79). Heed these warnings. It's also wise to plan your escape from a possible accident in case other drivers fail to obey these signs.

JDPX7605

Fig. 79 — Obey All Traffic Signs

Crossing Railroad Tracks

Railroad crossing accidents have claimed the lives of many farmers. It is usually a matter of not seeing or hearing the train or misjudging the speed of the train or the time it takes tractors and implements to cross the tracks.

Allow lots of time to cross railroad tracks. A tractor-implement combination 40 feet (12 m) long, crossing a rough railroad intersection at 5 mph (8 km/h) would take about 6 seconds to cross the tracks. From 500 feet (152 m) down the track, a train going 60 mph (97 km/h), or 88 feet per second (27 m/sec), would hit the implement before it crossed the track. Also, the tractor could stall, or a wide implement could get hung up on a rough crossing. You could be killed. Don't take that chance. Let the train pass before you drive across the track.

Towing Equipment

Extra care is necessary when towing equipment because you are controlling both the tractor and the implement behind it.

Most of the precautions for towing equipment have already been discussed in this chapter. Let's look at these again, along with a few others. First, we'll turn our attention to the precautions necessary for towing machines, implements, and wagons. Then we'll look at those that apply to towing tractors and self-propelled machines.

Towing Machines, Implements, or Wagons

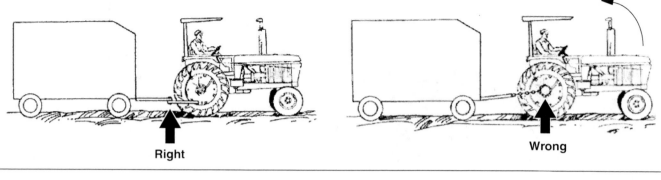

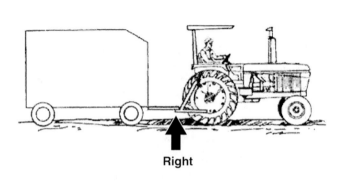
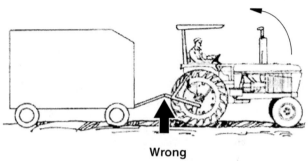

JDPX7870, JDPX7871, JDPX7872, JDPX7873

Fig. 80 — Prevent Upsets — Tow Only From the Drawbar or Three-Point-Hitch Drawbar Attachment, Set at Recommended Drawbar Height

1. Hitch only to the drawbar. Prevent tractor upsets by hitching pull-type equipment only to the drawbar or to the three-point-hitch drawbar attachment (Fig. 80). Never pull from any other point on the tractor. This aligns the tractor's center of pull with the load, helps to keep the front wheels of the tractor moving straight ahead, and provides extra steering control over the towed equipment.

2. Adjust drawbar height as recommended in your operator's manual. Do not position a three-point-hitch drawbar attachment higher than the position recommended for the regular drawbar. Install stay braces, if available, to prevent accidental raising of the draft arms. Remember, the higher the hitch point, the more susceptible your tractor is to upsets.

3. Use safety hitch pins. Prevent the possibility of any drawbar-connected load coming unhitched from the tractor. Pins with different types of locking devices are available. Some tractors and implements come equipped with hitch pins and locking devices incorporated into the design of the hitch. Keep these pins with their machines and always use the locking devices provided. Before highway travel, make sure the safety locks are in place. They can be knocked out by trash in field work.

NOTE: *Locking devices are also essential to keep implements connected to the three-point hitch in place. If an implement pin in one of the lower draft links pulls out, one side of the implement will drop to the ground. If the implement drags sideways behind the tractor, it could be caught by the tractor rear wheel and carried up to the operator's platform. To avoid this, make sure that each lower draft link implement pin is locked in place with a positive locking device.*

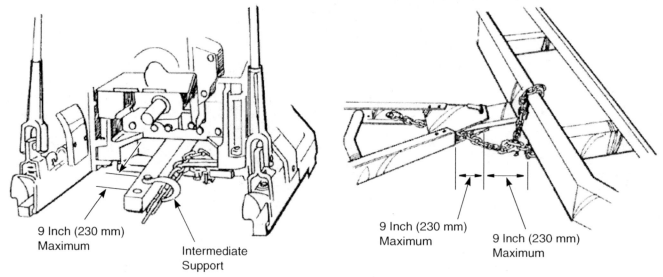

9 Inch (230 mm) Maximum

Intermediate Support

9 Inch (230 mm) Maximum

9 Inch (230 mm) Maximum

JDPX7874, JDPX7875

Fig. 81 — Install ANSI/ASABE Safety Chain for Highway Travel

4. Use safety chains on public roads (Fig. 81). They prevent trailed equipment from rolling into ditches or into the path of oncoming traffic if the hitch uncouples.

Safety chains defined by ANSI/ASABE Standard 5338.2 are available from dealers. Follow recommendations for installation and use, including:

a. Select adequate and proper attaching points:

- On towing unit, strength to support its gross weight.

- On towed unit, strength to support its gross weight.

- Within 9 inches (230 mm) vertically and laterally of the point where the hitchpin couples the units.

- Close enough to the primary attaching points so no more than 23.6 inches (600 mm) of chain is required for fastening on the towed machine, and no more than 35.4 inches (900 mm) on the towing vehicle.

b. Select a suitable chain:

- Strong enough to support the gross weight of the towed machines.

- Long enough to attach properly and to permit unrestricted turns, according to manufacturer's instructions.

c. Fasten chains to the towed and towing units with safety devices that can't be opened accidentally.

d. Provide intermediate support for the chain if the point of attachment is more than 9 inches (230 mm) in front of the hitch point on the towing unit, or over 9 inches (230 mm) behind the hitch pin on the towed unit (Fig. 81). intermediate support points should be at least half as strong as chain capacity, applied in any direction. (Follow manufacturer's instructions for ANSI/ASABE standard chains.)

5. Use transport links. Expensive damage and serious injury could occur if an implement or machine lowers and hits an obstacle or drops unexpectedly during travel. To prevent this, some equipment is provided with one or more transport links to lock it in transport position. Use these links when transporting. If a fold-up extension wing drops during transport, severe damage to the implement could occur, you could lose control of the tractor, and the wing could tear up fences, strike obstructions along the roadway, or collide with another vehicle.

6. Use safety warning equipment on public roads (Fig. 82). Do not drive on public roads unless safety warning equipment is in place and functioning properly (see highway safety equipment, Fig. 72 through 75, in this chapter.

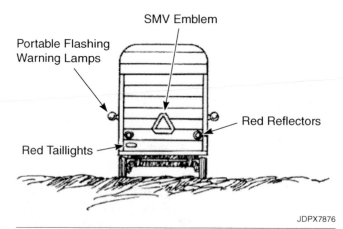

SMV Emblem

Portable Flashing Warning Lamps

Red Reflectors

Red Taillights

JDPX7876

Fig. 82 — Portable Warning Lamps Can Be Mounted on Equipment

Mount safety lights, reflectors, and SMV emblem on the towed equipment as described earlier. The lights should be at least 39 inches (1 meter) above ground level. Portable safety lighting and marking equipment is available.

7. Do not carry passengers on towed equipment. This can be as hazardous as carrying them on tractors (Fig. 83). Accident data show that farm wagons, for example, are involved in a significant number of personal injury accidents, and that falls account for many of these accidents. Avoid carrying passengers on wagons if possible. If you must, insist that they remain seated. Start and stop slowly. Maintain a slow travel speed, watch for bumps and obstructions, and avoid quick turns. Use hand or electric turn signals for the benefit of passengers, so unexpected movements won't catch them off guard.

JDPX7877

Fig. 83 — Implements Are Not Safe to Ride On, Especially on Public Roads

8. Maintain safe travel speeds. The dangers of travelling too fast include tractor upsets, collisions, and running off the road or into obstacles on the shoulder. Maintain a safe towing speed so you are in control at all times. If equipment fishtails, slow down (Fig. 84).

JDPX7878

Fig. 84 — Slow Down if Equipment Fishtails

9. Do not rely too much on your brakes (Fig. 85). Stopping distance increases with speed and weight of towed loads, and on slopes. Tractors must be heavy and powerful enough with adequate braking power for the towed load. Reduce speed if towed loads do not have brakes and weigh more than the tractor. For towed equipment without brakes, observe these ASABE recommended maximum road speeds:

Weight Ratio: Total Towed Load vs. Towing Machine	Maximum Road Speed
1 to 1, or less	Up to 20 mph (32 km/h)
2 to 1, or less	Up to 10 mph (16 km/h)
More than 2 to 1	**DO NOT TOW**

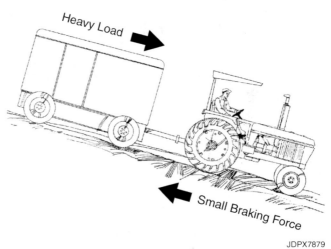

JDPX7879

Fig. 85 — Towed Loads Make It More DIfficult to Stop, Especially on Slopes

10. Put wide machines in transport position. Many machines have hitches that can be adjusted so the equipment is towed more in line behind the tractor (Fig. 86). Placed in transport position, this type of hitch reduces the width of the machine for safer towing. Do not drive half on and half off the road. If traffic backs up, pull off and stop to let traffic pass.

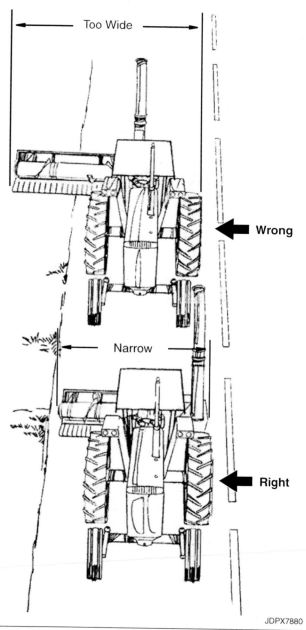

Fig. 86 — Place Equipment in Narrow Transport Position When Towing, if Possible

11. Watch out for overhead obstructions. Remember that towed equipment has height as well as width. Always watch for overhead obstructions like tree limbs and power lines.

Towing Tractors and Self-Propelled Machines

Driving tractors and self-propelled machines or hauling them on a truck is usually more desirable than towing them. Towing doesn't save travel time, since tractors and self-propelled machines should never be towed faster than they could move under their own power. There are other problems, too. Finding strong and conveniently located hitch points is not always easy. Damage to the transmission, differential, or final drive of the towed unit can occur if the recommendations outlined in the operator's manual are not followed. And careful coordination of movement between the towing vehicle and the towed unit is necessary to prevent a collision between them. If it is necessary to tow a tractor or a self-propelled machine, follow these practices:

1. Prepare the unit for towing. Follow instructions given in the operator's manual (Fig. 87). This is the only way to make sure that lack of lubrication will not damage bearings and gears in the transmission, differential, or final drive, and to prevent other damage.

Fig. 87 — Follow Instructions in the Operator's Manual to Prepare the Unit for Towing

    a. Place all transmission shift levers in neutral or tow position and disengage the PTO drive.

    b. Disengage the rear wheels from the transmission driveline, if recommended in the operator's manual. This may involve the simple shifting of a lever, or removing the drive shafts from some self-propelled machines.

    c. Run the engine to obtain hydraulic pressure for the power steering and brakes and to lubricate the transmission (on some tractors).

CAUTION: Machines that have power brakes or steering should not be towed except with a rigid tow bar unless someone is in the operator's seat and the engine is running to provide hydraulic power. Check the operator's manual for details.

2. Hitch safely. Find a hitch point strong enough for towing and located as close as possible to the centerline. Avoid hitching to any linkage of the steering mechanism, or to the front wheel knuckles of utility-type tractors. Do not pull from the front pedestals of tricycle-type tractors. These pedestals are not designed for towing and may break if subjected to shock loading. When attaching chains to self-propelled machines, avoid bending sheet metal. Use a rigid tow bar, if possible, to control speed and steering of the towed unit. If you use a chain, select one strong enough and long enough to give the driver of the towed unit adequate stopping time and distance to avoid a rear-end collision with the towing vehicle (Fig. 88).

Never use a nylon rope. Nylon stretches and produces a slingshot effect, which jerks the towed vehicle suddenly. Hooks or chains attached to ends of nylon rope will break before stronger nylon. Then, the rope "slingshots" the hook or chain, injuring or killing people in its way.

Always attach tow chains with hooks up. Then if the hook spreads under load, it flies down, not up.

Use Long Tow Chain

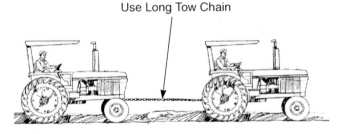

JDPX7881

Fig. 88 — When Towing, Use a Chain Long Enough to Prevent a Rear-End Collision. Do Not Use Nylon Rope

3. Maintain a safe towing speed. Always tow tractors and self-propelled implements more slowly than they would travel under their own power. Towing faster could result in machine damage or personal tragedy including:

• Loss of control of the towing vehicle due to the speed and weight of the towed unit — especially when going downhill

• Loss of control of the towed unit caused by liquid ballast rotating and surging in the tires (Fig. 89)

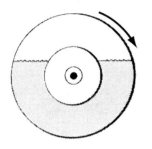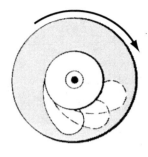

Liquid Ballast Maintains Its Own Level... at Normal Speeds

Liquid Rotates With Tire... Excessive Speed Makes Control Difficult

JDPX7882

Fig. 89 — Liquid Ballast Rotates With the Tire at Higher Speeds, Making Control of the Tractor Difficult

• Heating and seizing of the brakes on the towed unit that could make it swerve off the road

• Damage to driveline gears and bearings because of lack of lubrication at high rotating speeds (Fig. 90)

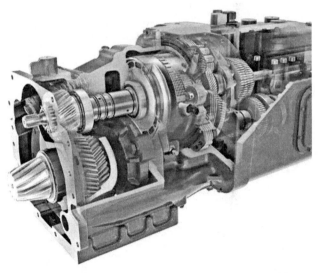

JDPX8093

Fig. 90 — Upper Gears and Bearings in Some Transmissions Are Not Lubricated When the Tractor Is Towed

IMPORTANT: Some tractors should not be towed faster than 5 mph (8 km/h). Be sure to consult your operator's manual.

Safety Feature Checklist for Tractors

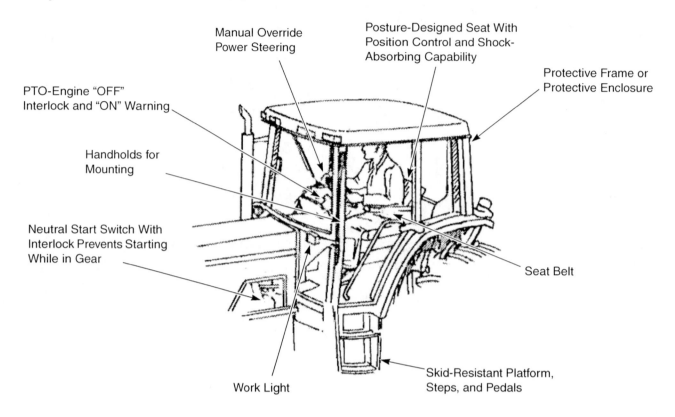

Manual Override
Power Steering

Posture-Designed Seat With
Position Control and Shock-
Absorbing Capability

Protective Frame or
Protective Enclosure

PTO-Engine "OFF"
Interlock and "ON" Warning

Handholds for
Mounting

Neutral Start Switch With
Interlock Prevents Starting
While in Gear

Seat Belt

Work Light

Skid-Resistant Platform,
Steps, and Pedals

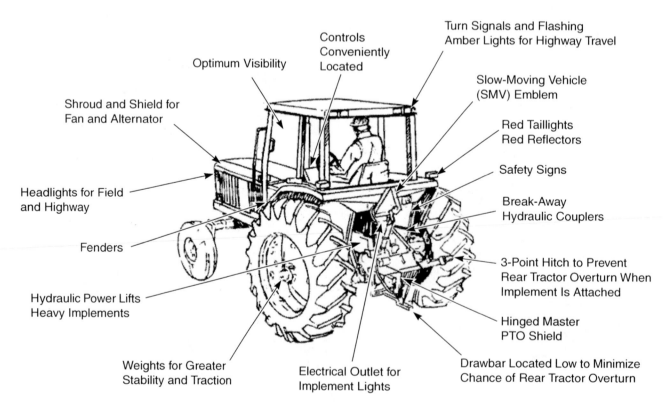

Controls
Conveniently
Located

Turn Signals and Flashing
Amber Lights for Highway Travel

Optimum Visibility

Slow-Moving Vehicle
(SMV) Emblem

Shroud and Shield for
Fan and Alternator

Red Taillights
Red Reflectors

Safety Signs

Headlights for Field
and Highway

Break-Away
Hydraulic Couplers

Fenders

3-Point Hitch to Prevent
Rear Tractor Overturn When
Implement Is Attached

Hydraulic Power Lifts
Heavy Implements

Hinged Master
PTO Shield

Weights for Greater
Stability and Traction

Electrical Outlet for
Implement Lights

Drawbar Located Low to Minimize
Chance of Rear Tractor Overturn

JDPX7883, JDPX7884

Fig. 91 — Safety Features: Keep Them Working and Use Them!

Tractor manufacturers have made impressive and effective improvements to make tractors easier and safer to operate (Fig. 91). Most of these improvements fall within four classifications:

- Visibility and recognition
- Operator comfort and ease of operation
- Protection from operational hazards
- Increased stability

Here is a list of safety features within each classification.

Visibility and Recognition

- Headlights, taillights, reflectors
- Work Lights
- Flashing amber warning lights
- Turn signals
- Electrical terminal outlet for implement lights
- SMV emblems
- Windshield wipers, washers, and tinted glass
- Rear-view mirrors and horns

Operator Comfort and Ease of Operation

- Adjustable, shock-absorbing seats
- Steps and handholds for mounting
- Adjustable steering wheels
- Hydraulic power and coupling devices to control heavy implements
- Fans, filters, air conditioners, and heaters (in cabs)
- Weather shields
- Color- and shape-coded controls and instruments using universal symbols
- Noise reduction (in cabs)

Protection From Operational Hazards

- Rollover protective structures (ROPS)
- Seat belts (used with ROPS)
- Fenders
- Steps and handholds for mounting
- Skid-resistant platforms and steps
- Shields for rotating parts
- Powered, self-adjusting brakes
- Key starting and shutoff switch
- Transmission park position or parking brake

- Power steering
- Neutral-start switches
- PTO-engine shutdown interlock and "ON" warning
- Tachometers and speedometers
- Self-cancelling remote hydraulic control valves
- Standardized controls
- Safety signs, symbols, and pictorials

Increased Stability

- Low-mounted drawbars
- Front-end weights
- Rear-wheel weights
- Tire ballast
- Easily adjusted wheel spacing
- Interconnected brake pedals

These design features, and others, are intended to make tractor operation easier and safer. To benefit from them:

1. Know what they are.
2. Know how they function.
3. Keep them in good condition.
4. Use them, and make sure others do also.

Your tractor may not have all of these features. If not, and if you feel that some are essential for safety in your particular operation, ask your farm equipment dealer if they are available and can be installed on your tractor.

Self-Propelled Implements

Self-propelled implements are mobile power units that incorporate harvesting, planting, spraying, and other working unit functions into their design. Thus, they are associated with two types or sources of hazards — those involving a self-propelled implement when driven as a mobile unit, and those related to the operation of the working unit.

To learn how to avoid accidents involving a working unit or a mobile unit in field operation, study Chapter 3, "Recognizing Common Machine Hazards". It addresses basic types of hazards such as moving and rotating components. Also read other chapters about self-propelled implements, including Chapter 10, "Hay and Forage Equipment" and Chapter 11, "Grain and Cotton Harvesting Equipment".

Let's look now at important recommendations for driving self-propelled machines safely as mobile power units:

1. Follow the recommendations in this chapter for safe tractor and implement operation. Many of them also apply to the operation of self-propelled machines. These are especially important:

- Make pre-operational checks

- Refuel safely to prevent fires and explosions

- Start and stop cautiously

- Maintain safe travel speeds

- Prevent upsets, collisions, falls, and hitching accidents

- Drive safely on public roads

- Perform maintenance according to operator's manual

2. Prepare the machine for safe transport.

   a. Empty grain tanks and hoppers. This lowers the center of gravity of most machines and makes them more stable. It makes the machine easier to stop, and it also relieves the machine of the strain of carrying the weight of the harvested product.

   b. Place all unloading augers and elevators in the transport position. This decreases the width of the machine and reduces the chances of hitting obstructions (Fig. 92).

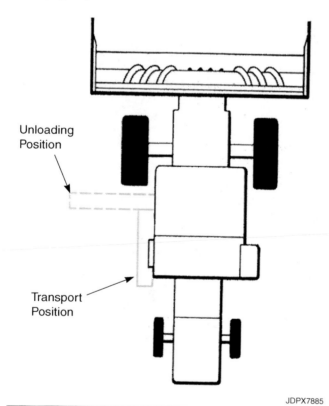

Unloading Position

Transport Position

JDPX7885

Fig. 92 — Place the Unloading Augers in Transport Position Before Transporting a Combine

c. For good visibility, lower the header to a height that will clear obstructions on the ground or roadway. Headers are heavy, and the higher they are carried, the more they shift the center of gravity forward, making the rear wheels lighter (Fig. 93). You need as much weight as possible on the rear wheels for effective steering.

Wrong:
Header Raised Too High

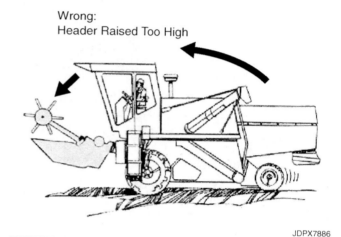

JDPX7886

Fig. 93 — Headers Carried Higher Than Necessary Shift Machine Weight off the Rear Wheels, Decreasing Steering Control

It is often desirable to remove the header from the machine and transport it by truck or implement carrier. If possible, follow this procedure if the machine is to be transported more than a short distance, if the header is much wider than a normal traffic lane, if traffic might be hazardous, or if the shoulders of the road are narrow and in poor condition.

JDPX8092

Fig. 94 — Transport Wide Headers and Combines on Flatbed Carriers for Highway Safety – For You And Motorists

Instructions for carrying the header in transport position will vary with the kind and make of machine. Check your operator's manual for instructions.

d. Lock both brake pedals together for highway transport (see Fig. 32), if a lock is provided, to maintain equal braking on the drive wheels. Self-propelled machines can swerve and go out of control just like tractors.

e. Inspect the safety warning devices before driving on public roads. Check to see that all warning lights are in place and functioning. Display an SMV emblem, making sure that it is clean, and that reflectors are, too.

3. Look before moving the machine. Do not run the risk of running over someone. You have good long-range vision from the operator's platform but, on many self-propelled machines, the area immediately surrounding the machine is not within the operator's line of sight (Fig. 95). After an individual has been run over by a self-propelled machine, the operator often explains that he "didn't see him." Do not get yourself into this situation. Before you start off, make sure you know exactly where all other persons are. Sound the horn before starting the engine, and clear people away from the machine.

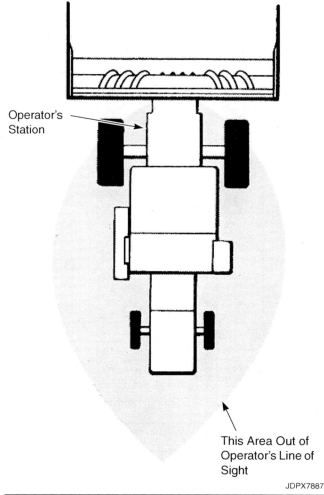

Operator's Station

This Area Out of Operator's Line of Sight

JDPX7887

Fig. 95 — You Can't See the Ground Near Some Self-Propelled Machines. Clear People Away Before Starting Engine

4. Maintain safe travel speeds. Self-propelled machines are heavy, big, wide, high, and bulky. Do not drive them too fast. Travel slow enough to maintain steering control, to prevent bumping and jarring of the machine on rough ground, and to give you time to stop.

Other safety precautions:

a. Slow down when passing obstructions on either side. Have a helper guide you if you're not absolutely sure that you can pass safely.

b. Slow down gradually. Reduce engine speed and travel speed before applying the brakes. Remember that sudden slowing down shifts machine weight forward, decreases the weight on the rear wheels, and makes steering control less effective.

c. Avoid panic stops by watching your driving environment in all directions. There are two good reasons for this: you might not be able to stop in time to prevent striking some obstacle and, during panic stops, you may lose all steering control. A self-propelled machine with rear-wheel steering will not turn if the drive wheels are braked solidly, or if the rear wheels are lifted off the ground.

5. Avoid collisions on the highway. The extra width and limited rearward visibility on some self-propelled machines can lead to collisions with other vehicles. Machines with wide platforms of pickup attachments, and with grain tanks that obstruct rearward vision, should be driven with special care. If possible, drive on public roads when traffic is the lightest. Drive within your own traffic lane to keep the passing lane clear for traffic to pass. If motor vehicles can't pass from behind, turn off the road at your first opportunity and let them go by (Fig. 96). However, before you pull over to the shoulder, make sure it is wide, firm, clear of obstacles, and smooth enough for your machine. If it is not, stay on the roadway, but stay out of the approaching traffic lane. If you must travel on narrow roads, arrange to transport the header on a flatbed carrier, if at all possible. Also, consider the need for escort vehicles. Check highway regulations.

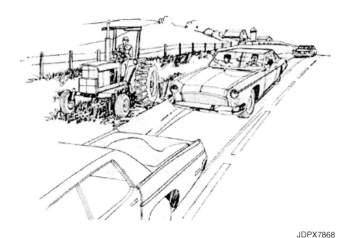

JDPX7868

Fig. 96 — Pull Off Road So Others Can Pass, but Only if Shoulder Is Wide and Safe to Drive On

Towing Self-Propelled Implements

Avoid towing self-propelled implements, especially combines, except in an emergency. Each type of self-propelled implement has some special requirements and instructions for preventing personal injury accidents, and for preventing serious damage to the machine. Follow the towing instructions in the machine operator's manual.

Operator's Manuals and Safety Signs

The operator's manuals and the safety signs on the machines are essential reading for anyone who operates self-propelled implements or any other farm equipment. Each machine has special instructions for proper operation and maintenance. They also have specific safety instructions to help you avoid serious injury or death to yourself and to others. Read and follow these instructions and safety signs. They can make your work more productive, and they help you avoid tragedy.

Test Yourself

Questions

1. (T/F) If a safety concern shows up in a pre-operational check, you should get the necessary repairs during your next trip to your farm equipment dealer.

2. Of the following, the safest way to start a tractor engine is:

   a. With the transmission in neutral

   b. With your foot disengaging the clutch

   c. With the operator in the seat, the transmission in neutral, the clutch disengaged, and the brakes applied.

   d. None of the above

3. Safety cans for storing gasoline should be painted _____, while those for diesel fuel should be painted _____.

   a. Blue / yellow

   b. Black / green

   c. Red / green

   d. Red / yellow

4. Turning too sharp at high speeds can cause an overturn due to:

   a. Breaking an axle

   b. Centripetal force

   c. Centrifugal force

   d. Gravity

5. Most tractor rollover incidents involve:

   a. Sideways rollovers

   b. Rear overturns

   c. Rollovers to the front

   d. None of the above

6. When driving down a hill with a tractor, the best means of maintaining proper speed is:

   a. Use a lower gear and take advantage of engine braking

   b. Ride the brake pedals (which are locked together)

   c. Drive in a zig-zag pattern

   d. Back down the hill

7. (T/F) Carbon monoxide has an easily recognized odor.

8. What is the most common type of fatal tractor accident?

   a. Being run over

   b. PTO entanglements

   c. Tractor overturns

   d. Collisions with automobiles

9. While pulling a large object, such as a log, where should the chain or cable be attached to the tractor?

   a. Wrapped around the rear axle

   b. Through the three-point hitch arms

   c. Around the base of the seat

   d. To the drawbar

10. (T/F) When stuck in the mud, it is best to try to drive out forward first.

11. The safest place for a child to ride while Grandpa is driving the tractor is:

    a. On Grandpa's lap

    b. Standing on the drawbar

    c. On the left fender

    d. In the pickup truck with Grandma

12. (T/F) Rollover protective structures and rollover protective enclosures are designed to prevent tractor overturns.

13. (T/F) Shorting across starter terminals is acceptable if the tractor is designed to start only in neutral.

14. How can you tell if a tractor cab is designed to provide rollover protection?

    a. By looking at the size of the metal frame

    b. If the cab was original equipment, it provides rollover protection

    c. The frame will have a label from ASABE, SAE, state, or federal government stating it meets approved standards

    d. None of the above

15. (T/F) You should avoid wearing loose clothing with dangling strings or shoelaces when working around PTO-powered equipment.

16. Which of the following are the two standard PTO speeds found on tractors today?

   a. 250 rpm and 540 rpm

   b. 500 rpm and 1000 rpm

   c. 540 rpm and 1000 rpm

   d. 1040 rpm and 1500 rpm

17. (T/F) Heavy clothing will protect an individual in the event of a PTO entanglement.

18. The PTO master shield is designed to:

   a. Completely enclose a rotating PTO shaft

   b. Protect the connection of the tractor stub shaft and the front universal joint of the attached machine's driveline

   c. Protect the connection of the driveline to the machine's gearbox

   d. None of the above

19. (T/F) Tractors and other self-propelled machines can be towed safely at highway speeds.

20. (T/F) Injury data shows that farm wagons are involved in many personal injuries.

References

1. ROPS Directory — A Guide to Tractor Roll Bars and Other Rollover Protective Structures. 1990. Wisconsin Rural Health Research Center, 1000 North Oak Ave., Marshfield, WI 54449-5790.

2. Lighting and Marking of Agricultural Field Equipment on Highways (ANSI/ASABE S279.9/SAE J137). July 1993. American National Standards Institute, New York, New York, 10036; American Society of Agricultural Engineers, St. Joseph, Michigan 49085; Society of Automotive Engineers, Warrendale, Pennsylvania 15096.

3. Safety Chain For Towed Equipment (ANSI/ASABE S338.2). 1990. (See ANSI and ASABE sources above.)

4. Slow-Moving Vehicle Identification Emblem (ANSI/ASABE S276.3/SAE J943). 1991. (See ANSI, ASABE, SAE sources above.)

Tillage and Planting Equipment

JDPX7669

8

Introduction

Most tillage and planting equipment does not have a lot of powered rotating parts as you expect on harvesting machines. However, while you might think safety isn't an issue, many people are seriously injured in tillage and planting operations (Fig. 1).

Some of the implements are long, wide, and heavy, and they may extend high above tractors. They require special care for hitching and unhitching, loading fertilizer and seed, maneuvering, changing from field to transport mode, and for transporting them on public roads.

JDPX8051

Fig. 1 — Tillage and Planting Operations May Seem Safe, but They Require Special Safety Practices

Understanding the Hazards

You can avoid serious injury if you understand the potential hazards and how to avoid them. There are several chapters in this book that address hazards and safe practices related to tillage and planting equipment operations. You should become familiar with each of the preceding chapters. However, for tillage and planting safety, you should give special attention to these chapters:

Chapter 3, "Recognizing Common Machine Hazards" page 3-1.

Chapter 7, "Tractors and Implements" page 7-1.

Chapter 9, "Chemical Safety" page 9-1.

John Deere FMO manual titled "Tillage and Planting" also includes safety information.

Operator's Manuals and Safety Signs

Each type of tillage and planting implement requires some unique safety practices. Read and follow the safety instructions in the operator's manuals and safety signs for the tractors and implements you operate and service. They are the most important safety information sources available.

Hazards in Tillage and Planting

As noted above, many of the hazards in tillage and planting are covered in other chapters. This chapter will address several safety concerns specifically related to tillage and planting equipment, including:

• Matching equipment to tractor

• Hydraulic connections

• Hitching and unhitching

• Raising and lowering wings and markers

• Machine operation

• Transporting Implements

• Environmental hazards

Matching Equipment to Tractor

Excessive power can damage an implement, and too large an implement can damage the tractor or cause personal injury accidents. Proper matching of tillage and planting equipment with tractors is generally quite obvious, but it deserves attention.

Operators are occasionally tempted to operate or tow implements with tractors that are too small to operate or control them, "just for a short time." That can cause accidents. A tractor that is too small for the equipment may become unstable because extra weight on the three-point hitch shifts the center of gravity. The tractor may not be able to safely steer or stop an excessively heavy load. The tractor and implement could upset or collide with a vehicle on the highway. Before attaching an implement to a tractor, check the operator's manuals for minimum and maximum power or size requirements (Fig. 2).

Fig. 2 — Match the Size of the Tractor and Implements as Recommended by the Manufacturers

CAUTION: When hydraulic oil escapes under pressure, it can cause serious injury. High-pressure jet streams can penetrate skin and flesh. (See previous chapters: Chapter 3, "Recognizing Common Machine Hazards", and Chapter 7, "Tractors and Implements") To prevent that possibility, relieve the hydraulic system pressure before attaching implement hydraulic hoses to tractor couplers. Shut off the tractor engine and move remote hydraulic control levers back and forth to relieve pressure. Follow the same procedure to disconnect hoses.

Fig. 3 — Color-Code Painting of Mating Parts Is One Way To Prevent Mismatching

Hydraulic Connections

It is possible to connect some hydraulic couplings that aren't intended to mate. A common error is to interchange hose ends on the auxiliary cylinder so the control valve operates in reverse. Instead of the implement being raised when the control is pulled for raising, it goes down. This can be hazardous because the implement motion is opposite of what is expected. This problem is discussed in more detail in Chapter 3, "Recognizing Common Machine Hazards" page 3-1.

After you attach hydraulic hoses between a tractor and implement, cautiously test to see if you get the proper result. If not, correct it at once.

If hydraulic couplers on your tractors and implements are not coded to prevent mismatching, consider color-coding them by painting the mating parts (Fig. 3). Identification kits are available from some manufacturers. Check with your dealer.

Hitching and Unhitching

Basic principles for hitching and unhitching implements and tractors are addressed in these chapters: Chapter 3, "Recognizing Common Machine Hazards", and Chapter 7, "Tractors and Implements". In addition, you should give special attention to the following recommendations for hitching and unhitching tillage and planting implements.

People Between Tractor and Implement

When you have helpers, make sure they are not between the tractor and implement when the tractor is backing up or when the hitch is being moved up or down for alignment (Fig. 4).

Fig. 4 — *Don't Allow Anyone Between the Tractor and Implement When Backing Tractor or Adjusting Hitch*

Some tractors have remote controls in the hitch area so the operator can move the hitch for proper alignment. Be sure the transmission is in park position, stay clear of potential interference or pinch points, and keep others out of the area before raising or lowering the three-point hitch.

Implement Support Stands

Tillage and planting implements are big and heavy. If some machines are not properly supported before hitching or unhitching, they can tip unexpectedly and quickly. You or others could be seriously injured while hitching or unhitching, or while the machine is stored or otherwise not attached to a tractor.

Many implements are front-heavy. They must be supported in front before unhitching from the tractor to prevent them from tipping. Others may tend to tip rearward or sideways if not properly supported.

Some implements are decidedly heavy to the rear. If you attempt to pull the hitch pins without first lowering the rear support stands, the implement tongue can whip upward, striking you with great force and causing serious injury.

Don't take dangerous shortcuts. Lower support stands or jacks and lock them in place before attempting to unhitch implements from tractors (Fig. 5). Each type of implement has some special procedures for safe hitching and unhitching, so follow the instructions in your tractor and implement operator's manuals.

Fig. 5 — *Pin Support Stands or Jacks in Place Before Unhitching Implements*

Machine Operation

Good tillage and planting jobs begin with adequately sized, properly maintained equipment and safe operating practices. Tillage equipment that is in bad shape or improperly adjusted pulls harder and wears faster, resulting in early wearout, failure, frequent plugging, and other problems that can lead to accidents.

Size, shape, and slope of fields affect the pattern of tillage and planting movement. Avoid dangerous working situations like extremely tight turns where the tractor tire may catch the implement tongue or frame, causing damage or forcing equipment up onto the operator's station (Fig. 6). Working on steep slopes where a sideways or rearward upset is likely or getting too close to fences and other obstructions is also risky.

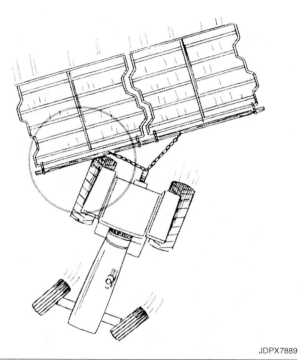

Fig. 6 — Avoid Tight Turns During Tillage Operations

Machine Adjustment

Safe and efficient machine operation demands that the machine be properly adjusted. For example, improperly hitched plows can cause poor steering control and penetration.

Improperly set coulters can cause a plow to plug. And trip beams that fail to trip may result in a bent beam or broken bottom. Such difficulties may cause you to become frustrated or angry, and this could lead to accidents. The chapter "Human Factors" discusses how emotions contribute to accidents.

Always lower the implement to the ground or use jacks, blocks, transport links, or lock pins when not in use or when you're working on the machine (Fig. 7). That is also important when you replace plowshares, shovels, or other ground-engaging tools. If there is a failure in the hydraulic system, the unsupported implement or wings can suddenly lower, crushing anyone beneath them.

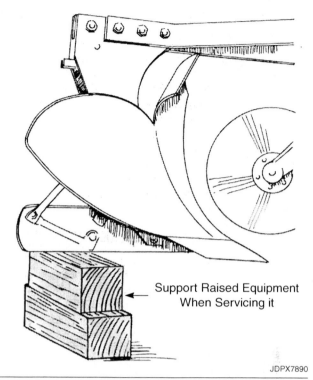

Support Raised Equipment
When Servicing it

Fig. 7 — If a Machine Must Be Serviced in the Raised Position, Use Transport Locks and Block It Up

Check outrigger locking mechanisms on fold-up tillage equipment. The mechanisms can fail and let the equipment fall.

Raising and Lowering Wings and Markers

Many tillage and planting machines are designed so side wings (outriggers) and markers can be folded upward or forward to be narrow enough to travel through gates and on public roads. Raised wings or markers may be 10 to 18 feet (3 to 5.5 m) high. Implement wings may weigh hundreds or thousands of pounds (kilograms).

Occasionally mechanical linkages or hydraulic systems fail or do not function exactly as expected, especially if they are not installed, maintained, or operated properly. If an implement wing or marker unexpectedly falls, anyone in the path can be critically or fatally injured (Fig. 8).

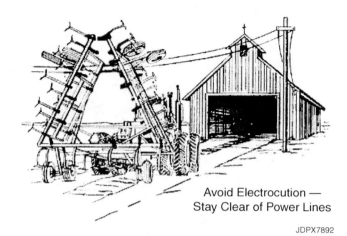

Avoid Electrocution —
Stay Clear of Power Lines

JDPX7892

Fig. 9 — Be Aware of Overhead Power Lines When Raising, Lowering, or Transporting Implements

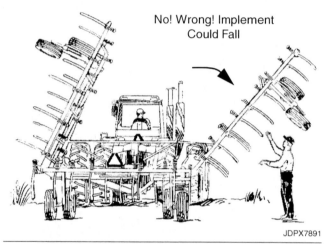

No! Wrong! Implement
Could Fall

JDPX7891

Fig. 8 — Keep Everyone Clear When Raising or Lowering Wings and Markers. They Can Fall Without Warning

You must exercise caution when raising, lowering, or folding implement wings or markers, and when working around them. Follow these safety practices:

1. Locate the tractor and implement on level ground for raising or lowering implement wings or markers.

2. Be aware of overhead power lines. If raised wings or markers contact power lines, you or others could be electrocuted. Know the transport height of your implement with wings or markers raised (Fig. 9).

3. Clear all people from the area before raising or folding the implement wings or markers. Mechanical or hydraulic failure can result in tragedy.

4. Be sure the hydraulic cylinders and lines are fully charged with oil before raising or lowering implement wings or markers. If there is air in the system, wings or markers can fall without warning. Refer to your operator's manual for instructions on how to check it.

5. Don't raise or lower wings with the machine in motion.

6. Lock wings or markers securely in folded or upright position before transporting or storing the machine. Make sure that the wings can't fall and that markers can't fall or swing outward.

7. Never walk or work under wings when they are in the raised position.

8. If hydraulic wing cylinders or transport locks must be removed or adjusted when implement wings are folded up, lock the wings, chain them together, or otherwise secure them, so they can't fall accidentally.

9. Before lowering wings or markers, clear all people from the area. They could be seriously injured by a wing or marker being lowered or falling unexpectedly.

10. For implements that fold forward or rearward for transport, keep people out of the folding area. People could be injured by folding or unfolding wings.

Transporting Implements

It is reported that each year about 30,000 accidents in the U.S.A. involving farm implements on rural roads, including many fatalities. Many of the accidents involve tillage and planting equipment.

Review the transport safety information in previous chapters. They discuss safety chains, lighting, and various safe practices for avoiding transport accidents. Follow operator's manuals instruction. Also, note the following concerns specifically related to tillage and planting equipment:

1. Use a tractor sufficiently large, heavy, and ballasted for towing, steering, and braking control.

2. Put the implement in its narrowest transport configuration (Fig. 10).

Fig. 11 — Platforms Are for Easier Filling of Hoppers. Don't Allow Them to Be Used for Riders

Fig. 10 — Put Implements in Narrow Position Before Travelling on Public Roads

3. Use transport links or locks to support the implements. Don't depend on the hydraulic system.

4. Don't allow anyone to ride on implements. Some planting implements have steps and platforms to aid in filling hoppers, but don't let anyone risk injury by riding in transport or field operations (Fig. 11).

5. Beware of overhead electrical lines. Make sure raised wings or markers won't contact them.

6. Check with state police or sheriff's office for height and width regulations.

7. Use safety chains, and display the proper lights, reflectors, and SMV emblem to alert motorists.

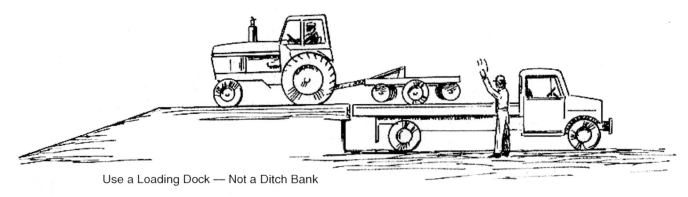

Use a Loading Dock — Not a Ditch Bank

JDPX7893

Fig. 12 — Use a Loading Dock or Ramp to Load Equipment — Not a Ditch Bank

When equipment must be moved some distance, loading it onto a trailer or truck is often the best thing to do. Follow these safe practices:

1. Use an adequate loading ramp or a loading dock, if possible (Fig. 12). Trying to drive onto a trailer from a ditch bank is risky. It invites tractor upsets. Many operators have been injured or killed and pieces of equipment have been damaged or tipped over as a result of this method of loading. To avoid upsets, back the equipment onto the trailer or truck, and drive forward for unloading (Fig. 12).

2. Secure the load with chain binders. Be sure they are tight. A pipe extender may help pull the binder tight. Wire the handles to prevent loosening in transit. If chain binders are not available, use rope, wire, blocks, or a winch cable. Check the load after traveling a few miles (kilometers), and frequently during transport to make sure the binders don't loosen. Also, check it after rough road bumps.

3. Display proper safety lighting equipment, and follow state and local height and width regulations.

Environmental Hazards

The environment can present challenges and problems in tillage and planting operations. The chapter "Human Factors" discusses fatigue, temperature, and other human performance-type issues. Also, the "Tractor and Implement" chapter addresses field-related conditions, getting stuck, tractor upsets, and other environmental conditions. Study them with regard to tillage and planting operations. In addition, note the following concerns:

Rough Ground

High speed and rough ground lead to fatigue, accidents, and damaged equipment. Occasionally, the operator loses control of a machine and may be bounced off due to a combination of speed and rough ground, especially if the operator is not wearing a seat belt.

Look ahead for rough ground. Anticipate bumps and slow down. Mark sharp breaks, gullies, and ditches ahead of tilling and planting, and stay away from them. You can't run harvest equipment near these areas without risking an upset. Leave a grass-covered margin around these gullies and ditches for safety and soil conservation.

Stones and Other Obstacles

Large stones and buried stumps or logs can damage equipment. Plow points can be broken if you hit large stones, buried roots, or other obstacles. Beams and bottoms may be bent, and disk blades and bearings are frequently broken. Similar damage can happen to planters. Be alert so you can avoid such hazards.

Temperature

Occasionally, tillage and planting must be done when the temperature, although not extreme, is far from the comfort zone. Above 80°F (27°C) and below 40°F (4°C), human performance decreases. When the temperature is cold, a person will usually be uncomfortable and will voluntarily stop work to get warmed up or put on warmer clothes before any bodily damage is done. Heat stress, however, can occur without a person recognizing it.

Darkness

The danger in working at night is in not being able to see or be seen clearly. The major nighttime hazard in operating tillage or planting equipment in a field is limited sight of obstructions or the equipment itself. Good lights on the tractor, both in front and to the rear, will help (Fig. 13). Without adequate working lights, a rock, fence, stump, or ditch can be hit before it's seen. Equipment may malfunction, and you won't realize what has happened until it's too late,

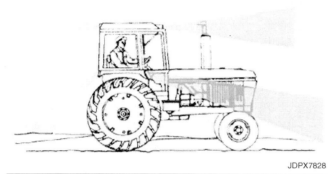

JDPX7828

Fig. 13 — Make Sure Your Tractor and Implement Have Adequate Lights for Field Operation

Dry Fields

Dust is a problem in dry fields. Pulling a disk harrow or drag for several hours in a cloud of dust can be annoying and unhealthy.

Sometimes it's possible to work a field so that the wind blows the eye- and lung-irritating dust away from you If this isn't possible, a particulate respirator may help. In hot weather, filters can be almost as annoying as the dust, but they are healthier.

The ideal solution is an air-conditioned tractor cab that filters and cools the air and lets you work in comfort.

Test Yourself

Questions

1. Safety concerns related to tillage and planting equipment include:

   a. Matching equipment to tractor

   b. Hitching and unhitching

   c. Transporting implements

   d. All of the above

2. Towing or operating equipment with a tractor that is too small affects:

   a. The stability of the tractor by changing the center of gravity

   b. The steering of the tractor

   c. The ability to stop

   d. All of the above

3. (T/F) Color-coding your hydraulic hoses and connections can help prevent injuries.

4. When being assisted in attaching an implement, the person helping should be:

   a. Standing on the drawbar to help guide as you back up

   b. Standing on the implement hitch to help guide as you back up

   c. Not between the tractor and implement

   d. In the cab with you where the person will be safe

5. (T/F) You should remove the hitch pin before setting the implement support stands.

6. Before raising, lowering, or folding implement wings the tractor should be:

   a. Facing uphill

   b. On level ground

   c. Facing downhill

   d. Moving slowly

7. Good safety guidelines for transporting tillage and planting equipment on public roads include:

   a. Equipping the machine with adequate warning lights and an SMV emblem

   b. Putting the machine in transport position

   c. Allowing no one to ride on the implement

   d. all of the above

8. (T/F) Loading equipment onto a trailer from a ditch bank is an acceptable alternative if you do not have a loading ramp.

9. (T/F) Rough ground can cause fatigue which causes the likelihood of accidents.

10. (T/F) The major nighttime hazard in operating tillage or planting equipment is fatigue.

Chemical Safety

9

JDPX8062

Introduction

Agriculture is chemistry. We use chemicals to feed plants; to control weeds, diseases, and insects; and to control fruit set on trees. Chemicals contribute greatly to the high productivity and profitability of agriculture (Fig. 1).

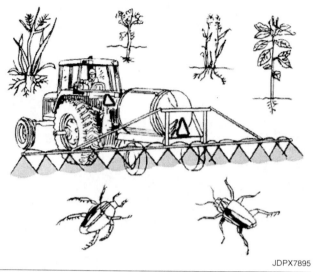

JDPX7895

Fig. 1 — Modern Agriculture Uses Many Chemicals

Chemicals can be harmful, however, if used, stored, or handled improperly. They can injure or kill people, domestic animals, wildlife, and desirable plants, as well as contaminate soil, water, and air.

Who has accidents with chemicals? There are two groups:

- People who are NOT aware of the hazards

- People who are aware of the hazards but, because of carelessness, hurry or for some other reason don't use safe practices

In this chapter, you will learn to recognize many of the hazards involved in working with chemicals. You will also learn how you can protect yourself and others from the dangers involved in chemical accidents and exposure.

Chemical Hazards

Chemicals used in agriculture are handled both by hand and by machine. Like other machinery, chemical application equipment can be hazardous if it is not operated properly. Hazards involved in chemical application machinery operation fall into three groups:

- Hazards common to all machines

- Hazards specific to chemical equipment

- Hazards involved in the use of some agriculture chemicals

Fertilizers

Fertilizers are materials used to supply plants with nutrients needed for growth. Chemical fertilizers are available in solid (dry), liquid, and gaseous forms.

Dry Chemical Fertilizer

Dry fertilizer is hygroscopic — it attracts moisture. It can draw water out of skin, causing skin burns. These skin burns usually leave the skin red and cause minor discomfort, but they can be painful to people with sensitive skin. Dry fertilizer also can get into and irritate the mouth, nose, and eyes. The scalp is another area that is often very sensitive to the effects of dry fertilizer.

Prevent fertilizer burns by keeping the material off your skin. Wear a long-sleeved shirt buttoned at the collar and a cap or hat to keep dust out of your hair (Fig. 2). Change clothes daily — more often if you are exposed to a lot of dust — and don't let fertilizer stay in contact with your skin. Wash your face, hands, arms, and other exposed skin areas several times a day with soap and water.

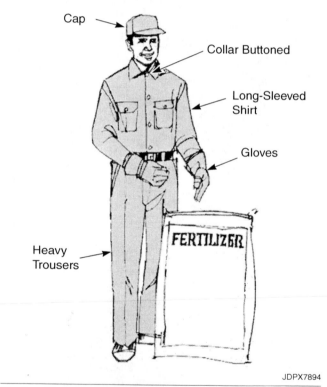

JDPX7894

Fig. 2 — Clothing Protects Against Fertilizer Burn

Let the wind blow dust away from you. Stand upwind when filling hoppers (Fig. 3). Drive crosswind in the field, if possible, so dust is blown off to one side (Fig. 4). If you can't stay out of the dust, wear goggles to protect your eyes and use a particulate respirator to keep from breathing the dust, You should be able to purchase the goggles and dust respirator from a local bulk fertilizer dealer or from a well-equipped hardware store.

Fertilizer Spreaders

Centrifugal broadcast spreaders throw fertilizer particles at high speeds. These particles can be painful if they hit you, and they can get into your eyes, ears, or mouth. Stay away from the rear of the machine when the spreader is operating (Fig. 5).

JDPX7896

Fig. 3 — Wind Can Keep Dust Away From You — Stand Upwind When Filling Hoppers

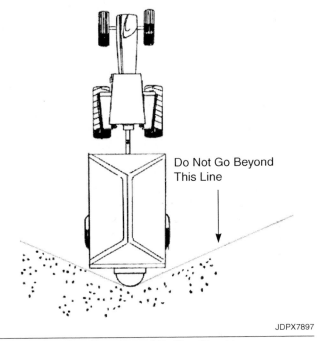

JDPX7897

Fig. 5 — Stay Away From Rear of Fertilizer Spreaders

If the spreader plugs, it's tempting to jump off and go back to see what is wrong. Don't do it! Besides getting hit by flying particles, you could get tangled in the spinner mechanism. Always stop the machine, shut off the engine, and set the brake before doing any inspection or maintenance work.

Use a safe speed when pulling any spreader. Spreaders are heavy. If you drive too fast, the unit can go out of control while going downhill or around corners. Spreaders are not normally equipped with brakes, so the tractor brakes alone must do the stopping.

Fertilizer equipment is often cleaned with diesel fuel. The fumes can ignite and burn. Always work outdoors or in a well-ventilated area, and stay away from flames. Do not allow smoking in the work area.

JDPX7647

Fig. 4 — Wind Can Keep Dust Away From You — Drive Crosswind in The Field, if Possible, so Dust Is Blown off to One Side

Anhydrous Ammonia

Anhydrous ammonia (NH_3) is a chemical made up of one part nitrogen and three parts hydrogen. The properties of this fertilizer make it one of the most potentially dangerous chemicals used in agriculture. Anhydrous means without water. Consequently, when anhydrous ammonia contacts moisture, it rapidly combines with the moisture. When it is injected into the soil, the liquid ammonia expands into a gas and is readily absorbed by the soil moisture.

Similarly, the liquid or gas that contacts the body tissue — especially the eyes, skin and respiratory tract — will cause dehydration, cell destruction and severe chemical burns. Victims exposed to even small amounts of ammonia require immediate treatment to avoid permanent injury (Fig. 6).

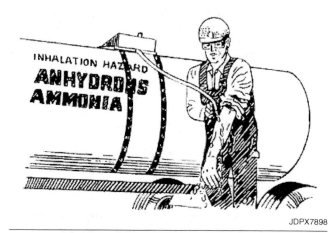

JDPX7898

Fig. 6 — Even a Little Anhydrous Ammonia Causes Major Injuries. Flush With Water at Once

Anhydrous ammonia is a colorless gas that has a built-in safety factor because of its sharp, penetrating odor — you can't stand to breathe it. No one can voluntarily remain in a concentration of anhydrous ammonia gas that is strong enough to damage the nose, throat, lungs, eyes, or skin. When people receive burns or eye damage from the product, it is because of a sudden release of it where the victim is unprotected and cannot escape.

Accident Situations

The accidental release of anhydrous ammonia can create a dangerous situation for both the handler and bystanders. The following situations are dangerous:

- Over filling the tank

- Handling hose by valve handle or hand wheel

- Faulty valves and deteriorated or out-of-date hoses

- Not using personal protective equipment

- Failure to bleed pressurized NH_3 from the hose before connecting or disconnecting

- Not enough water available

- Overturning an applicator tank

- External overheating of the storage container

An estimated 80% of reported accidents result from improper procedure, lack of knowledge or training, and failure to follow proper safety precautions. Accidents can be reduced if all individuals follow safety rules and maintain the equipment properly. It is essential that all equipment be in good operating condition. Only trained individuals should handle and apply anhydrous ammonia.

Operator Protection

Even with the best precautions, you may be involved with the accidental release of ammonia. Simple protection can prevent serious injury if used consistently (Fig. 7).

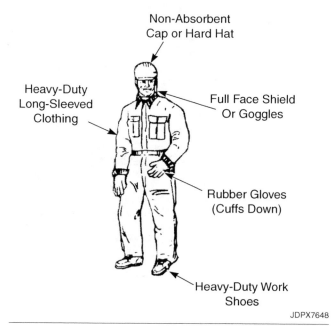

Non-Absorbent Cap or Hard Hat

Heavy-Duty Long-Sleeved Clothing

Full Face Shield Or Goggles

Rubber Gloves (Cuffs Down)

Heavy-Duty Work Shoes

JDPX7648

Fig. 7 — Personal Protective Clothing Can Prevent Serious Injury

Proper-fitting face shield or goggles, rubber gloves, and heavy-duty long-sleeved shirts are recommended as minimum protection for operators routinely handling ammonia. Wear rubber gloves and eye protection whenever you are handling hoses or working on or operating anhydrous equipment.

Regular glasses do not provide adequate protection. Never wear contact lenses when working with ammonia. The chemical might get under the lenses and cause permanent eye damage before you can remove the lenses and flush your eyes with water. The lenses also can trap the gas, causing them to freeze onto your eye.

Loose-fitting rubber gloves with an extended cuff are recommended for handling anhydrous ammonia. Turn the extended cuff down so anhydrous ammonia doesn't run down your sleeve when you raise your arms. You can remove gloves that fit loosely in case of an emergency.

You can further protect your arms from splashes by wearing heavy-duty clothing, such as coveralls or work shirts that cover your arms. Thin dress shirts or short sleeves do not provide satisfactory protection.

First Aid = Water + Water + Water

Seconds are critical when someone is sprayed with liquid ammonia or engulfed in concentrated vapors. Exposure to anhydrous ammonia can be harmful if it contacts the skin and eyes or if it is inhaled or swallowed. When ammonia contacts the skin or eyes, tissue damage occurs rapidly. Immediate use of water to flush the exposed body area(s) is crucial.

Water must be available for flushing the eyes and skin in case of exposure. Each vehicle used for anhydrous ammonia must carry a 5-gallon (19 L) container of clean water (Fig. 8). Anyone handling NH_3 should carry a 6- to 8-ounce (170 to 226 g) squeeze bottle of water in a shirt pocket for rapid emergency access (Fig. 9).

JDPX7651

Fig. 9 — *Keep a Container of Water in Your Pocket to Flush Ammonia From Eyes*

Washing with fresh, clean water is the only emergency measure to use when skin or eyes are exposed to anhydrous ammonia. Time is important! Get fresh water onto the exposed area of the skin or eyes immediately and FLUSH FOR AT LEAST 15 MINUTES. Contaminated clothing should be removed quickly but carefully. Thaw clothing frozen to the skin by running water over it before attempting removal. Wash the affected skin area with abundant amounts of water and do not apply anything except water for the first 24 hours. Stay warm and get to a physician immediately.

JDPX7649

Fig. 8 — *Anhydrous Ammonia Vehicles Must Carry Clean, Fresh Water for Emergency First Aid*

Container and System Requirements

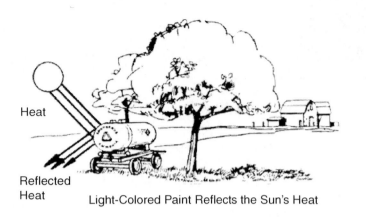

Heat

Reflected Heat

Light-Colored Paint Reflects the Sun's Heat

Put Tank in Shade When Possible

JDPX7650

Fig. 10 — White or Silver Paint and the Shade Help Keep Ammonia Tanks Cool in Hot Weather

The specially fabricated and designed pressurized equipment should meet the guidelines provided by American National Standards Institute "K61.1-1999, Safety Requirements for Storage and Handling of Anhydrous Ammonia." All parts and contact surfaces must be able to withstand a minimum working pressure of 250 psi (1723 kPa). This includes pressure welds, safety valves, gauges, fittings, hoses, and metering devices.

Ammonia is corrosive to certain metals and their alloys, such as copper and zinc. Galvanized pipe and brass fittings must not be used with equipment used for storing or applying ammonia. Containers should be made of high-strength steel or other suitable materials, and fittings should be made of black iron.

All containers used for storing ammonia must be painted white or silver. Light colors reflect heat and this helps keep the temperature and pressure down inside the tank during warm weather (Fig. 10).

LABELS, MARKINGS AND SAFETY SIGNS LEGAL REQUIREMENTS

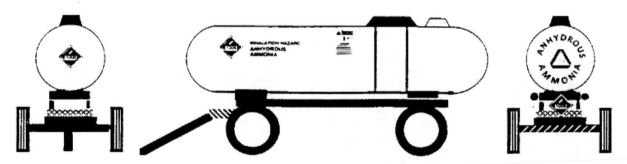

JDPX7604

Fig. 11 — Anhydrous Ammonia Tanks Must Be Properly Equipped and Identified on Highways

Nurse tanks must be labeled ANHYDROUS AMMONIA in 4-inch (101-mm) letters, on contrasting background, on the sides and rear of the tank. Federal DOT regulations require that the words INHALATION HAZARD, in association with the anhydrous ammonia label, be placed on both sides of the tank in 3-inch (76-mm) lettering. A Non-Flammable Gas placard with the numbers 1005 (identifying it as anhydrous ammonia) must be located on both sides and both ends of the tank (Fig. 11).

A slow-moving vehicle (SMV) emblem must be prominently displayed on the rear of the tank with the bottom of the sign at least 2 and not more than 6 feet (0.61 to 1.83 m) from the ground. The valves must be appropriately labeled by color or legend as vapor (Safety Yellow) or liquid (Omaha Orange). The letters of the legend must be at least 2 inches (51 mm) high on contrasting background and within 12 inches (305 mm) of the valves.

Care and Maintenance of NH$_3$ Equipment

Anhydrous ammonia can be handled and used safely. It is imperative that all equipment is properly maintained and checked daily. A regular, scheduled maintenance program will ensure that all the valves and the tank are safe for handling the high-pressure liquid and its vapor form.

Daily Inspection

Each day give the tank and hoses a brief inspection:

- Hoses — Look for cuts, soft spots, bulges, kinking, flattening, or slipping at the coupler.

- Tires — Inspect for proper inflation, cuts, weathering, wear, and tightness of lug bolts on wheels.

- Refill the emergency water tank with fresh, clean water.

Each time you fill the nurse tank, check the liquid level gauge and pressure gauge. The gauges should be working properly and be consistent in their readings. Don't use nurse tanks with faulty gauges.

Close all hand valves by hand only. Do not use a wrench, because you could break the stem of the valve or damage the seal. Either damage could allow ammonia to escape.

Immediate Inspections

Several situations are cause for immediate repair or replacement. Any leak in a liquid or vapor shutoff valve requires immediate repair or replacement of the valve. If an accident causes a dent, gouge, crack, or other damage to the tank that might result in failure, have tanks, valves, and hoses inspected before putting them back into service. An overturned tank or collision between the tank and other farm machinery is an example of cause for inspection. Any repairs should be performed by a qualified service technician. A certified welder must make any welding repairs.

Annual Inspection

At least once a year, inspect hoses carefully and repair or replace them as needed. Lay your hoses out straight and examine carefully for:

- Cuts exposing reinforcement fabric

- Soft spots or bulges

- Blistering or loose outer cover

- Unusual abuse, such as kinking or flattening by a vehicle

- Slippage of hose at any coupling

- Brass or copper fittings or water hose-type clamps

- Environmental damage resulting in checking and cracking

Immediately replace hoses that show any defects. Hoses exposed to anhydrous ammonia lose strength and should be replaced on a regular schedule, regardless of visible damage.

For a more detailed explanation, refer to "External Visual Inspection Guidelines for Anhydrous Ammonia Nurse Tanks, Applicators, and Tank Appurtenances," published by:

The Fertilizer Institute
820 First Street N. E., Suite 430
Washington, DC 20002

Highway Safety

Towing anhydrous tanks on public roads presents problems, because accidents can cause serious injuries to others, environmental damage, costly repairs, and possibly other liabilities.

Prior to operating a nurse tank on public roadways, make sure all running gear and tires are in safe working condition.

Running Gear — Inspect the wagon frame, tongue, reach poles, anchor devices, wheel bearings, steering, knuckles, ball joints, and pins for damage, cracks, excessive wear, and adjustments.

Tires — Check for proper inflation, cuts, badly worn spots, and signs of weathering. Ensure that lug nuts are tight.

Lubrication — Steering knuckles, wheel bearings, or other applicable wagon equipment should be lubricated at least once every year.

All nurse tanks and applicator equipment should be securely attached to the towing vehicle. Always use a safety hitch pin and safety chain.

Don't exceed 25 mph (40 km/h) when transporting anhydrous ammonia on public roads. For proper braking, it is strongly recommended that the vehicle used to tow the nurse tank be at least equal in weight to the gross weight of the nurse tank.

For proper safety lights, reflectors, and SMV emblem for transporting farm equipment on public roads, refer to "Accidents on Public Roads" in Chapter 7, "Tractors and Implements." Also, check with your local and state highway authorities.

Pesticides

Pesticide is a broad term for a large number of chemicals used to kill pests. A pest is anything that:

- Competes with humans, domestic animals, or crops for food, feed, or water

- Injures humans, animals, crops, structures, or possessions

- Spreads disease to humans, domestic animals, or crops

- Annoys humans or domestic animals

Some specific pesticides and the pests they control are: insecticides (insects), fungicides (fungi), herbicides (weeds), miticides (mites), rodenticides (rodents), avicides (birds), and bactericides (bacteria).

Pesticides are toxic or poisonous. They have to be ingested in order to kill the undesirable pest (Fig. 12). They also are toxic to desirable organisms, including humans. Exposure to a sufficient quantity of almost any pesticide can make a person ill or increase long-term health risks. Some pesticides are so toxic that very small quantities can kill a person. If you are handling pesticides, you need to be aware of the dangers. REMEMBER — PRACTICE SAFETY FIRST!

JDPX8098

Fig. 12 — Pesticides Are Poisonous. They Kill Insects and Weeds...and They Can Kill People

Accurate pest identification is the first step in an effective pest management program. Never attempt a pest control program until you can positively identify the pest. The more you know about the pest and the factors that influence its development and spread, the easier, more cost-effective, successful, and safer your pest control will be. Correct identification of a pest allows you to determine basic information about it, including its life cycle and the time that it is most susceptible to being controlled.

Regulations

As one who applies pesticides, you must be sure they are handled properly and safely. You must be familiar with all state and federal laws regulating pesticide use, storage, transportation, application, and disposal (Fig. 13).

Laws and Regulations Control Pesticides

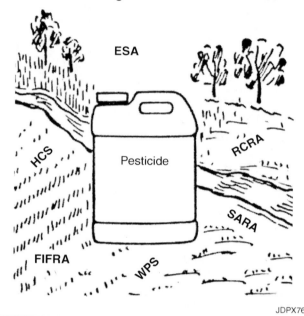

JDPX7631

Fig. 13 — Know What Laws and Regulations Require Before You Apply Pesticides

In the United States, federal laws and regulations set the standards for pesticide use. States have the right to be stricter than federal law, but not more lax. Some of the major federal pesticide laws and regulations are:

- Federal Insecticide, Fungicide and Rodenticide Act (FIFRA)

- Endangered Species Act (ESA)

- Title III of the Superfund Amendments and Authorization Act of 1986 (SARA Title III)

- Resource Conservation and Recovery Act (RCRA)

- Worker Protection Standard (WPS)

FIFRA — It governs the registration of pesticides. No pesticide may be marketed in the United States until the Environmental Protection Agency (EPA) approves the registration request from the chemical company (registrant) wishing to market it.

Some key provisions of FIFRA:

- Require EPA to register all pesticides, each use of each pesticide, and approval of the product label

- Require the classification of all registered pesticides as either "general use" or "restricted use"

- Require users of "restricted use" pesticides to be certified applicators or work under the direct supervision of a certified applicator

- Provide penalties for actions contrary to the Acts provisions

- Give EPA authority to develop regulations which are interpretations of the law and have the force of a law

ESA — Under the Endangered Species Act, it is a federal offense to use any pesticide in a manner that results in the death of a member of an endangered species. Prior to making a chemical application, the user must determine that endangered species are not located immediately adjacent to the site to be treated (Fig. 14).

JDPX7592

Fig. 14 — Don't Apply Pesticides Near Endangered Species Habitat; ESA Law Prevents It

If users are in doubt as to whether endangered species may be affected, they should contact the regional U. S. Fish and Wildlife Service Office or personnel of the State Fish and Game Office or Conservation Department.

HCS — The Hazard Communication Standard is a regulation under the Occupational Safety and Health Act (OSHA). The Standard requires employers to provide protection for their workers who may be exposed to hazardous chemicals (Fig. 15).

JDPX7593

Fig. 15 — OSHA-HCS Requires Employers to Inform Workers About Safe Use of Hazardous Chemicals

The Standard also requires employers to:

- Compile a list and obtain Material Safety Data Sheets (MSDS) for all hazardous chemicals in their workplace

- Ensure that all containers of hazardous chemicals are labeled at all times

- Train all workers about the hazardous materials in their workplace and how to safely handle them

- Keep a file (including MSDS) on the hazardous chemicals and make it available to all workers

SARA Title III — Some pesticide applicators may be affected by Title III of the Superfund Amendments and Reauthorization Act of 1986, also known as SARA Title III, administered by the EPA. The following sections of SARA Title III specifically relate to storage of pesticides:

- **Emergency Planning and Notification** — Under certain conditions, the law requires you to notify state and local officials about the location and amount of hazardous chemicals on site.

- **Material Safety Data Sheet** — Reporting copies of each pesticide's MSDS must be provided to the local fire department and the local Emergency Planning Committee (EPC).

- **Inventory** — All regulated facilities must submit an annual chemical inventory report to their local fire department, LEPC, and State Emergency Response Commission.

- **Reporting** — The Act requires the reporting of any accident that results in a "reportable quantity" spill of any extremely hazardous substance and that spill creates an off-site exposure.

Certain agricultural procedures are exempt from these sections. For specific requirements, contact your local Emergency Planning Commission, fire department, or county Extension Center.

RCRA — The EPA regulates waste under the Resource Conservation and Recovery Act. EPA issues a list of materials that are considered hazardous. However, there are certain flammable, corrosive, reactive, or toxic wastes that RCRA applies to even if they are not on the list. Therefore, some other pesticides could be "regulated hazardous wastes" under RCRA. To find out if a pesticide is listed in RCRA, call the EPA-RCRA Hotline: 1-800-424-9346 (Fig. 16).

JDPX7594

Fig. 16 — Call EPA to Determine if a Substance is Considered a "Regulated Hazardous Waste"

WPS — The Worker Protection Standard (WPS) is intended to reduce the risks of illness or injury among pesticide handlers and agricultural workers resulting from occupational exposures to pesticides used in production of agricultural plants on farms and in nurseries, greenhouses, and forests.

Major provisions of the standard (WPS) call for safety training, new labeling requirements, posting requirements, restricted entry requirements to treated areas, personal protective equipment, decontamination facilities, and emergency transportation, to name just a few (Fig. 17).

JDPX7899

Fig. 17 — Worker Protection Standards (WPS) Require Signs to Warn of Pesticides in Fields

More information — For more detailed information on any of these regulations or any specific state or local regulations, contact your local Extension Center for specific requirements applicable to your area and/or state.

Acute and Chronic Exposure

If someone is exposed to a pesticide or swallows some pesticide and immediately loses consciousness, then there is no doubt about the cause of the illness, and proper medical treatment can be administered. "One-time" cases like this are examples of acute exposure.

Chronic exposure is a low-level exposure over a longer period of time. The effect is prolonged, slow poisoning. The symptoms are less apparent than in acute cases, and are often difficult to isolate as having been caused by pesticides. Types of chronic effects include tumors and cancers, reproductive problems, damage to the nervous system, damage to internal organs, and allergic sensitization to a particular pesticide.

Chronic exposures can be every bit as dangerous as acute exposures. The greatest risk when working with a pesticide is a combination of the two types. If you've been receiving daily exposure from a contaminated hat or contaminated clothing, your body will be less able to deal with an acute exposure such as spilling pesticide on yourself.

Types of Exposure

Before a pesticide can harm you, it must be taken into the body. Toxic materials can be taken into the body in four ways (Fig. 18):

- **Oral Exposure** — when you swallow or ingest a pesticide

- **Inhalation Exposure** — when you inhale a pesticide

- **Dermal Exposure** — when you get a pesticide on your skin

- **Ocular Exposure** — when you get a pesticide in your eyes

PESTICIDE EXPOSURE

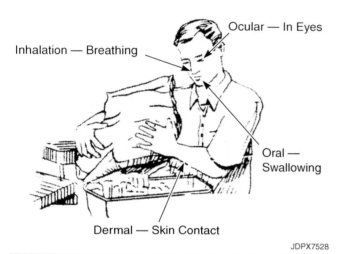

JDPX7528

Fig. 18 — Pesticides Can Enter Your Body From Four Types of Exposure

Causes of Exposure

One of the best ways to avoid pesticide exposures is to avoid situations and practices where exposures commonly occur.

Oral exposures or ingestions are caused by:

- Not washing hands before eating, drinking, smoking, or chewing tobacco

- Mistaking a pesticide for food or drink

- Accidentally applying pesticides to food

- Splashing pesticide into the mouth through carelessness or accident

Inhalation exposures often are caused by:

- Prolonged contact with pesticides in closed or poorly ventilated spaces

- Breathing vapors from fumigants and other pesticides

- Breathing vapors, dust, or mist while handling pesticides without appropriate protective equipment

- Inhaling vapors immediately after a pesticide is applied, for example, from chemical drift or from reentering a treated area too soon

- Using the wrong respirator or an improperly-fitted respirator, or using filters, cartridges, or canisters that are "full" of chemicals, dust, etc.

Dermal exposures and absorption often are caused by:

- Not washing hands after handling pesticides or their containers

- Splashing or spraying pesticides on unprotected skin

- Wearing pesticide-contaminated clothing (including boots and gloves)

- Applying pesticides (or flagging) in windy weather

- Wearing inadequate personal protective equipment while handling pesticides

- Touching pesticide-treated surfaces

Ocular exposures are caused by:

- Splashing or spraying pesticides in eyes

- Applying pesticides in windy weather without eye protection

- Rubbing eyes or forehead with contaminated gloves or hands

- Pouring dust, granules, or powder formulations without eye protection

Toxicity

Toxicity is a measure of the ability of a pesticide to cause a harmful effect. Oral and dermal toxicity are expressed by the LD_{50} value for the particular compound. The LD_{50} value is the number of milligrams (mg) of pesticides per kilogram (kg) of body weight that will kill 50% of a test animal population (usually rats or rabbits). The smaller the LD_{50} value, the more toxic (dangerous) the pesticide. For example, a pesticide with an oral LD_{50} of 500 would be much less toxic than a pesticide with an LD_{50} of 5 (Fig. 20).

LD_{50} values based on animal studies should not be interpreted as exact values for human toxicity. However, most experts agree that the LD_{50} value can be used as a relative guide for estimating acute human toxicity exposures. It should be noted that LD_{50} values provide little or no information about the possible cumulative effects of pesticide exposure.

What does the LD_{50} value mean in terms of danger to humans? If you know the LD_{50} of a material you're working with, you can find out how much of it would be likely to cause death by using this formula:

$$\frac{LD_{50} \times 0.0016 \times \text{body weight (lb)(kg)}}{100} = \text{Weight (oz)(g)}$$

of pesticide likely to cause death.

As an example, let's look at Sevin® 80S (Fig. 20). The LD_{50} of 281 milligrams per kilogram (i.e. 1000 grams) multiplied by 0.0016 equals 0.45 ounce (12.76 g) per 100 pounds (45 kg) of body weight. So if you weigh 100 pounds (45 kg) and swallow 0.45 ounces (12.76 g) of Sevin® 80S, you could expect to have about a 50% chance of dying from the exposure to the material. A person who weighs twice as much (200 pounds) (90 kg) might tolerate twice as much material (about 1 ounce) (28 g) and have the same (50%) chance of dying. Similarly, children would be killed by smaller pesticide doses than adults (Fig. 19).

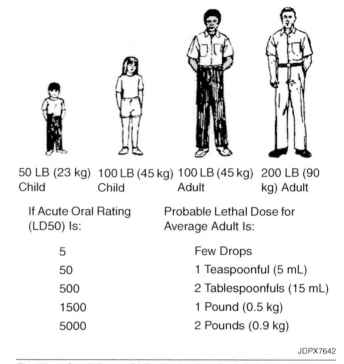

50 LB (23 kg) Child	100 LB (45 kg) Child	100 LB (45 kg) Adult	200 LB (90 kg) Adult

If Acute Oral Rating (LD50) Is:	Probable Lethal Dose for Average Adult Is:
5	Few Drops
50	1 Teaspoonful (5 mL)
500	2 Tablespoonfuls (15 mL)
1500	1 Pound (0.5 kg)
5000	2 Pounds (0.9 kg)

JDPX7642

Fig. 19 — Small Doses Of Pesticides Can Be Lethal

Pesticides are also rated on a relative toxicity scale. The ratings range from very low toxicity to extremely toxic depending on the LD_{50} rating of the material. The hazard or toxicity ratings and the amounts of some materials that probably would kill people are shown in Fig. 20 (oral) and Fig. 21 (dermal).

Inhalation Toxicity

Inhalation toxicity is expressed as an LC_{50}. LC_{50} is the concentration of an active ingredient in the air that is expected to cause death to 50% of the test animals treated. It is generally expressed as micrograms per liter for a dust or mist, but in the case of gas or vapor, as parts per million (ppm).

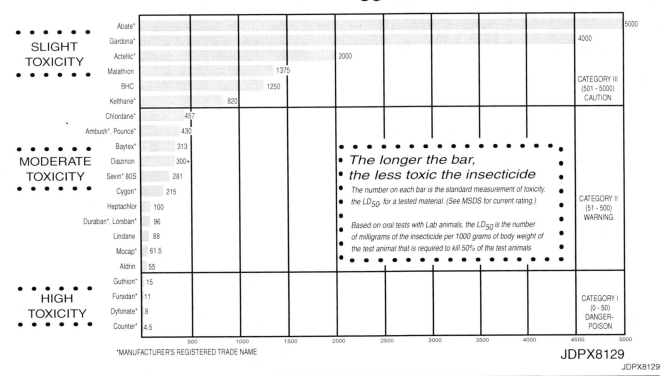

Fig. 20 — *Relative Oral Toxicity Of Some Common Pesticides (When Swallowed)*

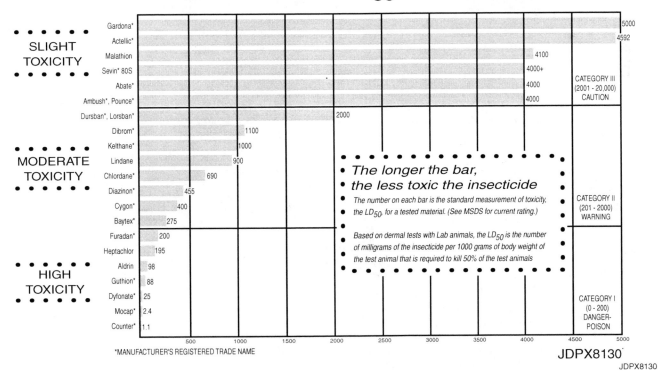

Fig. 21 — *Relative Dermal Toxicity Of Some Common Pesticides (From Skin Contact)*

Avoiding Exposure

Avoiding and reducing exposures to pesticides will reduce the harmful effects from pesticides. You can reduce exposures by using safety systems such as closed handling systems; by wearing appropriate personal protective equipment (PPE) and keeping it clean and in good operating condition; and by washing exposed body areas often (Fig. 22).

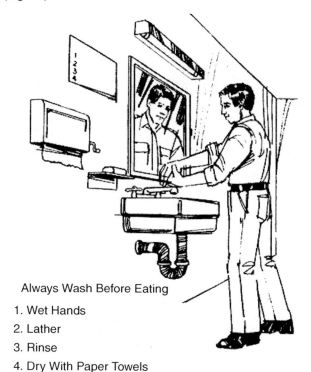

Always Wash Before Eating

1. Wet Hands
2. Lather
3. Rinse
4. Dry With Paper Towels

JDPX7900

Fig. 22 — Wash Carefully After Working With Pesticides

In most pesticide-handling situations, the skin is the part of the body that is most likely to receive exposure. Evidence indicates that about 97% of all body exposure that occurs during pesticide spraying is the result of dermal exposure or by contact with the skin.

However, inhalation can be a greater hazard than skin contact when you are working in a poorly ventilated, enclosed space and are using a fumigant or other pesticide that is highly toxic through inhalation.

The amount of pesticide that is absorbed through your skin (and eyes) and into your body depends on:

- The pesticide itself and the material used to dilute the pesticide. Emulsifiable concentrates, oil-based liquid pesticides, and oil-based diluents (such as xylene) are, in general, absorbed most readily. Water-based pesticides and dilutions (such as wettable and soluble powders and dry flowables) usually are absorbed less readily than the oil-based liquid formulations but more readily than dry formulations. Dusts, granules, and other dry formulations are not absorbed as readily as liquids.

- The area of the body exposed. The genital area tends to be the most absorptive (Fig. 23). The scalp, ear canal, and forehead also are highly absorptive.

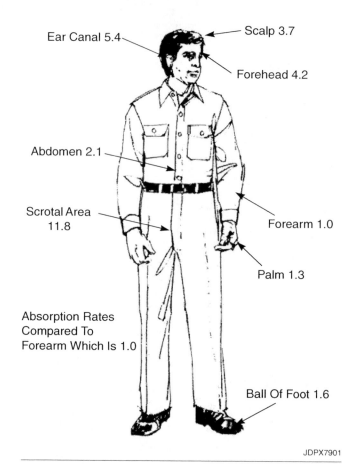

Ear Canal 5.4
Scalp 3.7
Forehead 4.2
Abdomen 2.1
Scrotal Area 11.8
Forearm 1.0
Palm 1.3
Ball Of Foot 1.6

Absorption Rates Compared To Forearm Which Is 1.0

JDPX7901

Fig. 23 — Dermal Absorption Rates Are Higher for Some Parts of the Body Than Others

- The condition of the skin exposed. Cuts, abrasions, and skin rashes allow absorption more readily than intact skin. Hot, sweaty skin will absorb more pesticide than dry, cool skin.

Labels

Pesticide product labeling is the main method of communication between a pesticide manufacturer and pesticide users. The information printed on or attached to the pesticide container is the label. Labeling includes the label itself plus all other information you receive from the manufacturer about the product when it is purchased.

Pesticide labeling gives instructions on how to use the product safely and correctly.

Pesticide users are required by law to comply with all the instructions and directions for the pesticide label.

The most important rule to follow when using pesticides is: READ THE LABEL BEFORE YOU BUY OR USE A PESTICIDE (Fig. 24).

JDPX7632

Fig. 24 — Read and Follow Label Information and Instructions on Pesticide Container

Before you buy a pesticide, determine:

- Whether it is the pesticide you need for the job

- Whether the pesticide can be used safely under the application conditions

- Whether the label has changed since you last purchased the product. Check for new restrictions, re-entry periods, notifications of others, new personal protective equipment requirements, etc.

Before you mix a pesticide, determine:

- What specific personal protective equipment you should use

- What the pesticide can be mixed with (compatibility)

- How much pesticide to use

- The mixing procedure

Before you apply a pesticide, determine:

- What safety measures you should follow

- Where the pesticide can be used

- When to apply the pesticide (including re-entry periods, application and harvest restrictions, etc.)

- How to apply the pesticide

- Whether there are any restrictions for use of the pesticide

- Who must be notified of application

- Posting requirements

Before you store or dispose of excess pesticide, pesticide rinsates, or pesticide container, determine:

- Where and how to safely store the pesticide

- How to safely dispose of the pesticide container

- Where to dispose of or safely use the surplus pesticides or rinsates

Read, follow and understand all instructions given on the pesticide label.

Labels do not generally report the toxicity of a pesticide as an LD_{50} value. This information can be found on the Material Safety Data Sheet (MSDS) available from your pesticide supplier. However, the label does include one of three signal words that express how toxic the chemical is to humans: Caution, Warning, or Danger.

Signal words are shown on every label and should be used as a guide to the toxicity of the material and as a signal to the possible hazards when mixing, loading, and applying the pesticide. See label safety warnings for different EPA pesticide toxicity levels in figures 25, 26, and 27.

DANGER ☠ POISON

KEEP OUT OF REACH OF CHILDREN
POISONOUS IF SWALLOWED
OR ABSORBED THROUGH SKIN
FLAMMABLE • VAPOR HARMFUL
Do Not Get in Eyes on Skin or on Clothing
Do Not Breathe Vapors or Spray Mist
Do Not Take Internally
Keep Away from Heat and Open Flame
Keep Container Closed

JDPX8131

Fig. 25 — Category 1 Toxicity: Highly Hazardous. Information Subject to Change

WARNING

IRRITATING TO SKIN AND EYES
CAUSES BURNS ON PROLONGED CONTACT
HARMFUL IF SWALLOWED

Avoid Breathing Spray Mists
Keep Out of the Reach of Children
Do Not Get in Eyes on Skin or on Clothing or Shoes

JDPX8132

Fig. 26 — Category II Toxicity: Moderately Hazardous. Information Subject to Change

CAUTION

KEEP OUT OF REACH OF CHILDREN
ABSORBED THROUGH SKIN
CAUSES EYE IRRITATION
HARMFUL IF SWALLOWED
FLAMMABLE LIQUID

JDPX8133

Fig. 27 — Category III Toxicity: Slightly Hazardous. Information Subject to Change

Material Safety Data Sheets

The Material Safety Data Sheet (MSDS) provides specific details on pesticides, including physical and health hazards, safety procedures, and emergency response techniques. Get an MSDS from your pesticide supplier for each product.

Read the MSDS before you start any pesticide job so you will know exactly what the risks are and how to do the job safely (Fig. 28). Then follow the label and MSDS procedures, and use the recommended personal protective equipment. That way, pesticides will be less hazardous to you, your family, your co-workers and the environment.

JDPX7902

Fig. 28 — Read MSDS (Material Safety Data Sheet) of Each Pesticide for Hazards and Safe Procedures

NOTE: For detailed information on various types of liquid and dry pesticide formulations and their characteristics, see "Pesticide Formulations" at the end of this chapter.

Personal Protective Equipment

Personal protective equipment (PPE) consists of clothing and devices that are worn to protect the human body from contact with pesticides and their residues. Personal protective equipment includes such items as coveralls, protective suits or aprons, footwear, gloves, respirators, eyewear, and headgear (Fig. 29).

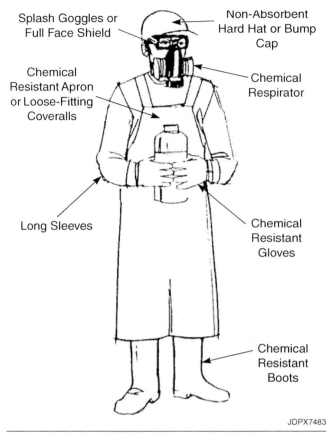

Splash Goggles or Full Face Shield

Chemical Resistant Apron or Loose-Fitting Coveralls

Long Sleeves

Non-Absorbent Hard Hat or Bump Cap

Chemical Respirator

Chemical Resistant Gloves

Chemical Resistant Boots

JDPX7483

Fig. 29 — Use Personal Protective Equipment (PPE) When Working With Pesticides. Follow The Label

Ordinary shirts, pants, shoes, and other regular work clothing usually are not considered personal protective equipment, although the pesticide label may require you to wear specific items of work clothing.

Pesticide labeling lists the minimum personal protective equipment you must wear while handling or applying the pesticide. Sometimes the label lists different requirements for different activities. For example, more personal protective equipment may be required for mixing and loading than for application.

Exposure to pesticides can cause harmful effects. To prevent or reduce exposure, you need to wear correct, clean, properly fitted personal protective equipment. You are legally required to follow all personal protective equipment instructions that appear on the label or labeling.

Gloves — Always wear unlined, elbow-length, chemical-resistant (liquid-proof) gloves when handling any pesticide concentrate or chemicals that carry the signal word Danger, Poison, or Warning. When spraying overhead, wear shirt sleeves inside gloves. At all other times, wear shirt sleeves on the outside to prevent chemicals from entering gloves at the cuff. Pay special attention to labels when determining what type of chemical-resistant gloves to wear, since some fumigants and pesticides are absorbed or trapped readily by the gloves.

Never wear cotton or leather gloves. They absorb the pesticide and provide constant dermal exposure to the chemical, which can be more hazardous than not wearing gloves at all.

Always check gloves carefully for leaks before wearing them. Fill the gloves with water and squeeze. If leaks appear, discard the gloves. Before removing the gloves, wash them with detergent and water to prevent contaminating your hands.

Boots — When handling or applying chemicals, wear unlined, lightweight rubber vinyl boots which cover your ankles. Wear trouser legs on the outside so pesticides cannot drain down into the boots. The boots should be washed daily and dried thoroughly inside and out to remove pesticide residues. As with gloves, especially check the label to determine what types of shoes or boots can be worn when using fumigants.

Goggles or Face Shields — Wear tight-fitting, non-fogging chemical splash goggles or a full face shield when pouring, mixing, or applying pesticides. Clean the equipment often and make sure the sweatband on the face shield is cleaned, since some materials used in face shield sweatbands absorb and hold chemicals. If possible, wear the sweatband under the head covering. Eyewear should meet or exceed the current requirements of ANSI (American National Standards Institute; Z87.1, 1968)

Head and Neck Coverings — The hair and skin on your neck and head should be protected. Several available types of headgear, such as waterproof rain hats or washable, wide-brimmed hard hats or bump caps can be used. Waterproof or repellent parkas will protect the neck and head at the same time. The headgear, including the sweatband, should be cleaned often to remove any chemical residues. Avoid cotton or felt hats since they also will absorb pesticides.

Respirators — Correct, properly fitted respirators prevent inhalation of toxic chemicals. Wear a respirator when the label calls for it. Respirators are especially necessary when handling concentrated, highly toxic pesticides.

Specific cartridges and canisters protect against specific chemical gases and vapors. Be sure to choose the type made to protect you against the pesticides you will use. The respirator must fit properly to ensure a good seal. Long sideburns, a beard, or glasses may prevent a good seal.

Chemical Cartridge Respirators — Usually recommended: a half-face mask, containing one or two cartridges, which covers the nose and mouth only.

Full facepiece respirators: provide both eye and respiratory protection. The inhaled air that enters the cartridge is pulled through a filter pad and a filtering medium such as activated charcoal. Use chemical cartridge respirators either for relatively short periods of exposure to high concentrations of toxic chemicals or for long exposure to low concentrations of toxic chemicals. This respirator should never be used in areas where the oxygen level is too low to support life.

Chemical canister respirator (gas masks): are designed to protect for a longer period of time than cartridge respirators. A gas mask usually protects the face better than the cartridge respirator since it covers the entire face (that is, it protects your eyes, nose, and mouth). Use a gas mask when you are exposed to toxic fumes in high concentrations or for a long period of time. Also, wear a gas mask when applying pesticides in enclosed or poorly ventilated areas. As with the chemical cartridge respirator, a gas mask should never be used in areas where the oxygen level is too low to support life.

Air-supplied respirators: should be worn in areas where the oxygen supply is low or where there are extremely high concentrations of very toxic pesticides in enclosed areas. Fresh air is pumped by a blower through a hose to the face mask from an uncontaminated area or from a backpack.

Care and Maintenance of Respirator. If breathing becomes difficult during spraying, get to fresh, clean air and change the respirator cartridges or canisters. Cartridges should be changed after 8 hours of use or sooner if you detect pesticide odor while wearing the respirator.

The facepiece should be washed with soap and water, rinsed, dried with a clean cloth, and stored in a clean, dry place away from pesticides. A tightly closed plastic bag works well for storage. The manufacturer's instructions on the use and care of a respirator and its parts should be read carefully before the respirator is used. Use only respirators approved by the National Institute for Occupational Safety and Health (NIOSH) or the Bureau of Mines.

General Care

Wear clean clothing each day you spray pesticides (Fig. 30). If pesticide solutions get on your clothing, change immediately.

JDPX7484

Fig. 30 — Change Clothes Daily for Pesticide Work, and at Once if Pesticide Gets on Clothes

Do not store or wash contaminated clothing with the family laundry, because your pesticide-contaminated clothing could contaminate theirs. Additionally, contaminated clothing may require special attention to ensure that it is thoroughly cleaned.

If your clothes have been heavily contaminated by pesticides, discard them. You may not be able to clean and safely wear them again.

All personal protective equipment should be washed daily with soap and water, rinsed, dried with a clean cloth, and stored in a clean dry place away from all pesticides.

Avoid Heat Stress

Several factors work together to cause heat stress. High temperatures, high humidity, direct sunlight, and personal protective equipment are just some of the factors that cause heat-related disabilities. Heat stress is one such illness, and it occurs when the body is subjected to more heat than it can handle. Personal protective equipment and clothing used in pesticide work in hot weather can lead to heat stress (Fig. 31).

JDPX7643

Fig. 31 — Personal Protective Equipment Can Be Hot to Work In. Use Caution. Avoid Heat Stress

Heat stress can be reduced by:

*   Using fans and ventilation systems and providing sun shade whenever possible

*   Allowing time to adjust to the heat factors and workload

*   Scheduling frequent breaks

*   Choosing personal protective equipment that is designed to be as cool as possible or will provide cooling (powered air-purifying respirators, back-vent coveralls, and cooling vests)

*   Drinking plenty of water

*   Scheduling work that requires personal protective equipment during the cool part of the day

First Aid for Heat Stress

It is not always easy to tell the difference between heat stress and pesticide poisoning. The signs and symptoms are similar. Don't waste time trying to decide what is causing the illness. Get medical help.

First aid measures for heat stress victims are similar to those for persons who are overexposed to pesticides:

*   Get the victim into a shaded or cool area

*   Cool victim as rapidly as possible by sponging or splashing skin with cool water, especially face, neck, hands, and forearms, or by immersing the victim in cool water when possible

*   Carefully remove all personal protective equipment and any other clothing that may be making the victim too warm

*   If the victim is conscious, have him or her drink as much cool water as possible

*   Keep the victim quiet until help arrives

Tractor and Sprayer Cabs

Tractor and sprayer manufacturers advise that positive-pressure cabs equipped with filtered ventilation systems aren't considered suitable alternatives to personal protective equipment (PPE).

EPA has approved the use of cabs that have properly functioning positive-pressure ventilation systems for pesticide application only if: 1) the cabs are used and maintained according to the manufacturer's written operating instruction; and 2) if the operator wears a chemical respirator as specified on the pesticide label.

NOTE: Contact your state EPA or Department of Agriculture for current personal protection requirements.

Before leaving the cab during field application, wear all the label-specified personal protective equipment. When you re-enter the cab, remove the protective equipment and store it outside the cab or in a closed, sealable container, or store it inside the cab in a pesticide-resistant container, such as a plastic bag. Decontaminate all PPE and clothing before re-entering the cab so chemicals won't accumulate inside the cab.

If the pesticide label requires a respiratory protection device, (e.g. chemical cartridge, gas mask and air supplier), then that device MUST be worn within the enclosed cab (Fig. 32).

JDPX8098

Fig. 32 — Even Inside a Cab, Use Respiratory Protection if Pesticide Label Calls For It

Preparing the Sprayer

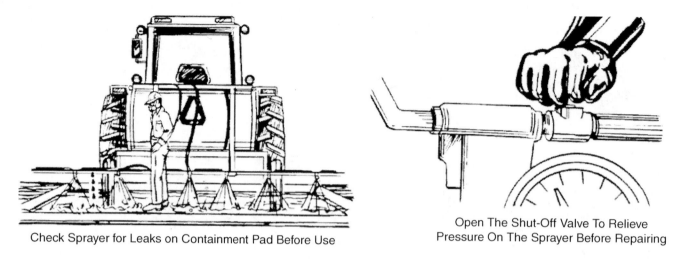

Check Sprayer for Leaks on Containment Pad Before Use

Open The Shut-Off Valve To Relieve Pressure On The Sprayer Before Repairing

JDPX7644

Fig. 33 — Before Use, Rinse Sprayer With Clean Water on a Containment Pad and Check for Leaks

Before a spraying operation is started, rinse out the sprayer on a containment pad and remove and clean all nozzles, nozzle screens, and strainers (Fig. 33). (Note: This also should be done after each use and before the sprayer is stored for the season.) All nozzles should be of the same type, size and fan angle.

Check all lines, valves, seals and the tank after filling the sprayer with water and during calibration to be sure there are no leaks.

Calibrate your sprayer frequently to be sure it is working properly and is accurately applying the pesticides. Follow the equipment operator's manual. Your local Cooperative Extension office may also have detailed instructions.

Always wear chemical resistant gloves to protect your hands from any pesticide residue. Replace weather-cracked or worn hoses. Adjust the nozzle height and spacing as suggested by the nozzle manufacturer or on the pesticide label.

Only water that is clean enough to drink should be used in the sprayer. Just a small amount of silt or sand in the water will rapidly wear pumps and other parts of the sprayer system.

Operation

Operating a pump dry or with a restricted inlet may damage it. Don't operate pumps at speeds or pressures above those recommended by the manufacturer. PTO pumps should be restrained from rotating by chains or torque bars. Be sure to keep all shields in place. Don't use ground speeds too fast for field conditions. Booms bouncing up and down or back and forth cause application rates to vary by 50%. Bouncing can damage the sprayer booms or frame.

Always follow equipment operator's manuals. Also, review Chapter 7, "Tractors and Implements," for equipment safety precautions.

Handling and Mixing

Take precautions to avoid exposure when pesticides are added to the sprayer tank, because air will be forced out, carrying some of the pesticide particles with it. Be sure the mixing operation is performed in an area with adequate ventilation. Always use the appropriate personal protective equipment.

Many types of solvents, some of which are chlorinated, are used in the pesticide formulation process. Vapors of chlorinated solvents are very dangerous to breathe. They can cause a "high," dizziness, or even unconsciousness. They also can cause permanent damage to kidneys, liver, and nervous system in workers exposed to vapors for a prolonged time.

Be sure to contain any spills. Never leave a sprayer unattended during filling; and never overfill. Always keep the water fill hose outside the tank to avoid back-siphoning during filling (Fig. 34). Back-siphoning carries chemicals back to the water supply. Check local and state regulations to determine if back-siphoning protection is required.

PREVENT BACK-SIPHONING

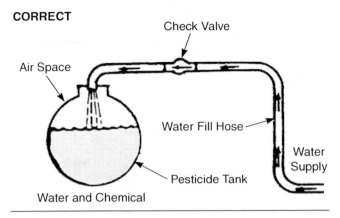

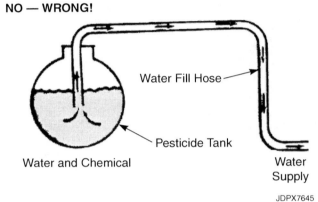

JDPX7645

Fig. 34 — When Filling Tank, Be Sure Pesticide and Water Can't Back-Siphon Into Water Supply

NOTE: See "Pesticide Formulation" at the end of this chapter for ingredients used in liquid and dry pesticides and for mixing considerations.

Closed Systems

A closed mixing and loading system is a system designed to reduce pesticide contact by handlers or others during mixing and loading. There are two primary types of closed mixing systems. One type of closed system monitors and transfers pesticide products from the shipping container to a mixing or applicator tank, or directly injects chemicals into the line carrying water to the booms and often rinses the emptied containers.

Another type of mechanical device is a lid-fill closed handling system (Fig. 35). When the transfer valve on the lid of the returnable chemical container is coupled to the mating valve on the hopper lid, both valves open automatically and allow the pesticide to flow into the hopper. When the container is removed, both valves close, resealing the container and the hopper, thus greatly reducing operator exposure to chemical dust or granules.

There is also a closed mixing system called soluble packaging. Soluble bags or containers are used by putting the entire package (pesticide and container) into the spray tank. The container dissolves in the tank and releases the pesticide into solution.

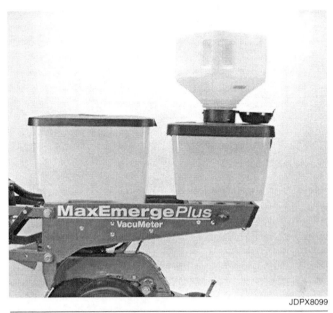

JDPX8099

Fig. 35 — Lid-Fill Closed Handling Systems Reduce Exposure to Pesticide Dust and Granules

Field Application

Always be aware of the meteorological conditions existing during pesticide application. Spray drift from treated areas increases during high winds or low humidity.

Avoid spraying near beehives, lakes, streams, pastures, houses, schools, playgrounds, hospitals, or sensitive crops when possible. If these areas must be sprayed, don't spray during windy or low humidity conditions. Always spray downwind from the sensitive area. Check to see if a permit is required before application.

Begin spraying where drift will be blown away from the next area to be treated. Likewise, with airblast sprayers, direct the air blast from the sprayer with the prevailing wind and away from the next area to be treated (Fig. 36). These two procedures will minimize the amount of pesticide that will be blown onto the user. Remember, these are only guidelines and are intended to supplement the good judgment of the user.

Fig. 36 — *Direct the Air Blast With the Wind So Pesticide Spray Won't Drift Back Onto You*

To minimize drift hazards, use the lowest pressure possible, the lowest boom height, and drift-reducing tips, and add thickeners (if the pesticide label permits) in areas where drift is likely to be particularly hazardous.

Be alert for nozzle clogging and changes in nozzle patterns. If nozzles clog or other troubles occur in the field, shut the sprayer off and move to an unsprayed area before dismounting from the sprayer to work on it. Use the appropriate personal protective equipment — rubber gloves and goggles are the minimum. Always carry extra spray tips on the sprayer so that plugged tips can be replaced with clean ones. Never try to unclog a nozzle by blowing through it (Fig. 37). Don't touch your mouth with pesticide-contaminated objects.

Check the pesticide label for re-entry and pre-harvest intervals. The re-entry interval is the elapsed time after a pesticide application before workers can safely re-enter a field. The pre-harvest interval is the elapsed time between a pesticide application and harvest of the crop. New labeling requirements involve posting treated areas and notifying others of treatment and re-entry intervals (Fig. 38). CHECK THE LABEL!

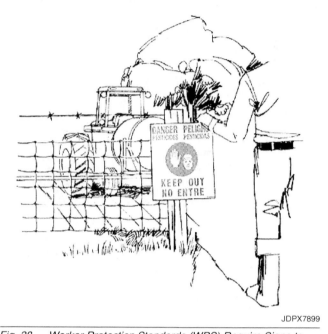

Fig. 38 — *Worker Protection Standards (WPS) Require Signs to Warn of Pesticides in Fields*

Fig. 37 — *Never Blow Through Nozzles With Your Mouth to Clean Them. Use an Air Hose*

Equipment Cleaning

All spray equipment should be thoroughly cleaned inside and outside immediately after it is used (Fig. 39). When cleaning equipment, you should always wear proper protective clothing, including rubber boots, a rubber apron, goggles, and possibly a respirator.

A specific area should be designated for cleaning operations. Use a rack or concrete containment apron with a well-designed sump and curbing to catch contaminated wash water and pesticides. Clean-up is important because many chemicals rapidly corrode some metals and may react with succeeding chemicals, possibly causing loss of effectiveness.

The most important step in machinery decontamination is a thorough washing with soap (or a mild detergent) and water, followed by a complete rinse with plenty of water. A high-pressure washer can be used if available. Compacted chemical deposits can usually be removed by wetting and scrubbing with a stiff-bristled brush. Pay particular attention to tires and hoses when cleaning sprayers. Dry material can be removed conveniently with a vacuum cleaner.

Use a strong soap-and-water solution to clean the inside of a sprayer if you have been using carbonates or organophosphates. If organochlorines have been used, the wash water should be mixed with acetic acid (vinegar). Do not use soap with organochlorines.

Sprayer parts should be washed in the same mixture as recommended for washing the inside of the sprayer.

Wash Outside of Sprayer With a Long-Handled Brush and Lots of Soapy Water, Wear an Apron Boots, Gloves, Hat, and Goggles

Hose Off the Machine

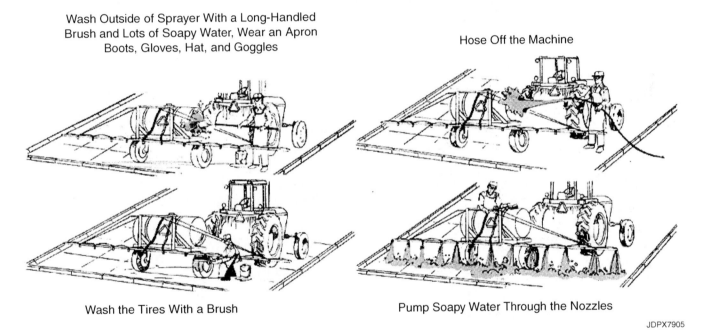

Wash the Tires With a Brush

Pump Soapy Water Through the Nozzles

JDPX7905

Fig. 39 — Decontaminate Chemical Application Machinery on Containment Pad. Wear Personal Protective Equipment

Pesticide Storage and Transport

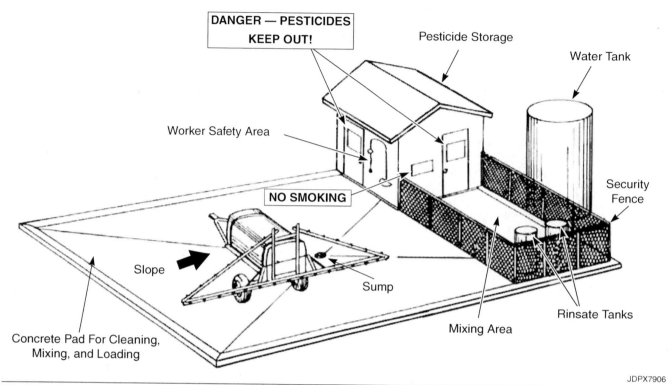

DANGER — PESTICIDES KEEP OUT!

Pesticide Storage

Water Tank

Worker Safety Area

NO SMOKING

Security Fence

Slope

Sump

Rinsate Tanks

Mixing Area

Concrete Pad For Cleaning, Mixing, and Loading

JDPX7906

Fig. 40 — Store Pesticides in a Separate Building With a Lock and Danger Signs

Pesticides are usually purchased before they are needed, and they must be stored on the farm between purchase and use. Existing buildings or areas within existing buildings often are used for this storage. It is best to build a special building just for pesticide needs. It could include storage, mixing, and containment pad facilities (Fig. 40).

Contact your local Cooperative Extension office or MidWest Plan Service for a storage plan. (Note: Check to see if there are special building codes for such facilities in your area.) Never store pesticides in a garage, basement, or home.

Establish a Storage Site

A suitable storage site:

- Protects people and animals from accidental exposure

- Protects the environment from accidental contamination

- Prevents damage to pesticides from temperature extremes and excess moisture

- Protects pesticides from theft, vandalism, and unauthorized use

- Reduces the likelihood of an accident, thereby reducing liability

Secure the Site

Keeping unauthorized people out is an important function of the storage site. Whether the storage site is as small as a cabinet or closet or as large as an entire room or building, keep it securely locked. Post weatherproof signs on doors and windows: DANGER — PESTICIDES. KEEP OUT! Post NO SMOKING signs, also.

Prevent Water Damage

Choose a storage site where water damage is unlikely to occur (Fig. 41). Water from burst pipes, spills, overflows, flooding streams, or excess rain or irrigation can damage pesticide containers and pesticides. Water or excess moisture can cause:

- Metal containers to rust

- Paper and cardboard containers to split or crumble

- Pesticide labeling to peel, smear, run, or otherwise become unreadable

- Dry pesticides to clump, degrade, or dissolve

- Slow-release products to release their pesticide

- Pesticides to move from the storage site into other areas

Keep Chemicals off the Floor

JDPX7633

Fig. 41 — Store Pesticide Containers off the Floor so They Will Stay Dry

Prevent Runoff

Inspect the storage site to determine the likely path of pesticides in case of spills, leaks, drainage of equipment wash water, or heavy pesticide runoff from fire fighting or floods. Pesticide movement away from the storage site could contaminate sensitive areas, including surface water or ground water, and ultimately well water. If you store large amounts of pesticides, regulations may require a collection pad to contain pesticide runoff. Check with your Cooperative Extension office or chemical supplier.

Use Non-Porous Materials

The floor of the storage site should be made of sealed concrete or another easily cleaned material. Wood, soil, gravel, or other absorbent floors are difficult or impossible to decontaminate in case of a leak or spill. For ease in cleaning, shelves and pallets should be made of non-absorbent materials such as plastic or metal.

Control the Temperature

The storage site should be indoors. Choose a cool, well-ventilated room or building that is insulated or temperature-controlled to prevent freezing or overheating.

Provide Adequate Lighting

The storage site should be well lighted. Pesticide handlers using the facility must be able to see well enough to:

- Read pesticide container labeling

- Notice whether containers are leaking, corroding, or otherwise disintegrating

- Clean up spills or leaks

Provide Clean Water

Each storage site must have an immediate supply of clean water. Keep an eyewash dispenser immediately available for emergencies.

Maintain the Storage Site

Prevent Contamination

Store only pesticides, pesticide containers, pesticide equipment, and a spill clean-up kit at the storage site. Do not keep food, drinks, tobacco, feed, medical or veterinary supplies, medication, seeds, clothing, or personal protective equipment (other than personal protective equipment necessary for emergency response) at the site.

Keep Labels Legible

Store pesticide containers with the label in plain sight. Labels should always be legible. If the label is destroyed or damaged, request a replacement from the pesticide dealer or the pesticide formulator immediately.

Keep Containers Closed

Keep pesticide containers securely closed whenever they are being stored. Tightly closed containers help protect against:

- Spills

- Cross-contamination with other stored products

- Evaporation of liquid pesticides or solvent

- Clumping or caking of dry pesticides in humid conditions

- Dust, dirt, and other contaminants getting into the pesticide, making it unusable

Store Volatile Products Separately

Volatile pesticides, such as type 2, 4-D, should be stored apart from other types of pesticides and other chemicals. The labeling of volatile herbicides usually will direct you to store them separately from seeds, fertilizers, and other types of pesticides.

Isolate Waste Products

If you have pesticides and pesticide containers that are being held for disposal, store them in a special section of the storage site. Clearly mark containers that have been triple-rinsed or cleaned by an equivalent method. They are more easily disposed of than unrinsed containers.

Know Your Inventory

Keep an up-to-date inventory of the pesticides you have in storage (Fig. 42). Each time a pesticide is added to or removed from the storage site, update the inventory list. Regulations require it for some storage sites.

JDPX7907

Fig. 42 — Maintain an Inventory of Pesticides in Your Storage

Pesticide Disposal

Use extreme care when disposing of pesticides. They are hazardous wastes. Do not contaminate ground water, surface water, or air.

Plan for the disposal of these kinds of materials:

* Leftover chemicals
* Spilled materials
* Empty containers

Your pesticide dealer or local Cooperative Extension office can help with disposal plans.

Dilute Solutions and Rinsates

Leftover spray solution and waste solution from washing equipment, rinsing tanks and booms, surplus tank mixtures, and spilled pesticides must be disposed of with minimal impact to the environment.

Follow these guidelines to minimize the amount of pesticide waste solution:

* Mix only enough pesticide for the acreage to be treated.

* If you mix too much pesticide, try to find other areas with the same pest problem and use up any extra tank mix or rinse water on these areas. Small amounts of surplus mixtures can be diluted and reapplied to the treated area. Be sure that the total application rate does not exceed the maximum rate for which the pesticide is labeled.

* Properly calibrate your sprayer.

* Pour the rinsate into the spray tank while triple rinsing containers.

* Read the product label carefully before combining two or more pesticides, as some pesticides are incompatible.

* Check out the job carefully before selecting a pesticide to make sure you do not mix the wrong pesticide.

Spilled Materials

Even when proper procedures are followed, pesticide spills can occur. Knowing what steps to take in the event of a pesticide spill will allow you to respond quickly and properly. Remember, always be sure to wear proper protective clothing when dealing with pesticide spills and to clean up your equipment and clothing when you are finished.

* Control the spill. Immediately after a spill has occurred, make sure that the source of the spill has been identified and controlled, preventing further spilling.

  − If a small container is leaking, you should immediately place the leaking container into a larger, watertight container.

  − Keep people and animals at least 30 feet (9.1 m) upwind from the spill. Avoid coming into contact with any fumes.

  − Call authorities as soon as possible for help and information on controlling the spill.

* Contain the spill (Fig. 43). Use appropriate diking materials. Pesticide dealers can advise about type and source. Never allow any chemical to get into any body of water, including storm sewers or sanitary sewers. Never hose down spills, because this will spread the chemical.

Fig. 43 — *Quickly Contain Spilled Pesticides so They Can't Spread. Use Good Diking Materials*

- Clean up the spill (Fig. 44). Use an absorbent material such as dirt, sand, clay, or kitty litter to soak up the spill. Shovel all contaminated material into a leakproof container for proper disposal.

Fig. 44 — *Scoop Contaminated Materials Into a Leakproof Container for Safe Disposal*

Once the spill has been cleaned up, it may be necessary to decontaminate the area. Common household bleach is usually effective. However, you should read the label for specific decontamination directions. For example, synthetic pyrethroid labels will state that bleach should not be used for clean-up.

Additional information can be obtained by dialing the emergency telephone number listed on the label or by calling:

CHEMTREC: 1-800-424-9300

- Call the authorities. Authorities may need to be notified in case of any spill, regardless of size, or only for major spills, or spills on public road with streets, or in case of food or water contamination. Check with local health, fire, or police department; your chemical dealer; or Cooperative Extension office to determine the requirements for your area.

Empty Containers

Empty container into spray tank and drain 30 seconds in vertical position.

Fill container 1/4 to 1/3 full with rinse water or other dilutent.

Rinse container, pour contents into tank, and drain 30 seconds. Add fluid to bring tank to level.

Do not reuse pesticide containers.

JDPX7640

Fig. 45 — Rinse and Drain Pesticide Containers Before Disposal. EPA Recommends Triple Rinsing

Chemical containers are never really empty. They always contain a residue of hazardous material. Keep them in a locked storage area and dispose of them at least once a year in a way that poses no hazard to humans, animals, or the environment.

Use a rinse-and-drain procedure to clean containers (Fig. 45). EPA classifies triple-rinsed containers as non-hazardous waste for disposal purposes.

Triple-rinsed containers are usually returned to the dealer or local container recycling program. Do not cut up a drum with a cutting torch — t could cause a fire or explosion. And do not convert empty pesticide drums into feed troughs, water storage tanks, or raft floats. Pesticide residue could contaminate the water and endanger livestock and people.

Burning may not be permitted in your area. Check with your local or state air quality control board before burning any pesticides or pesticide containers. If you are not satisfied with the answer you get, check with the State Department of Agriculture or the EPA in Washington, D. C.

Transporting Pesticide Containers

Transport pesticides only in containers with intact, undamaged, and readable labels (Fig. 46). Inspect containers before loading to be sure all caps, plugs, and other openings are tightly closed and there are no pesticides on the outside of containers. Handle containers carefully to avoid rips or punctures.

JDPX7909

Fig. 46 — Transport Pesticides in Tightly Closed Containers, and Pack Securely to Prevent Damage

Anchor all containers securely to keep them from rolling or sliding. Packing or shipping containers will provide extra cushioning. Protect paper and cardboard containers from moisture since they get soggy and split easily when wet.

Protect pesticides from extreme temperatures during transport. Extremely hot or cold temperatures can damage pesticide containers by causing them to melt or become brittle. Such temperatures also may reduce the usefulness of the pesticides.

Pesticide Fires

In the event of a fire, call the fire department. Make it clear that it is a fire involving pesticides. Provide any information you have about the pesticides. Evacuate all people from the area to a safe distance upwind from smoke and fumes.

If you attempt to fight the fire yourself until the fire department arrives, do so with extreme caution; do not take unnecessary risks. Attack fire from a safe distance. Bottles, drums, and metal and aerosol cans are not vented and may explode.

Use as little water as possible and contain runoff. Contaminated runoff can be the most serious problem. Environmentally, it may make more sense to let a fire burn rather than using large amounts of water and creating large amounts of contaminated runoff. If a water source (e.g. lake or a well) becomes contaminated, inform the appropriate local or state authority.

Pesticide Poisoning

If you or anyone using pesticides shows any of the signs of poisoning, get to a doctor immediately (Fig. 47).

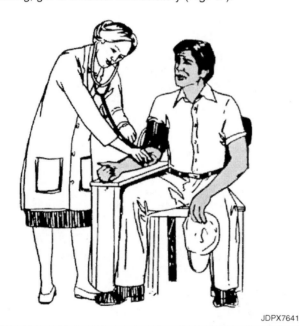

JDPX7641

Fig. 47 — Get Medical Attention Immediately if You Show Any Symptoms of Poisoning

Poisoning symptoms depend on the particular chemical. The label on the pesticide containers and the MSDS list the poisoning symptoms. Read the label and MSDS and know the symptoms to watch for.

First Aid for Pesticide Poisoning

First aid is the initial effort to help a victim while medical help is on the way. It is most important in pesticide emergencies to stop the source of pesticide exposure as quickly as possible. If you are alone with the victim, make sure the victim is breathing and is not being further exposed to the pesticide before you call for emergency help. Do not become exposed to the pesticide yourself while you are trying to help (Fig. 48).

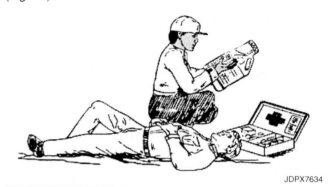

JDPX7634

Fig. 48 — Emergency First Aid for Poisoning Is Critical. Read Pesticide Label Instructions

Apply artificial respiration if the victim is not breathing. To help avoid exposure to communicable diseases, use a one-way protective airway device. It can prevent oral contact with the victim's blood, saliva, and vomitus.

Look at the pesticide label if possible. If it gives specific first aid instructions, follow those instructions carefully. If labeling instructions are not available, follow these general guidelines for first aid:

Pesticides on Skin

1. Drench skin and clothing with plenty of water (Fig. 49). Any source of relatively clean water will serve. If possible, immerse the person in a pond, creek, or other body of water. Even water in ditches or irrigation systems will do unless you think they may have pesticides in them.

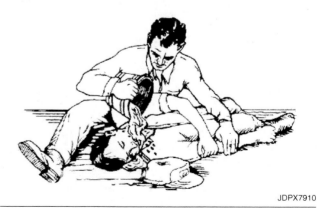

JDPX7910

Fig. 49 — For Pesticide Poisoning From Skin Contact, Drench Skin and Clothing With Water

2. Remove personal protective equipment and contaminated clothing.

3. Wash skin and hair thoroughly with a mild liquid detergent and water. If a shower is available, it is the best way to completely and thoroughly wash and rinse the entire body surface.

4. Dry victim and wrap in blanket or any clean clothing at hand. Do not allow victim to become chilled or overheated.

5. If skin is burned or otherwise injured, cover immediately with loose, clean, dry, soft cloth or bandage.

6. Do not apply ointments, greases, powders, or other drugs in first aid treatment of burns or injured skin.

Pesticides in Eyes

1. Wash eye quickly but gently. Use an eyewash dispenser if available. Otherwise, hold eyelid open and wash with a gentle drip of clean running water positioned so it flows across the eye rather than directly into the eye.

2. Rinse eye for 15 minutes or more.

3. Do not use chemicals or drugs in the rinse water. They may increase the injury.

Inhaled Pesticides

1. Get victim to fresh air immediately.

2. If other people are in or near the area, warn them of the danger.

3. Loosen tight clothing on victim that would constrict breathing.

4. Apply artificial respiration if breathing has stopped or if the victim's skin is blue. If pesticide or vomit is on the victim's mouth or face, avoid direct contact and use a shaped airway tube, if available, for mouth-to-mouth resuscitation.

Pesticides in Mouth or Swallowed

1. Rinse mouth with plenty of water.

2. Give victim large amounts (up to one quart) of milk or water to drink.

3. Induce vomiting only if instructions to do so are on the label.

Procedure to Induce Vomiting

1. Position victim face down or kneeling forward. Do not allow victim to lie on his or her back, because the vomit could enter the lungs and do additional damage.

2. Give syrup of ipecac.

3. Do not use salt solutions to induce vomiting.

Do Not Induce Vomiting:

- If the victim is unconscious or is having convulsions.

- If the victim has swallowed a corrosive poison, which is a strong acid or alkali that will burn the throat and mouth as severely coming up as it did going down. It also may get into the lungs and burn them, too.

- If the victim has swallowed an emulsifiable concentrate or oil solution. Emulsifiable concentrates and oil solutions may cause death if inhaled during vomiting.

Be Prepared for Any Pesticide Emergency

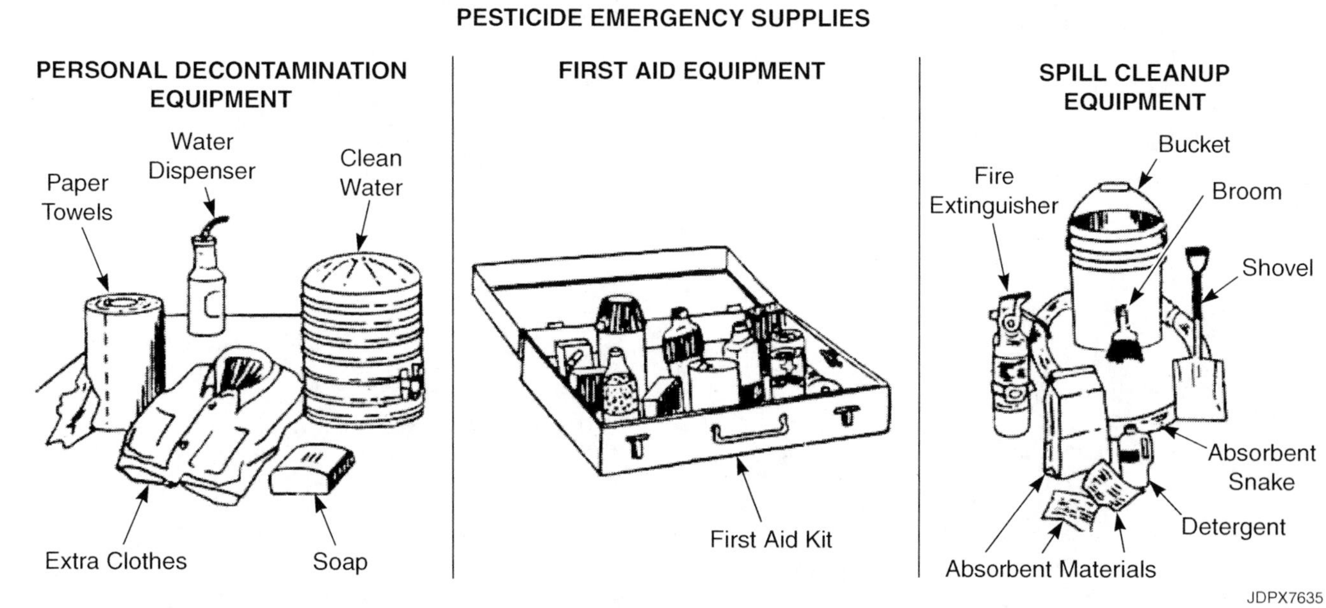

PESTICIDE EMERGENCY SUPPLIES

PERSONAL DECONTAMINATION EQUIPMENT

Paper Towels · Water Dispenser · Clean Water · Extra Clothes · Soap

FIRST AID EQUIPMENT

First Aid Kit

SPILL CLEANUP EQUIPMENT

Fire Extinguisher · Bucket · Broom · Shovel · Absorbent Snake · Detergent · Absorbent Materials

JDPX7635

Fig. 50 — Store Emergency Supplies Where You Can Get to Them Quickly for Pesticide Accidents

Before you begin any pesticide-handling activity, be sure you are prepared to deal with emergencies such as spills, injuries, and poisonings (Fig. 50). Your emergency supplies should include at least:

* **Personal decontamination equipment** — Keep plenty of clean water, detergent, and paper towels nearby in a protected container to allow for fast decontamination in an emergency. Have an extra coverall-type garment nearby in case clothing becomes soaked or saturated with pesticide and must be removed.

* **First aid equipment** — Have a well-stocked first aid kit on hand. It should include a plastic eyewash dispenser that has a gentle flushing action.

* **Spill cleanup equipment** — Keep a spill cleanup kit on hand at all times. The kit should contain not only all the items needed for prompt and complete spill cleanup, but also personal protective equipment to protect you while you are dealing with the spill.

Emergency phone numbers. Post a list of phone numbers near your telephone for emergency help and information (Fig. 51). Examples: physician, poison control center, hospital emergency room, fire department, and CHEMTREC 1-800-424-9300.

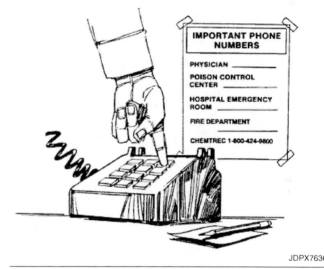

JDPX7636

Fig. 51 — Keep a List of Emergency Phone Numbers Near Your Telephone. They May Save a Life

Pesticide Formulations

The active ingredients in a pesticide are the chemicals that control the target pests. Most pesticide products you buy also have other ingredients called inert (inactive) ingredients. They are used to dilute the pesticide or make it safer, more effective, more convenient to handle, easier to measure, mix, and apply. This mixture of active and inactive ingredients is called a pesticide formulation.

Some formulations are ready to use. Others must be further diluted by the user with water, a petroleum solvent, or air before they are applied.

Liquid Formulations

Emulsifiable Concentrates (EC or E) — Liquid formulations containing the active ingredient, one or more solvent, and an emulsifier which allows mixing in water.

Solutions (S) — Some pesticide active ingredients dissolve readily in a liquid solvent such as water or a petroleum-based solvent. When mixed with the solvent, these ingredients form a solution that will not settle out or separate. Formulations of these pesticides usually contain the active ingredient, the solvent, and one or more other ingredients.

Ultra-Low-Volume Concentrates (ULV) — Liquid formulations which may be applied with specialized equipment as is or diluted with a small quantity of specified carrier. They are designed to apply only in ounces per acre.

Low-Concentrates (LS) — These formulations, usually solutions in petroleum solvents, contain small amounts (usually 1% or less) of active ingredients per gallon. They are designed to be used without further dilution for structural pests, space sprays in barns, mosquito control, etc.

Flowables (F or FL) — Liquid formulations consisting of a finely ground active ingredient suspended in a liquid. Flowables are mixed with water for application.

Aerosols — These formulations contain one or more active ingredients and a solvent. There are two types of aerosol formulations — the ready-to-use type and those made for use in smoke or fog generators.

Invert Emulsions — Water-soluble pesticides dispersed in an oil carrier. Large droplets are formed, which do not drift easily.

Fumigants — Pesticides that form poisonous gases when applied. Sometimes the active ingredients are gases that become liquids when packaged under high pressure. Other active ingredients are volatile liquids when enclosed in an ordinary container and so are not formulated under pressure. They also become gases during application. Others are solids that release gases when applied under conditions of high humidity or in the presence of water vapor.

Dry Formulations

Wettable Powders (WP or W) — Dry, finely ground formulations that look like dusts. The active ingredient is combined with a finely ground, dry carrier, usually mineral clay, along with other ingredients that enhance the ability of the powder to suspend in water. Mixed with water for application as a spray, wettable powders are probably the most widely used pesticide formulations.

Dry Flowables (DF or WDG) — Dry flowables, also known as water-dispersible granules, are like wettable powders except the active ingredient is formulated as a granule instead of a powder. They are easier to pour and mix than wettable powders because there is less dust.

Dusts (D) — Low percentage of an active ingredient on a very fine dry inert carrier like talc, chalk, or clay. Most are ready to use.

Baits — Active ingredient mixed with food or other attractive substance.

Granules (G) — Most often used for soil applications. The active ingredient is coated or absorbed onto coarse particles like clay, ground walnut shells, or ground corn cobs.

Pellets (P or PS) — Very similar to granules, although pellets are usually more uniform (of a specific weight and shape).

Microencapsulation — Particles of a pesticide, either liquid or dry, surrounded by a plastic coating. Mixed with water and applied as a spray. Encapsulation makes timed release possible.

Synergism — Synergism occurs when the combined action of two compounds produces an effect greater than the individual additive effects of the compounds. For example, pyrethrin alone would kill 10 insects, while piperonal butoxide alone would kill 5 insects. If these two chemicals were combined, you might expect them to kill 15 insects, but due to the synergistic effect, up to 50 insects could be killed.

This increased toxicity affects all three types of exposure. Use particular caution whenever you use combination sprays. Never make up your own mixtures. Use only those combinations recommended by your state cooperative Extension Service or some other reputable agency.

Test Yourself

Questions

1. (T/F) The accidental release of anhydrous ammonia is the most common cause of anhydrous ammonia injuries.

2. Recommended personal protective equipment when handling anhydrous ammonia includes:

    a. Face shields or goggles

    b. Rubber gloves

    c. Heavy-duty coveralls or long-sleeved shirts

    d. All of the above

3. Pesticides include:

    a. Herbicides

    b. Insecticides

    c. Fungicides

    d. All of the above

4. Which federal regulation governs the registration of pesticides?

    a. ESA (Endangered Species Act)

    b. HCS (Hazard Communication Standard)

    c. FIERA (Federal Insecticide, Fungicide, and Rodent-cide Act)

    d. SARA (Superfund Amendments and Authorization Act) Title III

5. What is the most common type of pesticide exposure?

    a. Dermal

    b. Oral

    c. Inhalation

    d. Ocular

6. The toxicity of agricultural pesticides is measured in terms of:

    a. mg/kg

    b. ppm

    c. LD_{50}

    d. mph

7. The least toxic chemicals would be labeled with what signal word?

    a. Danger

    b. Warning

    c. Caution

    d. Hazard

8. (T/F) Shop-type goggles provide adequate protection from chemical exposure.

9. Which type of respirator must be worn in an oxygen-limiting environment?

    a. Dust/mist

    b. Chemical cartridge

    c. Gas mask

    d. Air-supplied

10. (T/F) Leather gloves provide adequate protection from most agricultural pesticides.

11. (T/F) Triple-rinsed pesticide containers are safe to use for other purposes around the farm.

12. The most commonly used pesticide formulations are:

    a. Wettable powders

    b. Emulsifiable concentrates

    c. Baits

    d. Granules

References

1. Applying Pesticides Correctly — A Guide for Private and Commercial Applicators. United States Environmental Protection Agency Core Manual. EPA Regional Office.

2. Checklist for Safe Anhydrous Ammonia Equipment. 1992. Michigan State University, East Lansing. Doss, H.

3. Farm Chemical Handbook. Willoughby, OH: Meister Publishing Company. 1992.

4. First Aid for Pesticide Poisoning: MU Guide 1915. 1992. University of Missouri. Columbia. 1992. Baker, D. E.

5. Fundamentals Of Machine Operation, FMO-Crop Chemicals. Deere & Company. Moline, IL

6. Homeowner Chemical Safety: MU Guide 1918. 1991. University of Missouri, Columbia. Baker, D. E.

7. Personal Protective Equipment for Working with Pesticides, MU Guide 1917. 1992. University of Missouri, Columbia. Baker, D. E.

8. Pesticide Application Safety: MU Guide 1916. 1992. University of Missouri, Columbia. Baker, D. E.

9. Preventing Ammonia Burns and How to Treat Them. 1992. Michigan State University, East Lansing. Doss, H.

10. Safe Handling of Anhydrous Ammonia: HEX-394. 1991. Ohio Cooperative Extension Service. Columbus, OH. Bean, Thomas: Lawrence, Timothy; and Carpenter. Thomas.

11. The Safe And Effective Use Of Pesticides. 1988. Publication 3324. University of California. Div. of Agriculture and Natural Resources. Oakland, CA.

12. The Standard Pesticide User's Guide. 1990. New Jersey: Prentice Hall Inc. Bohrnont. Bert L.

Hay and Forage Equipment

10

JDPX8029

Introduction

Hay and forage crops comprise more acres of cropland in the United States than any other crop. There are three factors involved in hay and forage operations that are important to your safety:

- Timing of operations and crop conditions
- Field terrain
- Hay and forage equipment

Timing of Operations

Maturity and moisture level are more important with hay than with forage coming from row crops.

Weather is also an important factor in haying. Because of delays caused by rain or extremely fast drying conditions, part of the crop may be lost.

These factors sometimes rush operators into disregarding safety. Faster hay and forage harvesting is possible. But the operator must remember that safe operation of the machine is more important than hurried operation. An operator in a hurry, disregarding safe operating procedures for machinery, can make hay and forage harvesting hazardous (Fig. 1).

JDPX8100

Fig. 1 — Plant Maturity Stage and Weather Conditions Usually Rush Operators to Harvest Quickly — Increasing Chances for Injury

Field Terrain

In many areas, hay and forage crops are grown on ground too rough or hilly for row crops. In addition to the potential hazards of operating the machinery, rough or hilly terrain introduces additional hazards that deserve your continued attention. However, there are several precautions that can help reduce the hazards related to field conditions:

1. Remove stumps and stones or mark them clearly to prevent upsets, breakdowns, and dangerous driving conditions.

2. Inspect ditches for undercutting.

3. Plan harvesting so equipment travels downhill on steep slopes to avoid overturning.

4. Space tractor and equipment wheels as far apart as possible for stability.

5. Use tractors equipped with rollover protective structures (ROPS) and seat belts (Fig. 2).

JDPX7445

Fig. 2 — Be Alert for Obstacles and Terrain That Could Cause Upsets. Use Tractors With ROPS

Because hay and forage harvesting is strenuous and is frequently performed during hot weather, "field preparation" of the operator is also important. See Chapter 2, "Human Factors and Ergonomics" for guidelines for reducing the dangers of heat stress.

Hay and Forage Equipment

Farmers and ranchers in the United States harvest and feed millions of tons of forage crops each year. Hay and forage harvesting and handling equipment makes it possible. However, as in all machinery operations, there are potential hazards associated with the equipment.

Hay and forage equipment presents many of the potential hazards generally associated with tractors and implements as discussed in earlier chapters. See especially these two chapters:

- Chapter 3, "Recognizing Common Machine Hazards" — Basic types of hazards involving farm equipment.

- Chapter 7, "Tractors and Implements" — Safe operation including ROPS, hydraulics, PTO systems, starting, hitching, and transporting.

The John Deere publication, FMO —"Hay And Forage Harvesting" discusses machines and their functions in detail and provides some safety information.

Hay and forage equipment has powered components which cut, shear, chop, crush, crimp, and handle crop materials (Fig. 3). Those components can seriously injure you and others if you do not follow proper safety practices.

JDPX8101

Fig. 3 — Moving Components Are Potentially Dangerous to the Careless Operator

These general safety practices are particularly applicable to hay and forage equipment:

1. Study the machinery operator's manuals. Follow the safety instructions in the manuals and in safety signs on the machines.

2. Maintain machinery as recommended in operator's manuals to avoid frustrating breakdowns that lead to accidents.

3. Keep safety shields and guards in place and in good operating condition.

4. Do not attempt to unplug, service, or adjust a machine when it is running. Disengage the power, shut off the engine, and make sure all moving parts have stopped.

5. Keep others away from machinery operations unless they must be involved in the work.

6. Equip machines with the proper safety chains, lighting, reflectors, and SMV emblem for travel on roads. Check local regulations.

To avoid injury, you must understand and respect the equipment and operate it safely. The following pages of this chapter will address some of the primary safety concerns involving hay and forage equipment.

Types of Hay and Forage Equipment

Just as there are different methods and systems of storing hay and forage crops, there are a wide variety of machines to get the job done:

- Mowers and Mower-Conditioners

- Windrowers

- Small Rectangular Balers and Bale Ejectors

- Bale Handling Systems

- Large Rectangular Balers

- Large Round Balers

- Forage Harvesters

- Forage Wagons

Each type of machine will be discussed.

Mowers and Mower-Conditioners

These are the primary danger areas:

- Cutterbars

- Gathering Reels

- Conditioning Rolls and Impellers

Cutterbars

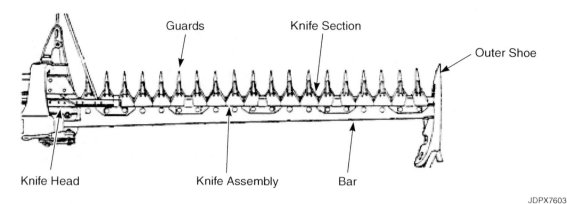

Fig. 4 — Sickle Cutterbar Components. Knife Assembly Moves Back and Forth to Shear Crop

Mowers and mower-conditioners have two popular types of cutterbars to cut standing crops: sickle cutterbars and rotary disk cutterbars.

Sickle Cutterbars

Sickle cutterbars are used not only on mowers and mower-conditioners, but also on windrowers, forage harvesters, and grain combines. They cut standing crops by shearing action. The knife assembly moves back and forth, shearing crop stems between the sharp knife sections and the stationary guards (Fig. 4).

Shearing Hazard

The primary hazard of sickle cutterbars is probably obvious to you: If reciprocating knife sections will cut or shear grass or stalks, they will (and they have) cut people or animals severely. Make sure there is no person or animal near the machine when it is operating (Fig. 5).

Fig. 5 — Sickle Cutterbar Windrower in Operation. Keep Others Away to Avoid Injury

Field Obstructions

When field obstructions are encountered, a safety spring release allows the cutterbar to swing back to avoid damage (Fig. 6). To relatch the cutterbar, disengage the PTO, leave cutterbar on the ground, and back the tractor until the latch catches. With some mowers it may be necessary to turn the front wheels to the left to provide more momentum to relatch the cutterbar. Make sure no one is standing nearby during this operation.

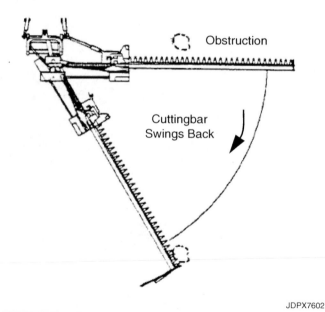

Fig. 6 — Breakaway Latch Permits Cutterbar to Swing Back to Prevent Damage if an Obstruction Is Hit

Transport Position

Transport position requires shutting off the tractor and manually lifting the cutterbar by the grass board or back side to a vertical position. Lock the cutterbar into place with a pin or other device provided. Always handle the cutterbar from the back, keeping hands away from the knife section. A small knife movement can easily cut your fingers.

Rotary Disk Cutterbars

Rotary disk mowers have gained popularity because they can be operated at fairly high field speeds without plugging in lodged or tangled crops and in fields with rodent mounds or uneven terrain.

Disk mowers have several disks mounted on a cutterbar (Fig. 7). The disks rotate horizontally at about 3000 rpm. There are two knives on each rotating disk that strike and cut the crop materials similar to the way a rotary lawn mower cuts grass.

JDPX8103

Fig. 7 — Rotary Disk Mowers Have Cutting Knives on Rotating Disks. Rapid Rotation Can Throw Objects

Cutting and Thrown-Object Hazards

The rotating disks and knives can severely injure people or animals by cutting them. Also, those rotating disks and knives can throw objects such as crop materials, rocks, or other debris causing serious injury. Follow these safety precautions and those in the operator's manuals:

1. Keep hands and feet away from the cutterbar when the machine is running. Always disengage and shut off power before unplugging, servicing, or folding the machine for transport.

2. Stand clear when mower is operating and do not operate near other people. The rotating disks can throw objects, which can cause serious injury.

3. Keep the mower safety curtain or cover in place over the cutterbar when operating the machine. It is provided to contain objects that can be thrown out by the rotating disks (Fig. 8).

JDPX8104

Fig. 8 — To Contain Thrown Objects, Keep Safety Curtain In Place When Operating Rotary Disk Mowers

4. For additional protection from thrown objects, use a tractor that has an operator enclosure.

5. Keep knives and attaching hardware in good condition to avoid having knives thrown from the machine. Follow operator's manual instructions for inspecting and maintaining rotating components.

Unplugging Cutterbars

Even though precautions are taken to ensure correct operation of mowers and mower-conditioners, plugging or trash buildup on cutterbars may still occur.

To safely unplug the cutterbar or remove trash, follow these steps:

1. Stop and disengage the PTO.

2. Raise the cutterbar and back up.

3. Shut off the engine and engage the parking brake or shift the transmission into park (or neutral).

4. Pull hay away from cutterbar.

5. Check the cutterbar for broken or damaged components.

6. Return safety curtain or cover to proper operating position.

7. Start engine and engage PTO at low speed and then increase it to rated speed.

8. Ease mower into standing hay and resume operation.

Gathering Reels

Most of the safety precautions applying to mowers and cutterbars apply to mower-conditioners. However, some mower-conditioners have reels located over the cutterbar to deliver the crop to the crimping rolls. The rotating reel is an extra reason, in addition to the cutterbar, to shut off the engine and disengage the PTO before going near it (Fig. 9).

Reels also have tines to grab the crop and direct it into the mower-conditioner. The tines can easily hook or grab your hand or clothing, and drag you into the cutterbar. Always make sure the engine is off and the PTO disengaged before unplugging or working on the cutterbar or reel.

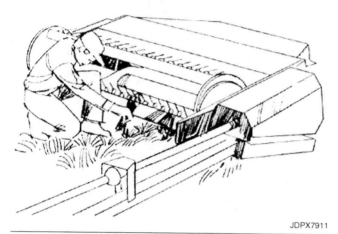

JDPX7911

Fig. 9 — Stay Clear of Reel and Other Components Unless Power Is Disengaged and Engine Is Shut Off

Conditioning Rolls and Impellers

Conditioners and mower-conditioners use crimping or crushing rolls to condition hay so it can dry faster. The conditioning rolls are PTO-powered. They pull the hay between them, and throw the hay out of the back of the machine. The rolls may pick up a stone or other object and throw it out of the machine (Fig. 10).

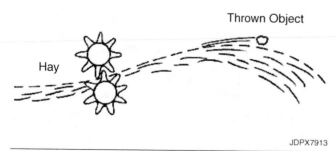

JDPX7913

Fig. 10 — Never Allow Anyone Behind Conditioning Rolls While Operating — An Object Could Be Thrown and Cause a Serious Injury

Some mower-conditioners have rotating steel impeller tines which drive or rub the plants against the conditioning hood and out the back of the machine. That action scuffs the waxy film from the plant stems, allowing the stems to dry more quickly. The PTO-powered impellers, rotating at 600 to 900 rpm, can throw solid objects out the back, along with crop materials (Fig. 11).

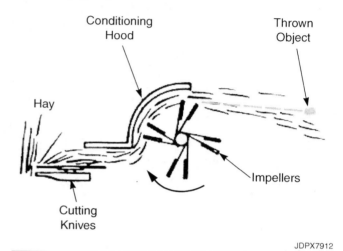

JDPX7912

Fig. 11 — Impellers Can Throw Solid Objects out the Back of Mower-Conditioners. Keep Everyone Away During Operation

Because of possible thrown objects, don't allow anyone to stand near the rear of a conditioner or mower-conditioner during operation. The crimping or crushing rolls, or the impellers, may pick up stones or other solid objects and throw them out the rear, causing very serious injury.

Never work around a conditioner or mower-conditioner unless the PTO is disengaged and the engine is off (Fig. 12). The rolls or impellers can grab your hand or clothing in an instant. Also, make sure machine hoods, covers, or shields are in place as manufacturers recommend during operation. Don't take a chance!

JDPX7914

Fig. 12 — Always Disengage Power and Stop the Engine Before Working on a Mower or Conditioner

Windrowers

JDPX8105

Fig. 13 — The Windrowers Rotating Auger and Reel, and Also the Cutterbar, Can Cause Serious Injury

Similar to mower-conditioners, windrowers can combine the functions of mowing and conditioning to convert standing forage crops into conditioned windrows for curing. Sometimes the terms "mower-conditioner" and "windrower" are used interchangeably. Windrowers are also available without conditioners for harvesting grain crops. A primary difference between mower-conditioners and windrowers is the type of crop handling mechanisms of the cutting platform.

Mower-conditioners move the crop directly rearward from the cutterbar into the conditioners. However, windrowers move crop materials laterally from both sides of the cutting platform to the center with an auger or a canvas draper (belt conveyor). Materials are then fed rearward into the conditioning system or dropped to the ground. Most windrowers used in haying operations have auger platforms and are self-propelled (Fig. 13).

Safety concerns for windrowers are similar to those for mowers and mower-conditioners. However, there are two additional concerns to be addressed:

• Auger Platforms

• Self-Propelled Windrower Operation

Auger Platforms

The auger platform poses potential safety hazards not only because it has moving parts but also because it may get plugged up by crops. Operators often get careless when unplugging machines.

The potential hazard of the auger is shearing and pinching action. The same precautions in unplugging, servicing, repairing, and inspecting mower-conditioners should be followed with windrowers. Never work on the platform (including cutterbar, reel, or auger) without first disengaging the PTO and shutting off the engine.

Don't work under a raised platform without installing the hydraulic cylinder stops or solid blocks. It could fall and crush you. Also, when you leave the windrower unattended, lower the platform to the ground or engage the cylinder stops so it can't fall on someone.

Self-Propelled Windrower Operation

As with any other piece of mechanized farm equipment, self-propelled windrowers require a responsible and mature operator. Become thoroughly familiar with the machine and controls before beginning field operations. The aspects of safe operation to be considered are:

- Pre-Operation

- Field Operation

- Dismounting Machine

Each phase of operation deserves special attention as to the safest methods and procedures. For general safety procedures on self-propelled equipment, refer to Chapter 7, "Tractors and Implements".

Pre-Operation

To avoid accidents, take these precautions before starting the windrower:

1. Be sure no one is near the windrower. The rear of the machine can move sideways if the steering wheel is turned while the engine is running.

2. Place the variable ground speed control lever in neutral, and put the steering wheel in center position. (Put steering levers in neutral on older machines.)

3. Engage the parking brake.

4. Disengage the platform drive control, and put the platform lift control in neutral.

By following these procedures, the operator can prevent sudden movement of the machine or parts of the machine during starting and avoid accidents.

Field Operation

Alertness to field and crop conditions is one key to operating windrowers safely. Also, a machine in good operating condition gets the job done faster and safer because of fewer operating problems or repairs. Proper operation includes:

- Safe windrower speed for conditions

- Correct operation of steering mechanisms

- Correct platform height

Windrower Speed

Windrower speed is determined by the terrain and the density of the crop. When operating over rough terrain or on hillsides, take care to avoid holes or obstacles that can tip a windrower or throw you from the machine.

Crop density also affects the speed at which you operate the windrower. In heavy crops, high operating speeds cause frequent clogging. The more often you must unplug the machine, the higher the chances for an accident. Also, remember that by taking the crop in at the proper speed you'll save time. Frequent stops to clear the machine almost always consume more time than harvesting at slower speed.

Never operate the windrower at transport speeds in the field, and do not operate in transport range without the platform in place. Transporting without the platform reduces your ability to steer and stop the windrower.

Travel at reasonable speeds on roads and use the proper safety lighting as described in Chapter 7, "Tractors and Implements".

Steering Systems

Steering characteristics of self-propelled windrowers are somewhat different from those of other machines (Fig. 14). They are driven and steered by the front wheels. The drive wheels may be powered by hydrostatic pumps or by variable-speed sheave and belt mechanisms. Turning the steering wheel slows or reverses one drive wheel relative to the other, and the rear end of the windrower swings wide in the direction of the turn.

Fig. 14 — Learn About the Controls and Steering Characteristics Before Operating

Some windrowers with mechanical drives may also have levers to control steering, forward speed, and sharp turns in the field.

Make turns on hillsides with care. Reduce speed and make turns with caution to avoid upsets.

Platform Height

The height at which you operate the platform will vary from crop to crop. For instance, if you windrow grain for combine harvest, the platform will usually be operated much higher than for forage crops. When harvesting hay crops, the platform may be operated very near the ground. Keep a close watch for rocks or other obstacles that could be hit or picked up by the platform. Irregularities in the ground also present safety hazards.

Manufacturers install skid plates and include other design characteristics on platforms to keep the platform from scalping or digging into the ground. However, if the windrower is operated on steep slopes over rough terrain or goes through a ground depression, the platform may dig into the ground. This may cause the windrower to veer, or the machine may stop suddenly. If this occurs, you may be thrown against the controls or completely off the operator's station. Keep these potential hazards in mind. Slow down and maintain maximum control.

Careful operators keep a watchful eye on the cutterbar and platform as well as watching where they are going.

Don't allow passengers to ride on the windrower. Sharp, fast turns and rough terrain can easily throw them off.

Dismounting Machine

Before dismounting the machine for any reason:

1. Move ground speed control lever to neutral.

2. Engage the parking brake.

3. Disengage the platform drive control.

4. Reduce throttle to idle and turn off ignition.

5. Lower the platform to the ground.

If you plan to leave the machine for any period of time, remove the ignition key. This will keep an unauthorized person from starting the machine.

Balers and Bale Handling Systems

Hay balers revolutionized hay making in the United States in the 1930s. Labor and cost reduction are becoming more and more important each year, so producers now rely increasingly on labor savers such as balers, bale ejectors, bale handling systems, and large-package hay systems.

As with any machine doing the job better and faster, the operator must be familiar with the mechanisms involved and with safety precautions. The machines we will be looking at from the safety standpoint in this section are:

• Balers and bale ejectors

• Bale handling systems

Though these machines greatly reduce the time and labor needed for haying, the operator must be more conscious of safety hazards. Baling usually puts pressure on the operator to get the job done quickly and efficiently (Fig. 15). Weather conditions can change rapidly and the value of the crop may be reduced when conditions are too wet, too cool, too hot, or too dry. But, no crop is worth injury or death. Slow down and work safely.

JDPX8108

Fig. 15 — Baling Places Stress on the Operator to Hurry to Beat the Weather

Small Rectangular Balers and Bale Ejectors

With the development of automatic twine and wire tying, baling has become more popular than any other hay packaging method in the United States. Small rectangular bales are commonly 14 x 18 x 36 inches (356 x 457 x 914 mm) and may weigh 50 to 85 pounds (23 to 39 kg). Most balers are tractor-drawn, but some are self-propelled.

Small Rectangular-Type Balers

Balers producing rectangular bales use a pickup head, auger, and other mechanisms to feed the hay to the bale chamber for compression and bale formation (Fig. 16). As the hay enters the bale chamber, it is sliced by knives. A plunger compresses the hay inside the baling chamber. Needles and tying mechanisms automatically tie the compressed or baled hay with twine or wire.

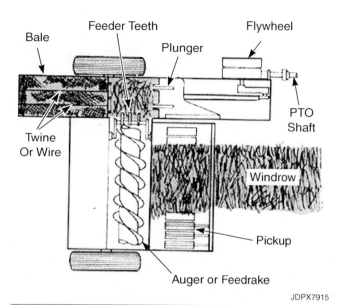

Bale

Feeder Teeth

Plunger

Flywheel

PTO Shaft

Twine Or Wire

Windrow

Pickup

Auger or Feedrake

JDPX7915

Fig. 16 — Moving Baler Parts Which Must Handle Tough Crops Can Injure Careless People

These and other components can injure you if you do not take all safety precautions during operation and servicing of balers. Make sure that all such parts are completely stopped and that stored energy is released or locked in place before you get near them.

Proper operation of the baler reduces the number of times it needs repair and therefore the number of potentially dangerous field repairs you must make. Repairs also take time away from field operation and increase the stress factor on you to get the job done.

Proper timing of all moving parts is essential for successful baler operation and for avoiding serious damage or breakage of moving parts. Timing is usually accomplished by means of chains or sprockets. Consult your operator's manual for specific timing instructions and other adjustments.

Correct ground speed is important to safe baler operation. If you travel too fast, the baler can become overloaded.

If the need should arise to adjust, inspect, or repair the baler, follow these procedures for your safety:

• Disengage all power.

• Shut off the engine.

• Wait for the flywheel and all other moving parts to stop.

If you need to test the bale knotter, the proper procedure is to disengage the PTO, shut off the tractor engine, remove the key, and then turn the flywheel by hand after tripping the knotter drive. With the flywheel turning slowly, you can observe the knotter going through a tying cycle in slow motion. But, be sure you just watch while the parts are moving. Keep hands away!

Other recommendations for safe baler operation are:

1. Never allow anyone to turn the flywheel while someone is working on the machine knives. Moving parts can easily injure someone.

2. Be sure bale twine or wire is properly spliced and threaded in the machine to avoid knotter problems.

3. Never pull anything out of the knotter while it's in operation. It can easily entangle you and cause a serious injury.

4. Keep all safety shields and guards in place and in good condition.

5. Don't hand feed material, such as broken bales or heavy windrows, into the machine when it is running. Spread the material on the ground so the machine can pick it up.

Bale Ejectors

Fig. 17 — Bale Ejectors Throw Bales With Considerable Force

Bale ejectors, or throwers, were introduced on small rectangular balers in the 1950s to eliminate the manual labor of loading bales from the ground or from the baler and stacking them onto wagons or trucks.

As each finished bale is pushed out the bale chute at the rear, the ejector throws it into a trailing wagon (Fig. 17). Bales are not stacked neatly in the wagon, but merely rest where they fall.

The two most common bale ejecting or throwing mechanisms are hydraulically powered, high-speed belts and bale-throwing frames (Fig. 18). Since those mechanisms can throw heavy bales of hay, they can seriously injure people.

You could be injured by the high-speed belts if you are careless. Also, you could be struck by a bale as it is ejected. Or, you could be struck a severe blow by the throwing frame and pan if you are too close.

Fig. 18 — Bale-Throwing Frame and Pan Swing Swiftly and Without Warning. Stand Clear to Avoid Injury

Whenever the baler and ejector are running, the throwing device can be activated. Even if the ejector is empty, someone checking the machine could accidentally trigger the ejector release mechanism. A bale-throwing frame swings swiftly and without warning.

Bale Ejector Safety

Follow the safety instructions in the operator's manual and in safety signs on the machine. Note these precautions:

1. Before inspecting, servicing, or adjusting a bale ejector, follow these steps:

    - Disengage all power.

    - Shut off the engine.

    - Wait until all moving parts have stopped.

    - Move the ejector lockout control into locked position.

2. Don't allow anyone to stand behind or work on the ejector while the PTO and engine are operating, or while a bale is in the ejector.

3. Keep all safety shields and guards in place.

4. Shut off the tractor engine, disengage the PTO, and engage the ejector lockout control before hitching or unhitching a wagon behind the ejector.

5. Don't allow anyone to ride in the bale wagon.

Manual Bale Loading

Manual bale loading on wagons or trucks can be done safely if it is done carefully. The nature of wagons and bale handling means the operator and bale handlers alike should take extra care. Be sure everyone is aware of these potential hazards:

1. Starts and stops can cause bale handlers to fall off the wagon or truck.

2. Accidentally stepping off the wagon or truck while loading bales can result in serious injuries.

3. Falls from the wagon or truck could mean getting run over as well as fractures, sprains, and concussions.

4. Tossing bales can easily knock someone off balance.

It is just as important to practice safety while transporting bales. Travel at a safe speed and do not allow riders on top of the hay or on the back of the wagon or truck.

Bale Handling Systems

In addition to bale ejectors discussed earlier, there are several other mechanical systems for handling small rectangular bales. They vary throughout the country according to farming and ranching practices. However, each system reduces the time and labor required to gather, transport, and store baled hay.

A number of bale handling systems and machines are explained in the John Deere publication FMO — "Hay and Forage Harvesting."

This section will address key safety concerns for two types of mechanical bale handling systems:

- Bale Accumulating and Handling Systems

- Automatic Bale Wagon Systems

As with other hay and forage machinery, these machines involve many of the safety considerations discussed in previous chapters of this book, especially Chapter 3, "Recognizing Common Machine Hazards", Chapter 5, "Equipment Service and Maintenance" and Chapter 7, "Tractors and Implements". Review those chapters and, most important, follow the machinery operator's manuals and safety signs.

Bale Accumulating and Handling Systems

Bale accumulating and handling systems generally consist of two pieces of equipment. The accumulator either is attached to or trails the baler or may be pulled independently (Fig. 19). It assembles the bales in groups of eight or ten. When the accumulator is filled, the cluster of bales is deposited on the ground without stopping the forward progress of the baler.

JDPX7917

Fig. 19 — Avoid Having Accumulator Deposit Bales on Steep Slopes. A Tractor Loading Bales Could Upset

The second phase of the system is usually stacking or loading the groups of bales with a special attachment for tractor loaders. Hydraulic powered steel tines are thrust into the bales to hold them together as a unit for stacking and loading.

The same tractor-mounted unit can usually be used to unload the bales and store them in ground-level storage, either in a building or in open stacks.

For safe operation of tractor-mounted front end loaders, see these chapters:

Chapter 7, "Tractors and Implements" (Tractor Upsets) and Chapter 12, "Material Handling Equipment" (Front End Loaders).

Special considerations for safe operation of bale accumulating and handling systems include these:

1. Carefully attach hydraulic hoses when hitching the accumulator behind a baler. The long hoses deserve special care. Keep them in good repair and check for leaks as discussed in Chapter 5, "Equipment Service and Maintenance".

2. Be alert to slick surfaces on top of bale accumulators and wagon or truck beds. Hay sliding over the surfaces makes them slippery, sometimes causing falls and serious injuries.

3. Be cautious when operating bale accumulating and handling equipment on steep slopes. Avoid dumping clusters of bales on steep slopes, because using a tractor loader to load bales where it is too steep could easily cause the tractor to tip over.

Automatic Bale Wagon Systems

Automatic bale wagons pick up, transport, and stack loads of hay bales in large haying operations. Both drawn and self-propelled wagons are available. They handle loads of 100 to 200 or more bales, and up to 7 tons (6.3 t, metric).

Automatic bale wagons gather individual bales into a flanged chute. Chain teeth grasp each bale and lift it onto a table. Bales are then mechanically placed into layers or tiers. After traveling to a storage site, the machine unloads its complete stack by tilting the hydraulically operated bale table. Some machines can unload bales one at a time, mechanically.

To safely operate or work around automatic bale wagons, read and follow the safety instructions in operator's manuals and machine safety signs. Here are some typical safety precautions:

1. Make sure everyone is clear before starting the engine, engaging the hydraulic system, or moving the bale wagon. Note that visibility is limited, and never allow passengers.

2. Disengage power and shut off the engine before climbing on bale tables, or before servicing or adjusting the machine. Also remove engine key.

3. Never work or reach under raised tables unless they are securely blocked, the engine is stopped and hydraulics are shut off. Falling tables can crush you.

4. Make sure the loader transport latch is engaged when transporting on roads or when parking with the loader raised.

5. Check overhead clearance when preparing to unload, especially near electrical lines. Locate stacks a safe distance from power lines.

6. Keep everyone clear when unloading bales. Bystanders could be crushed by falling stacks of bales.

7. Disengage the hydraulic system before inspecting or servicing. Check carefully for high-pressure leaks as noted in Chapter 5, "Equipment Service and Maintenance".

Large Rectangular Balers

The large rectangular balers were introduced in the 1970s. They serve custom or commercial operations and other large growers. They function essentially like the small rectangular balers described earlier. However, the machines are larger in order to produce larger bales (Fig. 20).

JDPX8109

Fig. 20 — Large-Type Mechanical Baler. Each Bale Weighs Several Hundred Pounds (Kilograms)

Common sizes for the large bales are 2 x 3 x 8 feet (0.6 x 0.9 x 2.4 m) and 3 x 4 x 8 feet (0.9 x 1.2 x 2.4 m). Such hay bales would weigh about 700 pounds (318 kg) and 1300 pounds (590 kg) or even more.

Instead of the two twine or wire ties that are common for the small bales. the large-type balers use four to six twine ties. The baler can drop each finished bale on the ground, or may be equipped with an accumulator that gathers three or four bales before dropping them in a group.

Since the machines function much like the smaller balers described earlier, you must exercise similar safety precautions. However, you should be aware there are some different hazards associated with handling larger and heavier bales.

If one of the large bales falls on you, the injury could be serious. Also, the larger, heavier bales can significantly affect the stability of tractor loaders or forks used to lift and handle them. Read, understand, and follow the safety instructions in the operator's manuals and in safety signs on the machines before moving large rectangular bales.

Large Round Balers

By the late 1980s, large round balers had become the dominant hay baling machines in the United States (Fig. 21). Large round bale systems require only about one-third to one-half as much labor as conventional small-package baling systems. One person can easily operate the entire system from baling to storage and feeding.

Fig. 21 — Large Round Baler. Never Attempt to Feed or Unplug the Machine While It Is Running

Large round balers can form bales from several different crops including alfalfa, grain straw, grasses, and corn stalks. Bale size varies up to about 5 feet (1.5 m) in diameter and 6 feet (1.8 m) wide. Weight may range from 750 to 2,000 pounds (340 to 900 kg). They are resistant to weather damage.

Round balers are driven by tractor PTO or hydraulic systems. Many of them use rotating tine pick-up units to lift crop from windrows as the machine moves forward. The crop is fed into a baling chamber where large belts, rolls, or other mechanisms roll and compress it into firm, round bales (Fig. 22).

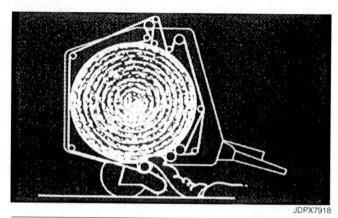

Fig. 22 — Round Bales Are Formed in the Baling Chamber by Belts, Rollers, or Other Mechanisms

When a bale has become large enough, twine is fed into the baling chamber with the hay. Forward machine travel is stopped. The machine wraps the bale with sisal or plastic twine, or with a mesh or plastic sheet. After wrapping, the rear gate is raised and the bale is discharged out the back onto the ground (Fig. 23).

Fig. 23 — Before Raising the Rear Gate to Discharge a Bale, Make Sure No One Is Nearby

Safety for Large Round Balers

Large round balers are powerful machines that produce large, heavy bales in varied crops and conditions. Operators have been seriously and fatally injured in large round baling operations. However, you can avoid injury by following safe practices.

Here are some of the safety precautions for large round baling operations:

1. Do not attempt to feed crop material into the intake area or unplug crop material from the intake area while the baler is running. The baler takes material in faster than you can release it, and it can pull you in, too. Don't take a chance! Disengage the tractor PTO and shut off the engine before dismounting to inspect, service, or adjust the machine, its bale-forming belts, or other moving parts.

2. Don't feed twine into the machine by hand, even with the tractor at idle speed. The twine is pulled in faster than you can react to let go.

3. Before raising the rear gate to unload a bale, be sure no one is nearby.

4. Discharge bales on level ground, crosswise on a slope, or in some other manner so it won't roll. A rolling bale can cause serious injury.

5. Always engage the rear gate lock before working on or around the gate in the raised position, or before working inside the baler. Engage the gate lock anytime the gate is open, and stand clear before unlocking it.

6. Close the gate whenever the baler is left unattended.

Moving Large Round Bales

There are tractor- or truck-towed bale movers that can load large round bales and transport several of them at a time. However, most operators move the bales with tractor-mounted loaders, one bale at a time, in short-distance situations.

Three-Point Hitch Bale Movers

There are several bale movers that can be attached to the three-point hitch of a tractor. They pick up an individual bale with steel tines that slide under a single bale, or they spear it. They are often called bale forks. Others grasp the bale at both ends (Fig. 24).

JDPX8111

Fig. 24 — Bale Mover on Tractor Three-Point Hitch

Front-End Loader Bale Movers

Tractor-mounted front-end loaders can be equipped with bale forks, or with special bale clamps or grapples (Fig. 25). Such devices hold a single bale in place so it can't fall or roll back onto the tractor operator as it is raised for loading.

JDPX8127

Fig. 25 — Tractor Front-End Loader With Grapple to Keep Bale From Falling or Rolling Down Loader Arms

Safety Precautions for Moving Large Round Bales

In handling or moving large round bales, you should recognize these basic characteristics:

- Large round bales are heavy.

- Large round bales will roll.

Note these safety precautions and also those in equipment operator's manuals and safety signs:

1. Use the right equipment. Tractor and bale movers must be properly matched, and tractors must have the capacity to handle the mover and bales. Check equipment specifications. Also, use a ROPS-equipped tractor for upset protection. (ROPS are not intended to protect against falling bales.)

2. Never move bales with a tractor loader unless it is equipped with a manufacturer's approved device, such as a bale fork, clamp, or grapple to hold bales in place (Fig. 25). People have been killed by falling bales and by bales rolling down loader arms onto the operator station when loaders were not properly equipped.

3. If bales are on a slope in the field, approach them from the downhill side. A slight bump can cause bales to roll downhill. Don't attempt to stop a rolling bale, not even with a tractor. They can damage property and kill.

4. When moving a bale with a tractor-mounted loader, carry it "low and slow" to avoid an upset.

JDPX8113

Fig. 27 — Rotating Cutterheads Have Sharp Knives to Chop Forage Crops Into Short Lengths

Forage Harvesters

Forage harvesters cut or pick up crops such as corn, alfalfa, or sorghum and chop them for livestock feed. The chopped material is collected in a trailing wagon or truck to be fed directly to livestock or stored in silos (Fig. 26).

All field and feeding operations can be mechanized.

JDPX8112

Fig. 26 — Forage Harvesters Have Parts That Rotate Several Minutes After Power Is Off. Be Alert

Forage harvesters may be self-propelled, tractor-drawn, or tractor-mounted. Most harvesters have sharp knives mounted on a rotating drum-type cutterhead, or on a flywheel (Fig. 27). Feed rolls move forage crops into position to be cut by the knives rotating past a shear bar (Fig. 28).

Fig. 28 — Header, Feed Rolls, Cutterhead, and Blower in a Cut-And-Blow Forage Harvester

Cut-and-throw harvesters have a cutterhead or a flywheel-type device that chops material and throws it directly into a truck or wagon. On cut-and-blow harvesters, a drum-type cutterhead chops the material. It is moved, directly or by an auger, to a blower that propels it into a wagon or truck (Fig. 28). Cut-and-blow machines are the most popular forage harvesters in the United States.

Flail Harvesters

On flail harvesters, cup-shaped or L-shaped flails or tines swing freely on bars of a horizontal rotor or reel-type mechanism (Fig. 29). The flails cut and chop a standing or windrowed crop. They throw the chopped material through a spout into a wagon or truck.

Fig. 29 — Curved Flails Swing on Horizontal Rotor to Cut and Chop Crops in Flail-Type Harvester

General Safety Precautions

General safety for operation of hay and forage equipment was discussed in the first pages of this chapter. In addition, review Chapter 11, "Grain and Cotton Harvesting Equipment". Many of those precautions also apply to safe forage harvester operation.

Consider the potential hazards of these moving parts: Headers that pull in crops; aggressive feed rolls, flails, and sharp knives; heavy cutterheads and blowers that freewheel after power is disengaged; and knife sharpeners that are operated while parts are moving. Note these and other precautions:

1. Never service, adjust, hand feed, or unplug the harvester when it is running. Always disengage the power, shut off the engine, and wait until all moving parts have stopped rotating.

2. Remember, the auger, fan, and cutterhead may continue to rotate for several minutes after power is shut off. To avoid injury, do not open doors or shields until these parts have stopped rotating.

3. Keep doors and shields latched tightly during operation. Objects are thrown from the cutter with great force. The shields must be in place to deflect these objects.

4. Never stand behind or under the discharge spout while the harvester is operating (Fig. 30). Also note this precaution when hitching wagons. Hard objects blown from spouts can be dangerous.

JDPX8116

Fig. 30 — Hard Objects Blown From Discharge Spout Can Be Dangerous. Stand Clear

5. Be sure everyone is clear of harvester and discharge spout before starting engine or engaging the power. Also check the swing clearance between discharge spout and wagon before field operation.

6. Keep tractors and self-propelled harvesters in gear when traveling downhill. Shift to a lower gear on steep downgrades.

7. Do not shift powered rear wheels of self-propelled harvesters on or off while transporting on slick roads. It changes drive wheel speeds and could cause loss of control.

8. Review these chapters: Chapter 3, "Recognizing Common Machine Hazards": Chapter 5, "Equipment Service and Maintenance"; Chapter 7, "Tractors and Implements"; Chapter 11, "Grain and Cotton Harvesting Equipment": Chapter 12, "Material Handling Equipment". Note especially: Freewheeling parts; field operation; hitching, towing and highway travel; corn heads; forage wagons.

Knife Sharpener Safety

Operating the built-in knife sharpener is an operation that must be performed while the harvester is running. Some machines allow the doors over the knife reel and auger to be closed during the sharpening process. Others require the door to be open so the sharpener can be operated. If your machine's doors must be open during the sharpening process, use extreme care. Make sure your feet are on dry, solid ground. Keep others clear and do not be distracted from your work. A moment's carelessness can easily result in a lost hand or arm.

Make sure no objects can fall into the knife reel. Because the blades tun at very high speeds, the object could be thrown a long ways. Wear eye protection to guard against particles of forage and steel thrown. Study the operator's manual for specific instructions and safety rules (Fig. 31).

JDPX7404

Fig. 31 — Follow Your Operator's Manual for Safe Knife Sharpening Practices

Here are general safety procedures for sharpening knives:

1. Park the tractor and harvester on level ground to prevent rolling. Set parking brake.

2. Disengage the harvesting unit and feed rolls, and wait until all components have stopped moving before opening the cutterhead door.

3. Make sure the self-sharpener is in correct position according to the operator's manual.

4. Close and latch auger and cutterhead doors before starting the engine if your machine is so designed.

5. After starting the engine, engage the power, adjust to proper rpm for sharpening, and then dismount to operate the sharpener.

6. Stay away from any moving parts.

7. When sharpening is completed, disengage the power and turn off the engine. Make sure the cutterhead has stopped rotating before inspecting the knives.

Knife Balance and Safety

If adding or removing knives, the primary concern is locating knives so the cutterhead or flail knives are in balance and properly adjusted. If not balanced and adjusted, excessive vibration and wear will result. Tighten cutterhead bolts with a torque wrench to manufacturer's specifications.

The cutterhead turns easily, and since it is heavy it has high inertia, it tends to keep on rotating. It can easily cut a hand or fingers.

Forage Wagons

Forage wagon safety is covered in Chapter 12, "Material Handling Equipment".

When hooking up wagons to balers or forage harvesters, always exercise great caution. Have your helper stand clear until you have stopped backing up. The helper can then use the telescoping tongue to hook up the wagon.

Forage Blowers

Safety precautions for forage blowers used to transfer chopped forage crops from wagons or trucks into silos are addressed in Chapter 12, "Material Handling Equipment".

Test Yourself

Questions

1. Safe hay and forage harvesting is affected by:

   a. Timing of operations and crop conditions

   b. Field terrain

   c. Condition of equipment

   d. All of the above

2. (T/F) To avoid overturning you should plan hay and forage operations so that equipment travels uphill on steep slopes.

3. (T/F) Gathering reels pose no real hazard to the quick operator.

4. Always handle cutterbars from the:

   a. Front

   b. Bottom

   c. Back

   d. Top

5. (T/F) Self-propelled windrowers steer and respond just like automobiles.

6. (T/F) Safely test the bale knotter by disengaging the PTO, shutting off the tractor engine, tripping the knotter drive, and then turning the flywheel by hand.

7. (T/F) Bale ejectors are safe to work on if there is no bale in the ejector.

8. Hazards of bale accumulating and handling systems include:

   a. Slick surfaces and steep slopes

   b. Gathering reels and cutterbars

   c. Knives and plungers

   d. Broken bales and bale ejectors

9. (T/F) Twine can safely be fed into a large round baler when the tractor is operating at idle speed.

10. Characteristics of large round bales that can lead to serious injury during handling are:

    a. Large round bales are heavy

    b. Large round bales will roll

    c. Both of the above

    d. None of the above

11. (T/F) Forage harvesters are dangerous even after power is shut off because the auger, fan, and cutterhead may continue to rotate for several minutes.

References

1. Fundamentals Of Machine Operation, FMO — "Hay and Forage Harvesting." 1993. Deere & Company. Service Publications FOS/FMO, Moline, IL, 61265.

Grain and Cotton Harvesting Equipment

11

JDPX7589

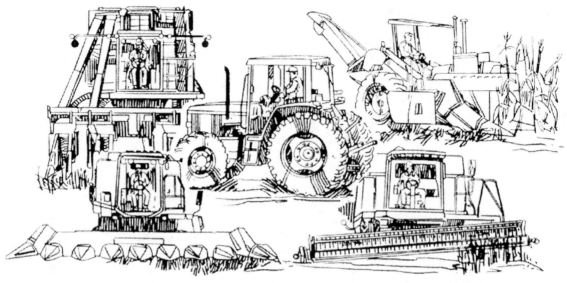

Fig. 1 — Operators Make or Break Profit When It Comes to Using Safe and Efficient Grain Harvesting Methods

Introduction

In the United States, most cereal grain crops and beans are harvested with grain combines. However, it is common in the northern U.S. and Canada for grain crops to be cut and windrowed for drying before combine harvesting. (Grain windrowers are similar to those used in hay crops. They are addressed Chapter 10, "Hay and Forage Equipment".

Some corn is harvested by mechanical pickers, which remove dry corn ears from standing stalks and load them into wagons.

Cotton is harvested by machines called pickers and strippers. The machines remove cotton from standing plants and convey it into a large "basket" mounted on the machine.

Safety in Grain and Cotton Harvesting

Harvesting machines perform many functions. They call for alert, careful operators. Engineers have designed many safety features into harvesting equipment. However, injury-free machinery operation will always require attention to safe practices (Fig. 1).

You should become familiar with several of the preceding chapters, especially these:

•   Chapter 3, "Recognizing Common Machine Hazards" — Basic hazards associated with farm equipment.

•   Chapter 7, "Tractors and Implements" — Safe operation and servicing, including refueling, starting, driving, PTO, hydraulics, hitching, towing, and highway safety.

The Deere & Company publication FMO — "Combine Harvesting" discusses combines in detail.

This chapter will address safe practices for grain combines, corn pickers, and cotton harvesters outlined as follows:

•   Machine Preparation

•   Field Preparation

•   Adjusting and Servicing Combines

•   Driving the Combine

•   Attaching Combine Headers

•   Combine Field Operation

•   Moving Combines on Public Roads

•   Header Attachments

•   Corn Heads

•   Corn Pickers

•   Cotton Harvesters

Machine Preparation

If harvesting machines are not properly adjusted or in good repair, they may plug unnecessarily, or they may break down causing frustration and accidents.

Start getting these machines ready for harvest in the off season or at least several weeks before you'll be using them. An early start is necessary because:

- It takes time to get these machines into efficient and safe operating condition.

- Lead time may be involved when ordering replacement parts.

- Other field operations near harvest time compete for time needed for machine preparation.

- Major adjustments may be necessary to convert a machine for use with a particular crop.

- Special attachments may have to be mounted.

Be sure to follow your operator's manual for proper service, adjustment, and repair.

Safety Preparation

Machine preparation is never complete without a safety check. All safety features should be in place and in good condition (Fig. 2). That includes shields, guards, field and highway lighting equipment, electric interlocks, brakes, steering, and every safety sign.

During harvest, you may be tempted to take shortcuts to get the job done on time. Don't take the risk of short cutting safety. Make sure safety features are ready before harvest.

Refer to "Making Pre-Operation Checks" in Chapter 7, "Tractors and Implements".

Operator's Manual and Safety Signs

The most important sources of safety instructions are the machinery operator's manuals and safety signs. The instructions were prepared to help you service and operate the machines efficiently and to help you avoid injury or death. Read them and follow them.

Field Preparation

Preparing a field for harvest begins long before the crop is mature. It's easy to see obstacles in a field during tillage and planting operations, but they might not be visible at harvest time (Fig. 3). For example, a stub post from an old fence line may not cause any trouble during planting, but might cause severe damage or personal injury later if hit by a combine. It could also cause costly down time and put extra stress on you, the operator, perhaps leading to personal injury.

JDPX7920

Fig. 3 — When Planting, Check Field for Obstacles That May Be Hard to See During Harvesting

JDPX7575

Fig. 2 — Adjust Equipment and Check All Safety Features Before the Harvest Season

Stones and Other Obstacles

Stones in the field can be hazardous during harvesting operations (Fig. 4) especially in:

- Direct-cut crops that must be cut low (soybeans, for example)

- Corn, where the row-gathering units may strike or pick up stones

- Windrowed crops, where stones can be taken into the machine with the windrow

Large stones or stumps can cause upsets, especially on slopes. Mark posts, stones, stumps, and other obstacles that can't be removed at planting time, with a tall pole or stake so their location can be seen clearly in the mature crop. You can then drive around the obstacles, avoiding damage to equipment and sudden stops or upsets that could cause personal injury.

JDPX8117

Fig. 4 — Stones Can Cause Safety Problems in Harvesting Crops That Must Be Cut Low

Ditches

Ditches cause special problems. What was a safe distance from ditches at planting time may not be safe during harvesting operations. Heavy rains can undercut banks or form new ditches. An eight-row planter may have been used, which would have kept the heavy tractor away from the bank. But if the crop is harvested with a combine with a four-row corn head, the combine's weight will be much closer to the bank than the tractor's was (Fig. 5). This problem is even more serious when the ditch runs at right angles to the rows, because turning at the end of some of the rows may require the combine to travel too near the edge of the bank. In general, allow at least one-fourth more turning space than that required by the largest piece of equipment you will be using (Fig. 6).

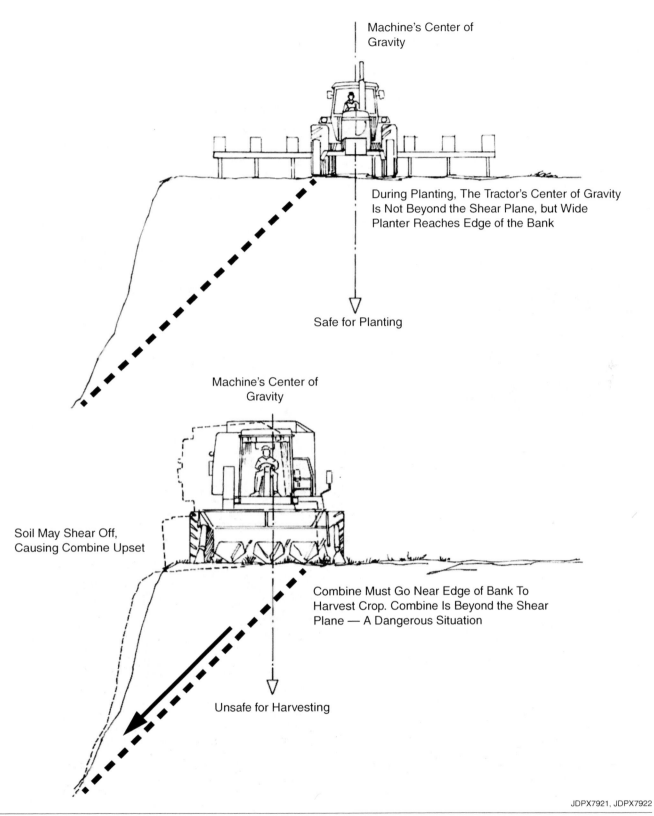

Machine's Center of Gravity

During Planting, The Tractor's Center of Gravity Is Not Beyond the Shear Plane, but Wide Planter Reaches Edge of the Bank

Safe for Planting

Machine's Center of Gravity

Soil May Shear Off, Causing Combine Upset

Combine Must Go Near Edge of Bank To Harvest Crop. Combine Is Beyond the Shear Plane — A Dangerous Situation

Unsafe for Harvesting

JDPX7921, JDPX7922

Fig. 5 — Combine Is Too Close to the Bank — A Dangerous Situation

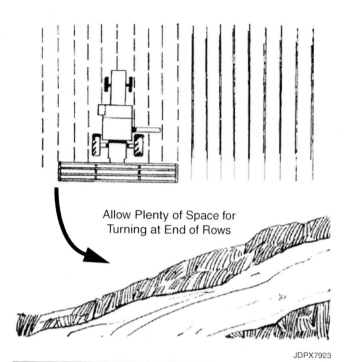

Allow Plenty of Space for
Turning at End of Rows

JDPX7923

Fig. 6 — When Planting, Allow One-Fourth More Turning Space
Than That Required By Your Largest Piece Of Equipment

Weeds

Weeds cause problems during harvesting operations. They
can wrap around rotating drives and plug machines, causing
costly delays (Fig. 7). You may become tired and be more
likely to take chances — like dismounting to unplug the
machine without shutting off the power. A sound weed
control program can give you a safer harvest and a higher
yield.

JDPX7590

Fig. 7 — Weeds Can Create Hazards for Operators. Use an
Effective Weed Control Program

Plan for Harvest

1. Remove as many posts, stones, stumps, and other
   obstacles from the fields as possible.

2. Mark any remaining obstacles clearly.

3. Plant the crop to allow extra turning space near ditches.

4. Check for undercut embankments and new ditches
   before harvesting.

5. Prepare the edges of the fields for smooth turning and
   maneuvering of harvesting equipment.

6. Plant crops with row spacing that matches row spacing
   of the gathering unit of the harvester.

7. Make sure the number of planter rows corresponds
   properly to the number of row units on the harvester, in
   whole-number multiples. That will help avoid plugging of
   the harvester's gathering units.

8. Use adequate weed control practices.

Adjusting and Servicing Combines

A self-propelled combine is large and may weigh up to 35,000 pounds (15,900 kg). It travels in fields and on roads. It cuts crops, and it separates grain from straw and chaff. A combine may carry 14,000 pounds (6,350 kg) of grain and unload it into a moving truck. It is complex and must be properly maintained.

Only a properly serviced machine is safe and efficient to operate. Damaged wiring can cause fires. A loose traction drive belt on a self-propelled machine can cause loss of control on a hill. Complete servicing and maintenance before starting work for the day, and as required during the day.

1. Adjust and service the combine before going to the field (Fig. 8).

JDPX8118

Fig. 8 — Inspect, Adjust, and Service Combine Before Going to the Field

2. Lubricate and fuel the machine after it cools in the evening. Rust may form on moving parts or moisture may condense in the fuel tank if a combine is left out overnight without servicing.

3. Block all parts before working under a header. You may need to raise the header to work on the cylinder and other parts. Do not rely on the hydraulic system. Be sure that the header lock or cylinder ram stop is fixed in place or that proper blocking is used (Fig. 9).

JDPX8119

Fig. 9 — Always Latch the Header Safety Stop or Put Blocks Under the Header When Working Under It

4. Shut down both the machine and the engine before inspecting or working on harvesting equipment. This is especially important for self-propelled combines. They have one or more main belt drives that turn whenever the engine is running, even when other components are shut off, like the variable-speed belt drive to the propulsion transmission shown in Fig. 10. Any moving parts can be hazardous for anyone servicing the machine. Follow the rule: Disengage all power, shut off the engine, and take the key.

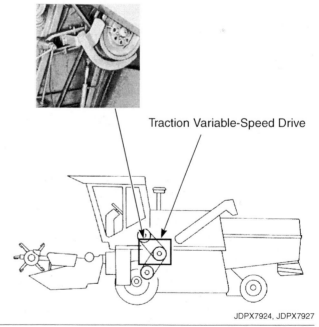

Traction Variable-Speed Drive

JDPX7924, JDPX7927

Fig. 10 — Shut Off the Machine and Engine When Working Around a Variable-Speed Propulsion Belt Drive — Except When Otherwise Specified In Your Operator's Manual

NOTE: It may be necessary to have some machines running to make certain adjustments, such as changing the speed of a combine cylinder or fan. Do it only as instructed in the operator's manual. If the manual doesn't specifically call for it, don't do it.

5. Don't try to remove weeds or other trash from a rotating shaft or from the header while the machine is running. The moving parts can pull the material faster than you can react to let go. You could be pulled in and seriously injured.

6. If two or more people are needed to position, adjust, or service a machine, make sure everyone knows what's being done and how to communicate with others. Have the person on the ground use standard hand signals when helping you position the combine (see Chapter 1, "Safe Farm Machinery Operation"), and be sure you know where everyone is before starting or moving the combine.

7. When working on a harvesting machine, always keep in mind that someone could show up unexpectedly. If you're working on the inside of the machine, or otherwise working on it where others can't see you, remove the ignition key and post a sign saying that the machine is being serviced. Keep everyone away from the controls while you're working on the combine unless you need someone's help.

8. Check carefully for hydraulic leaks. Pinhole leaks can inject oil into your flesh and cause serious injury. Use a piece of cardboard or wood. (For details, see Chapter 3, "Recognizing Common Machine Hazards" and Chapter 7, "Tractors and Implements".)

Driving the Combine

Safety Switch Interlock Systems

Electrical, safety-switch interlock systems are provided on some combines to help prevent serious injury to an operator. Some systems prevent starting the engine when certain machine functions are engaged or if the operator is not in the seat. Examples: Feeder house, header, unloading auger, or separator on a self-propelled combine. A seat switch may shut off one or more such functions if the operator leaves the seat.

Do not bypass seat switches or other safety interlock systems. Don't disconnect them or try to change the wiring. They vary by model and brand of machine, but they are all intended to help keep people from getting hurt or killed. Check them periodically according to the operator's manual. If they don't work, get them repaired immediately. (See also "Neutral Start Safety Switches" in Chapter 7, "Tractors and Implements".)

Operator's Seat

The operator must be comfortable and have the controls within easy reach. Adjust the seat to your height and reach.

All self-propelled combines have adjustable seats that can be raised and lowered and moved forward and rearward. Some can be folded up or moved out of the way so the operator can stand while driving. A typical seat mounting is shown in Fig. 11.

JDPX7584

Fig. 11 — Adjust the Seat for Your Comfort, and Don't Allow Any Joy Riders

Some harvesting equipment has seating space designed to accommodate a second person. Such extra seating area is intended to enable a person to briefly observe harvesting during instruction or supervision, such as in preparing to work the next operating shift.

If a manufacturer provides additional operator station space, it should NOT be used as a place for "extra riders," someone along for fun or baby sitting. An extra rider can cause accidents by distracting the operator, and riders can become injured. Also, persons who are not needed for the combining operation should not ride anywhere on the machine.

Steering

Most combines also have an adjustable steering column for individual arm lengths. Adjust the steering column by releasing the position latch and then moving the column to the desired position (Fig. 12). Don't adjust it when the machine is moving.

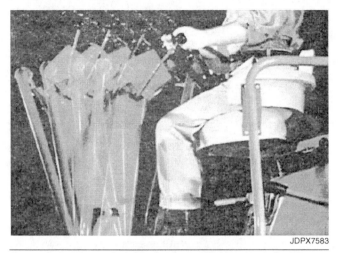

JDPX7583

Fig. 12 — Adjust the Steering Column to Fit Your Arm's Reach

The wheels that steer many combines and other self-propelled machines are mounted on the rear of the machine. Be careful when making a turn near obstacles. The rear of the combine may swing around and strike something.

Most combines have power steering, which makes guiding the combine down rows easy. However, the rear end swings around so far that the back of the machine can hit fence posts even when the header has plenty of clearance. Leave at least one-fourth more room for turning at the ends of fields than necessary for normal turns (Fig. 13 and Fig. 14).

Auger And Header Turning Radius

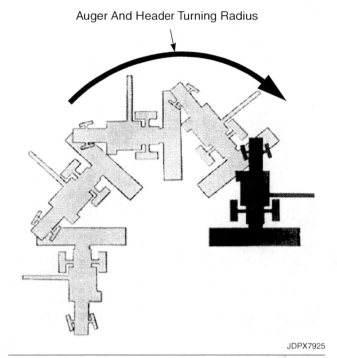

JDPX7925

Fig. 13 — Turning Machines With Rear Steering Requires Practice

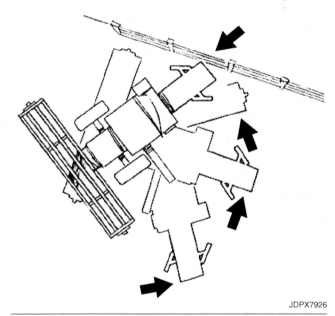

JDPX7926

Fig. 14 — Don't Turn Too Close to a Fence. Remember, the Rear of the Machine Swings Around Quickly During Turns

Brakes

Two brake pedals brake individual drive wheels (Fig. 15). When used separately, these pedals assist in turning. When used together, they stop the combine in a straight line.

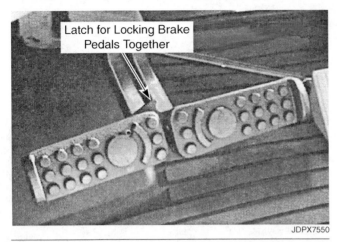

Fig. 15 — Two Brake Pedals Control Individual Drive Wheels. Lock Them Together for Road Travel

To stop the combine, press on both brake pedals. Uneven application of brakes will cause the combine to swerve to one side, especially at high speed on the road. This could result in an upset. Lock the brake pedals together for highway driving.

When using the brakes to make sharp turns, slow down to a safe speed. Start turning the steering wheel before applying the brake to assist turning. Otherwise, the rear wheels will skid sideways and the turn will be more difficult.

Always reduce travel speed before applying brakes. Quick stops can result in the combine nosing forward.

Parking Brake

Engage the parking brake when the engine is running and the combine is parked, and before leaving the machine with the engine turned off. Never attempt to move the machine with the parking brake engaged. It can damage brakes. Some combines have a warning light that tells when the parking brake is engaged.

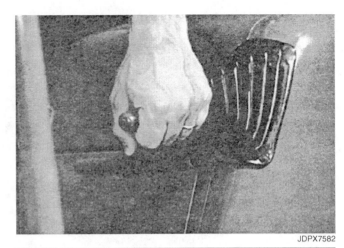

Fig. 16 — Using the Parking Brake

Combine Size

Know how big your combine actually is. The header is usually the widest and lowest part of the combine (Fig. 17). Don't ram it into rocks and stumps. And don't get too close to fences and trees.

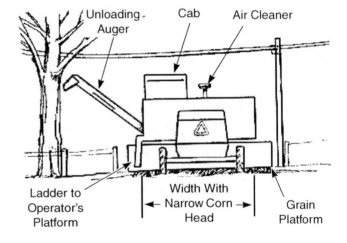

Fig. 17 — How Big Is Your Combine?

Look for parts on the sides of the combine that extend beyond the header, especially when you're using a narrow corn head. Know the position of the unloading auger and the ladder to the operator's platform.

The highest point on a combine is usually the cab or grain elevator, but could be the radiator screen, exhaust pipe, air cleaner, or CB antenna. Check to see which part extends the highest on your machine. Make sure there's plenty of clearance under power lines and trees.

Because of a combine's large size, visibility to the rear is usually limited (Fig. 18). Make sure everyone is clear before backing or turning the combine. Sound the horn before starting the engine.

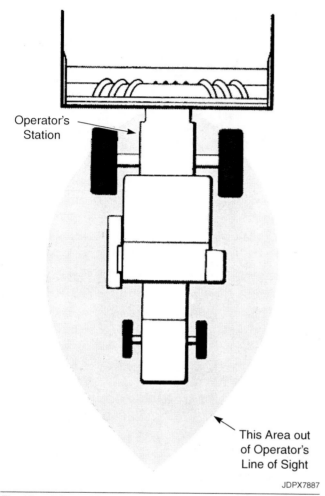

Operator's Station

This Area out of Operator's Line of Sight

JDPX7887

Fig. 18 — A Large Machine Blocks the Operator's View of Some Areas

Shielding

Grain harvesting equipment has many moving parts that have shields, guards, or cover plates to help prevent personal injury. Some people make a tragic mistake of not replacing shields or guards after working on machines.

To help ensure that safety shields or guards are kept in place, they may be permanently attached, for instance, by hinges or cables. Always do this:

1. Fasten all safety shields or guards securely in place after working on machines. Do not cut or remove cables or other devices intended to keep shielding with the machine.

2. Immediately repair or replace damaged safety shields or guards. Don't delay!

Ladders and Platforms

Ladders and platforms provide access to the operator's station and service areas of the combine. Keep the steps and walking surfaces free of grease and dirt. Use handrails for safe mounting and dismounting. Be careful when climbing steps or walking on platforms on frosty mornings. Frost-covered sheet metal is slippery. And don't use the operator's platform as a place for storing loose tools or lunch boxes.

The ladder leading to the operator's platform on some self-propelled combines may be hinged for lifting during operation or transport. Fasten it in place when lifted so that it doesn't fall on someone approaching the combine. Lock the ladder down to prevent falls when using the ladder.

Attaching Combine Headers

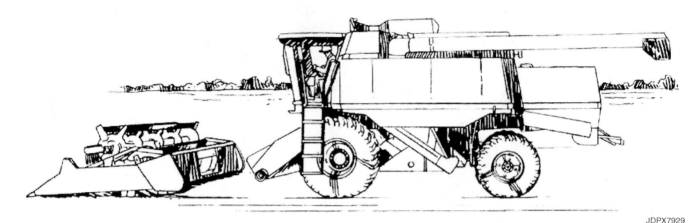

JDPX7929

Fig. 19 — Position Headers for Reattachment on Level, Firm Surface

The best way to have the header in the right position for attachment is to plan for reattachment when you're removing it. Here's how:

1. Position the header so that both the header and the combine are level and on a firm surface, preferably concrete (Fig. 19).

2. Allow enough room for the combine to be driven straight into the header.

3. Use blocks or stands to hold the header firmly in the correct position.

4. Release all pressure in hydraulic lines before disconnecting them.

5. Clean out any trash or dirt between the header and the combine header mounting.

Headers are heavy. If they move or fall while you make connections, you could be crushed between the header and combine or under the header. Follow the operator's manual for proper procedures and for height and angle of attachment. Note these precautions:

1. Add weight to the rear of the combine, if needed for stability. Check the operator's manual.

2. Align the combine and drive in slowly.

3. If you have to move the header to complete the connection, stop the combine and turn off the engine.

4. Make sure header can't move or fall. Raise it hydraulically and engage the hydraulic lift-cylinder safety stop (Fig. 20).

JDPX8119

Fig. 20 — Always Latch the Header Safety Stop or Put Blocks Under the Header When Working Under It

5. Locate pinch points and keep you hands away from them.

6. Before moving the combine, insert all connecting pins and bolts and connect all drives and hydraulic lines.

7. Raise all jacks and stands to their highest positions.

Combine Field Operation

Proper machine operation is important for safe and efficient harvesting. This means reducing harvesting problems and dealing safely and effectively with those that do occur.

Feeding

Uniform feeding of a crop in good condition provides a smooth flow through the combine and results in fewer breakdowns (Fig. 21). Non-uniform feeding can cause plugging.

For uniform feeding:

1. Use good tillage and planting practice to help produce a uniform crop.

2. Cut the crop evenly.

3. Make sure the header is the right size for the combine.

4. Select the proper header speed and ground speed so the crop will flow freely across the header toward the threshing cylinder.

The crop should be dry enough for good threshing without wrapping around the cylinder or other rotating parts. Early harvesting gives weeds less time to develop into a problem for the combine, but late harvest can sometimes be beneficial with weedy crops, especially where frost comes early enough to kill the weeds.

JDPX8120

Fig. 21 — Strive for Uniform Feeding to Prevent Plugging Problems

Combine Capacity

A combine's harvesting rate can be limited by separator capacity size, size of feeding and threshing mechanisms, available engine power, and ground conditions. I If you try to increase the harvesting rate above any of these limits, you could plug the machine or cause a breakdown.

Do not speed, especially on rough ground. Make sure your combine is under control at all times.

Performance Checks

Don't dip your hand into the bin when the machine is moving. There may be augers running below the surface of the grain. And never get into the bin while the combine is running.

To check tailings, use the inspection ports provided on most combines, where you can safely and conveniently observe the material being returned from the shoe to the threshing cylinder. Many self-propelled combines have inspection ports near the operator's station. Use the ports provided for that purpose. Never try to collect a tailings sample with the machine running by opening the door at the bottom of the tailings elevator where fingers could get into the elevator.

Checking grain losses by walking along at the rear of a moving combine to collect a sample is dangerous. It is also very inaccurate. Don't do it. You could be run over by the rear wheels, or struck by thrown objects if a straw spreader or chopper is operating. An electronic grain loss sensor system is more accurate and doesn't expose you to the hazards.

If you must check losses by hand, count the number of seeds in a given area on the ground after the machine has passed (Fig. 22). Follow operator's manual procedures and calculations.

Deere & Company publication FMO — "Combine Harvesting" also gives detailed instructions for checking and calculating losses per acre (hectare).

JDPX7578

Fig. 22 — Check Grain Losses On The Ground. Don't Collect Samples From Moving Machine

Plugging

Plugging the header, cylinder or rotor, or the separator causes increased stress and subjects an operator to the hazards of unplugging the machine. Cylinder plugging is usually the result of weeds or overloading.

You can sense the load on the cylinder or rotor by checking the tachometer. Load on the engine is a good indication of cylinder load. Slow down the combine and avoid taking in slugs of crops or weeds. Remember, the chance of plugging increases as cylinder or rotor speed decreases. After you've unplugged the machine a few times, you should be able to tell when it's about to overload.

Running the combine separator slower than recommended in the operator's manual will result in less effective harvesting and, ultimately, plugging of the separator and conveying mechanisms. Adjust and equip your combine properly for the particular crop being harvested to reduce plugging problems.

If the combine header or other mechanism becomes plugged, clear it using the reverser or other procedure described in the operator's manual. If it won't clear under power, proceed to unplug it by hand as instructed by the manufacturer.

Follow these precautions:

1. Disengage the power, shut off the engine, and remove the key.

2. Make sure all other persons are clear of the machine. Know where they are and what they are doing.

3. Remove the plugged material, using tools or hands protected by heavy gloves.

4. Get your hands and fingers out of the mechanism before anyone attempts to turn the drives.

5. Turn the plugged mechanism with tools that are adequate for that purpose. If special tools are provided, use them.

6. After the mechanism has been freed, remove all tools from the inside of the combine.

7. Replace all inspection or access doors and shields.

8. Be sure everyone is clear before starting the engine. Sound the horn to warn others.

JDPX7585

Fig. 23 — Disengage Power, Shut Off Engine, and Remove the Key Before Unplugging the Combine

Adverse Operating Conditions

Down Crops

Use the right equipment on the combine for down crops to avoid damage to the combine and reduce operator fatigue. Special attachments are available for down crops, including pickup reels, special dividers, and gathering devices.

Be especially alert for ditches, washouts, and holes in the field. A down crop covers them.

When handling down crops, reduce speed so the long straw will be taken in slowly and the header will feed evenly. Slower speed will keep the straw from tangling and bunching up and reduce the chances of plugging. You'll finish faster and safer if you can avoid plugging.

Muddy Fields or Weedy Crops

Avoid Wet Area. Return Later

JDPX7930

Fig. 24 — Skip Wet Areas in Field to Avoid Getting Stuck. Return Later When the Ground Is Dryer and More Firm

When you harvest, skip the muddy and weedy areas in your fields. Turn to them later when there's less pressure on you and the field is dryer and firmer (Fig. 24). See the chapter, "Tractors and Implements" for discussion of "Towing Equipment" precautions to follow when machines are mired.

Special equipment, such as powered rear wheels, wide flotation tires, and half tracks help harvest muddy fields. When installing these attachments, block the machine carefully and do the job on a hard surface.

Keep steps, platform, and controls clean when working in mud, snow, or ice. Reduce slipping hazards. Keep foot pedals and other controls clean, too.

Dry Crop Conditions and High Temperatures

Check for overheated bearings that could burn out and start a fire in dry chaff. Remove excess dirt and chaff from engine, particularly around the exhaust system (Fig. 25). Severely overloaded or slipping belts can cause fires. Watch speed indicators to catch slipping belts, and service all belt drives regularly.

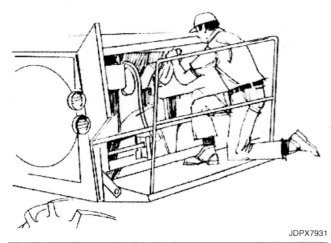

JDPX7931

Fig. 25 — Remove Dust and Chaff From Engine Area and Keep Fire Extinguisher Fully Charged

Operators who have experienced combine fires know the value of a good fire extinguisher. Just one fire costs more than all the fire extinguishers you will ever need. Mount large, multi-purpose ABC-type fire extinguishers on the combine. Locate one near the operator's station entrance and one near the engine. Keep them fully charged.

A low-hanging exhaust system or an engine backfire from a car or truck can start a fire in dry stubble fields. Take precaution if you drive a car or truck into a field. Catalytic converters start a lot of fires in wheat fields. Drive forward — never back up with a truck or car. The exhaust pipe can load up with straw, creating a fire hazard. And after a trip with heavy loads or at high speeds, let the vehicle cool for a few minutes before entering a dry field.

Operating on Hillsides

Operating any machine on a hillside demands good equipment and good operator control. The hillside combine is designed for harvest on very steep slopes, such as those in the northwestern United States. These machines level themselves automatically. Sidehill combines operate on less severe slopes as in Fig. 26. However, level-land combines may be operated safely on many minor slopes used for crop production if these practices are followed:

JDPX7589

Fig. 26 — Hillside and Sidehill Combines Automatically Level Themselves on Slopes

1. Operate the combine smoothly. Avoid quick changes in speed or direction, and don't brake or turn sharply, even at low speeds. Remember that combines have a high center of gravity and can be upset.

2. If space permits, make a large loop when turning to go up a hill (Fig. 27). Turning and braking to go uphill after following along a hillside can make the drive wheels spin out and cause side slipping and possible loss of control of the machine.

3. Don't attempt to shift gears when going up or down a hill, because most transmissions cannot be shifted back into gear while the machine is moving. You have much better continuous control when you change speed slowly with the variable-speed traction drive than when you brake or shift gears.

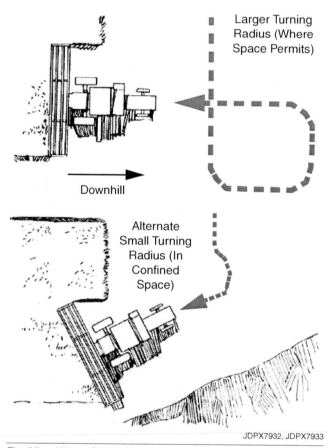

Larger Turning Radius (Where Space Permits)

Downhill

Alternate Small Turning Radius (In Confined Space)

JDPX7932, JDPX7933

Fig. 27 — Where Space Permits, Make a Large Loop When Turning on a Hill

4. Select a transmission gear to operate in mid-range when combining on steep slopes. If the drive slips on older machines when the variable-speed control is at its lowest speed position, increase the variable speed slightly to maintain control. Adjust the traction drive belt periodically, and recheck it if there is any indication of slippage. Check its tightness throughout the speed range.

5. Maintain a low center of gravity for machine stability on side hills. The more grain in the grain tank, the higher the center of gravity on most combines, so don't overfill the tank. And don't use a tank extension above that recommended by the manufacturer. Such extensions can make the equipment top-heavy, and the added weight could put too much strain on the combine axle.

6. Be alert for ditches and holes. If the wheel on the downhill side drops, it could cause an upset.

These practices apply to hillside and sidehill combines too. They can be used safely on much steeper slopes than level-land combines. Follow these rules for using the leveling system on hillside and sidehill combines:

1. Change slope slowly. Allow time for the hydraulic system to level the combine. Upsets can be caused by not allowing enough time for the leveling system to react to changes in slopes. Give it time. Slow down.

2. Don't operate on slopes steeper than the leveling system was designed for. Slopes that are too steep can cause upsets for hillside and sidehill combines just as for level-land machines.

3. Use the manual control wisely. Tilt the machine in the right direction for leveling. Prepare for changes in slope.

Unloading and Cleaning the Grain Tank

On most newer combines, the grain tank unloading augers are swung into unloading and transport position by hydraulics (Fig. 28). However, some machines still in operation require the auger to be positioned manually. Keep your hands away from pinch points in the connector during positioning. Also, keep the combine on level ground so the weight of the auger can be handled easily and safely.

JDPX8023

Fig. 28 — When Swinging Unloading Auger Into Position, Be Sure There Are No Obstructions

No one should be in the grain tank when the combine engine is running. If there's room for grain to get to the unloading auger, there's room to get a hand or foot caught in the auger.

If grain bridges during unloading, stop the auger and shut off the engine before trying to free the grain. Use a small shovel or stick to break the bridging.

Never use your hands or feet to remove trash or to push the last bit of grain into the unloading auger. You will get caught in the auger and be pulled in before you can react. Stop the machine and the engine, take the key, and use a broom.

Some combines have a sump-type auger system to unload the grain tank (Fig. 29). Grain is augured vertically from the sump, then out through the horizontal unloading auger to a truck or wagon.

Don't allow anyone to open the cleanout cover to clean the sump when the machine is running. The rotating vertical auger can cause severe injury. Before anyone begins cleaning the sump, make sure the machine won't be started. Disengage the power, shut of the engine, and remove the key. Be sure everyone is clear before starting the engine or engaging any part of the combine.

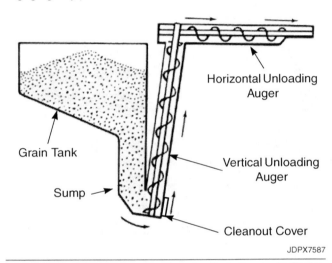

Fig. 29 — Don't Allow Anyone to Clean the Grain Sump with Engine Running

Unloading On-the-Go

If the combine grain tank is emptied on-the-go, cooperation between the combine operator and the hauler is essential to avoid accidents between the combine and the truck or wagon. The hauler is responsible for positioning the truck or wagon for unloading without getting too close to the combine. He or she must be prepared for unexpected stops of the combine and leave plenty of room for the combine to turn at the ends of the field. The combine operator must stop unloading in time for the hauler to turn corners and drive around obstacles safely.

Unloading on-the-go should only be used:

• When the combine is being operated at less than maximum capacity so that a constant ground speed can be maintained

• When the combine is traveling in a straight direction and the field is relatively flat

• When there is plenty of room for unloading before reaching the edge of the field, a fence, or other obstruction

Have a rearview mirror positioned so you can see the end of the unloading auger and the truck or wagon at the same time.

Straw Choppers and Spreaders

Straw, stalks, and other flying material thrown from choppers can injure nearby people. Make sure everyone is away from the discharge area of the machine before and during operation.

Discharge from straw choppers can be dangerous, because the material is thrown at high speed. Small stones and other heavy objects can be picked up and thrown by the machine. Corncobs can hurt, too.

Material from a chopper is discharged in a fan-shaped pattern (Fig. 30). The heaviest, most dangerous objects are thrown farthest. Don't allow anyone near the chopper when it is running.

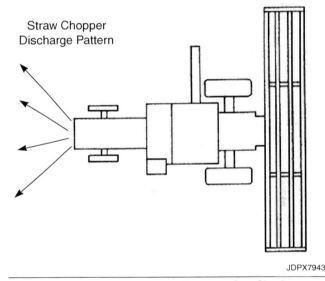

Fig. 30 — Straw Chopper Discharge Pattern — Stay Clear!

Chopper rotors are heavy and may continue running after the rest of the machine has been shut off. Disengage all power, shut off the engine, and allow plenty of time for the rotor to stop before removing covers, inspecting, or cleaning. Keep you hands away from the discharge area of the chopper whenever the combine is running.

Check and service choppers regularly to keep them properly balanced. Imbalance in the rotor causes severe vibration and may lead to blade breakup, damage to the combine, or serious injury to nearby persons. Maintain balance by:

• Removing excess dirt from the rotor

• Replacing defective blades

- Installing new blades in balanced pairs as directed in the operator's manual

Most straw spreaders throw material out from both sides and the back of the machine. The speed of the flying materials is less than that for a chopper but it still can be dangerous, especially to eyes. Stay away from the straw spreader discharge when it is operating (Fig. 31). Disconnect or remove the chopper or spreader if you have to be near the back of a combine for performance checks.

Even when you are far enough away to avoid being struck by dense thrown objects, there is often a lot of dust, dirt, and chaff in the vicinity. Eye protection is recommended.

JDPX7577

Fig. 31 — Stay Away From the Straw Spreader Discharge When It's Operating

Shutdown Procedure

When leaving a harvesting machine, even for a short time, make sure the header is down on the ground or locked in the up position with the safety latch. Shut off the engine, set the parking brake, and leave the drive in gear to keep the combine from moving. Remove the ignition key to keep children or unauthorized people from starting it up.

Moving Combines on Public Roads

Moving a self-propelled or pull-type combine on roads and highways requires special care, especially for larger machines. Anytime a combine is to be moved on roads, whether driven, towed, or hauled, it should be properly prepared (Fig. 32). To do this:

1. Empty the grain tank to reduce weight and lower the center of gravity.

2. Move the unloading auger to transport position.

3. When practical, remove the header if it is wider than the basic machine, and transport it on a truck or implement carrier.

4. Be sure the SMV emblem, all reflectors, and lights are in proper working order and that they comply with state laws. Check with the police or sheriff if you have any questions.

5. Measure the height and width of the machine and write this information near the operator's platform with paint or a wax pencil for quick reference.

Auger in Transport Position

JDPX7588

Fig. 32 — Auger in Position for Transport. It Must Not Block Safety Lights or Reflectors

Other preparation of the combine depends on road conditions, distance moved, and method of moving. Check the operator's manual for specific procedures.

Consult local or state officials for regulations on moving large machines on public roads.

When driving a self-propelled machine on the road, you need good visibility both to the front and to the rear. Use rearview mirrors.

Have a car or truck drive in front and behind the combine to warn traffic on public roads. Stay in your lane. Pull off. stop, and let traffic pass. Do not drive on the shoulder!

Because the wheels for steering are in the back, self-propelled machines often fishtail when turned quickly at transport spends. Steering the combine to the right will whip the rear of the combine to the left: and vice versa. If you steer suddenly to the right when meeting oncoming traffic, the back of the combine will swing out into the path of oncoming traffic (Fig. 33).

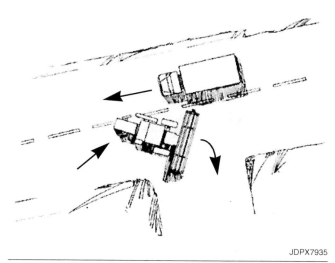

JDPX7935

Fig. 33 — As Combine Turns, Rear of Machine Can Swing Into Traffic Lane. Watch All Turns — Wait for Traffic to Clear at Intersections

If you slow or brake the combine too rapidly, you could lose some steering control (weight on rear wheels). This problem is most noticeable when driving a combine with a corn head or some other heavy header, or with the header raised high (Fig. 34). In this case, most of the combine's weight will be on the drive wheels. The following practices will reduce the problem:

1. Install rear wheel weights as recommended in the operator's manual.

2. Keep the header as low as possible.

3. Use the variable-speed drive or engine throttle to slow the machine.

4. Reduce speed before you to apply brakes. Always transport with the brake pedals locked together.

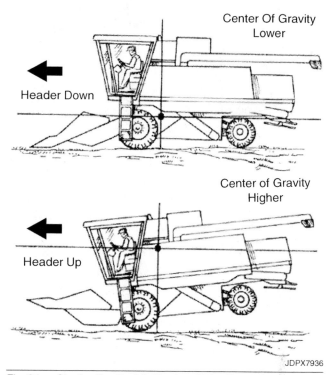

JDPX7936

Fig. 34 — Slowing Too Rapidly Tends to Raise Back Wheels When the Header Is High

Towing pull-type combines at transport speeds can be hazardous because of side forces on the tractor when stopping too quickly. Side forces from slowing a combine quickly may cause a tractor to skid, especially on loose gravel (Fig. 35). Slowing down while turning a corner can cause jackknifing. So, slow down before the corner so the towed combine doesn't get out of control.

aJDPX7937

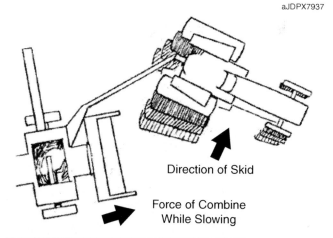

Fig. 35 — Side Forces From Slowing a Pull-Type Combine Can Make a Tractor Skid

Self-propelled combines should be towed on the road only if it's absolutely necessary. And even though it may be easier to tow from the rear, towing from the front is preferred because visibility is better and there are stronger structural members for fastening the tow chain. Follow Instructions in your operator's manual.

When hauling a combine on a truck or trailer, fasten the combine at each drive wheel and at the rear. Lock the brake pedals together. Use wheel chocks to block each side of the drive wheels. Some parts of the machine may have to be temporarily removed to comply with maximum transport height requirements. Low truck trailers are available for hauling with minimum overall height. When hauling a machine, be sure you know the height, width, and weight of the load, and check local and state, regulations. Watch for narrow culverts and bridges and low-clearance objects like power lines and bridges.

If the header is wide enough to cross the center line, a car should travel in front of and behind the combine, flashing its lights as a warning.

NOTE: For more detailed discussion of towing and travel on public roads, refer to Chapter 7, "Tractors and Implements". It addresses towed and self-propelled implements as well as tractors.

Header Attachments

Windrow Pickups

A problem with using a windrow pickup on a standard header is that it is usually quite a bit narrower than the header. This means you have two machine widths to watch — a narrow one for picking up the corp and a wider one that may hit obstacles along the windrows (Fig. 36). Keep an eye on the header to avoid fences and other obstructions along the edge of the field while concentrating on picking up the windrow.

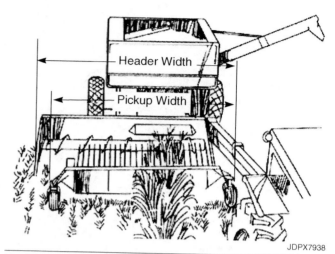

JDPX7938

Fig. 36 — Watch Both Machine Widths (Header and pickup) When Using a Windrow Pickup

Before traveling at road transport speeds, check pickup support chains to make sure they're securely hooked and in good condition. Having the pickup drop while moving at fairly high speed could damage the combine and injure you. Also, remember to add the amount of rear weight specified by the manufacturer before using or transporting the pickup.

Special Attachments

Several attachments can be used on combine headers to increase their effectiveness in gathering various crops. Many of these attachments require some modification of the grain header. In some cases, they're heavy and you'll need a hoist or some other lifting device to install them. For example, you should use a hoist to remove a bat reel and install a pickup reel. Using the right equipment saves time and reduces chances of injury.

Most attachments increase the weight on the front part of the header, so you may need additional weight at the rear of the combine for stability.

Correct installation of header devices is also important for proper operation of the rest of the hydraulic system, including steering, traction drive speed control, and other critical functions.

Corn Heads

Proper operation of snapping rolls, snapping bars, and gathering chains on corn heads is important for the smooth flow of materials into a combine. Improper functioning of these parts can cause uneven feeding, plugging, more downtime and increased risk for the operator. Trying to unclog snapping rolls without shutting off the machine is a leading cause of serious injuries during corn harvesting (Fig. 37).

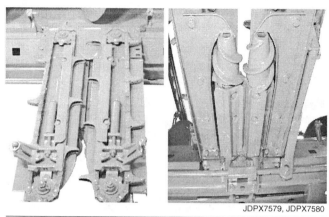

JDPX7579, JDPX7580

Fig. 37 — Properly Adjusted Snapping Bars and Rolls Reduce Plugging. That Reduces Accidents

Follow the recommendations in the operator's manual to adjust the corn head properly:

1. Tapered stalk rolls do not require adjustment to prevent plugging. To prevent plugging on corn heads with straight stalk rolls, the rolls must be set at a spacing that will aggressively feed the stalks through the rolls without breaking off the stalk.

2. Deck plates or snapping bars must be spaced far enough apart that stalks will pass through freely and at the same time prevent ears of corn from getting into stalk rolls.

3. Gathering chains must be timed and tensioned according to manufacturer's instructions in order to effectively move the stalks into the corn head.

Stalk roll speed must be comparable to ground speed. This is important for good stalk movement through the stalk rolls. If stalk roll speed is too fast, many stalks will be chewed off, which can cause plugging. If the stalk roll speed is too slow, stalks will be pulled out of the ground, plugging the stalk rolls.

If the corn head plugs:

1. Stop combine and header. Idle the engine.

2. Engage reverser to run header and feeder house backwards to free slug.

3. After slug is free, re-engage header and feeder house to run forward. Resume operation.

   (If combine does not have a reverser, after stopping and backing the combine, remove plugged material by hand. It may be necessary to cut tough materials with a heavy knife.)

4. If you must unplug the machine by hand disengage the power, shut off the engine, and remove the key before dismounting (Fig. 38).

   If plugging persists, snapping rolls or trash cutter may need adjusting or replacing. Follow operator's manual instructions.

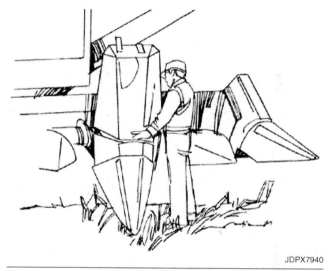

JDPX7940

Fig. 38 — Disengage Power and Shut Off Engine Before Removing Materials From Snapping Rolls

Replacing Snapping Rolls

Some rolls are made of cast iron, while others have replaceable steel bars. When cast iron rolls become worn, the entire roll has to be replaced. Replace snapping rolls while the corn head is on the combine — not while it's being stored. The combine holds the corn head rigid and provides power for turning corn head drives for safe and easy installation of the rolls. Block the header before working under it. Use the safety latch to hold it in the raised position. And keep your hands and fingers away from the many pinch points in the rolls and gathering chains.

Visibility in the Cornfield

Watch Out

Fig. 39 — Watch Out for People or Animals When Combining

Always be on the lookout for movement ahead of your harvester. It's hard to see people walking in a cornfield. Harvesting noise usually signals them to stay back, but children or animals may be attracted by the noise and run toward it without realizing how fast the combine is moving (Fig. 39). Keep children completely away from the fields when you are harvesting. Harvesting noise drowns out speech, and you may have to use hand signals to communicate with others (see Chapter 1, "Safe Farm Machinery Operation"). Make sure everyone knows your intentions and stands clear before the machine is moved.

Corn Pickers

The precautions for safe use of ear corn pickers are similar to those discussed earlier for combines and corn heads. They have many rotating parts. Therefore, safety shielding is very important, especially for tractor-mounted pickers where the operator sits close to gathering units and power drives (Fig. 40). Aggressive snapping and husking rolls are very dangerous if a person tries to clean them while they're running. Also, trash buildup on tractor-mounted pickers leads to fires.

Use Full Shielding
as Shown

Fig. 40 — Keep All Shields and Trash Screens in Place When Operating a Corn Picker

Fire Prevention

To prevent fires on mounted pickers:

1. Enlarge or raise the radiator screen for effective cooling.

2. Shield leaves from the hot exhaust manifold and muffler.

3. Clean leaves and trash from hot spots.

4. Mount a multi-purpose dry chemical fire extinguisher on the equipment to use in case of fire.

Operation

For effective and safe picker operation:

1. Maintain the ground speed for snapping ears midway along the rolls.

2. Adjust the snapping rolls and gathering chains for good stalk feeding without breaking the stalks.

3. Replace snapping rolls as required for good feeding action.

4. Keep all safety shields and safety signs in place and in good condition.

5. Don't attempt to inspect, clean, service, or adjust the picker while it is running. Disengage the power and shut off the engine.

Snapping Roll

Maintaining the right relationship between gathering chain speed and ground speed is important. When ground speed is too slow, stalks are fed too far back on the rolls and ears may catch in the stalk ejector and cause plugging at the back. When ground speed is too fast, stalks won't move back far enough, causing snapping and plugging near the front of the rolls and gathering chains. When these mechanisms get plugged, operators take chances and get hurt. Keep them operating smoothly.

Because spiraled rolls aren't as aggressive as fluted rolls, they often have more of a plugging problem. If snapping rolls plug, disengage the power and shut off the engine before trying to get stalks or weeds out. If there's a lot of plugged material, separate the rolls. You can do this from the operator's station on many mounted units. Remove as much material from the rolls as you can before starting the machine. Cleaning fluted rolls by running the machine is more effective than it is with spiraled rolls.

One hazard of snapping rolls is that a weed or stalk can catch between the rolls and stop. Then if it's moved slightly, the weed or stalk can feed through the rolls very quickly — faster than someone holding onto it 2 or 3 feet (600 to 900 mm) from the rolls could possibly release it (Fig. 41). This is how some operators are caught when they think they're reasonably safe. Never attempt to clean snapping rolls while the machine is running. Adjust rolls properly and use correct ground speed to reduce plugging problems.

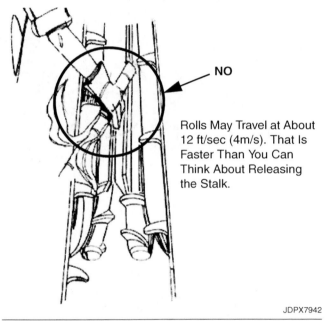

NO

Rolls May Travel at About 12 ft/sec (4m/s). That Is Faster Than You Can Think About Releasing the Stalk.

JDPX7942

Fig. 41 — Snapping Rolls Can Pull You In Faster Than You Can Let Go. Don't Try It!

Husking Mechanisms

Husking beds have pairs of rolls made of steel or rubber designed to grasp and pull the husks. These rolls can also catch gloves, fingers, or hands. Ear retarders above the rolls may consist of pressure wheels with rubber paddles or a chain-and-slat mechanism. Most of this equipment is open and can cause injury if you are careless. Disengage the power and shut off engine before doing any cleaning or adjusting around the husking bed.

Plugging can be as much of a problem with husking rolls as it is with snapping rolls. Damp or green material sometimes gets wrapped around them. If this happens, disengage the power, shut off the engine, cut the material free with a strong knife, and pull it out by hand. Wear well-fitting gloves, because corn husks and leaves have sharp edges that can cut, and weeds or trash wrapped around the rolls can injure bare hands.

Shelling Mechanisms

If inspection of the sheller is necessary, disengage the power, shut off engine, and see that the sheller has stopped completely before opening any covers or doors. If the cylinder plugs, use procedures similar to those recommended for unplugging combine cylinders. Using the right tools, protecting hands with heavy gloves, and avoiding pinch points are musts for unplugging a sheller safely.

Turning

A tractor hitched to a pull-type picker and wagon needs a lot of space for turning. If equipment is turned too sharply, the wagon's steering mechanism can be damaged or the hitch broken free from the picker. Someone could be seriously injured by an uncontrolled wagonload of corn. Sharp turns cause the wagon to create large side forces on the picker, and these can cause skidding of the picker and loss of control, particularly when turning on a down hill or side hill with a fully loaded wagon.

Under some conditions, the offset loading of a pull-type picker adds to the problem. Take special care when moving down steep slopes to avoid jackknifing the picker and wagon (Fig. 42). Give yourself plenty of room for turning at the end of the field.

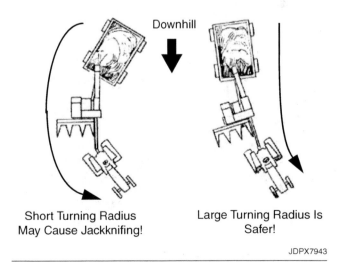

Downhill

Short Turning Radius May Cause Jackknifing!

Large Turning Radius Is Safer!

JDPX7943

Fig. 42 — Make Wide Turns on Downward Slopes to Avoid Jackknifing

Personal Protective Equipment

When harvesting conditions are especially dirty, eye and respiratory protection may be needed (Fig. 43). This may be of greatest concern when operating machines that place the operator near the crop-gathering devices. Noise from the many gears. chains, and other drives may also make ear protection desirable, especially when units are operated several hours a day.

JDPX7944

Fig. 43 — Use Protective Equipment When Operating a Corn Picker Under Moldy, Dirty, and Noisy Conditions

Riders

Keep riders out of wagons behind corn pickers. When ear corn is harvested, the ears can hit a rider (Fig. 44), or the elevator that carries the corn to the wagon could injure a rider, especially when turning or harvesting on rolling ground. And wagon fall-off accidents are frequent. Do not stand while operating corn pickers. You cannot adequately control the machine if a sudden emergency occurs.

JDPX7945

Fig. 44 — Riders in Wagons Behind Corn Pickers Can Be Hit by Ears Thrown by the Elevator and They Can Fall Off — Insist On No Riders

Cotton Harvesters

Mechanical cotton pickers and strippers are large, high-profile or tall machines. A self-propelled picker may weigh up to 30,500 pounds (13,835 kg), and it may be about 16 feet (5 m) high in field operation. Pickers selectively remove only the white cotton fibers and seeds from the bolls (pods). Mechanical strippers remove or "strip" the entire boll from the plant, including cotton fiber, seed, and trash.

Although most modern cotton harvesters are self-propelled pickers (Fig. 45), cotton strippers are still being marketed and used (Fig. 46). In earlier years, pull-type and tractor-mounted strippers were common. Some are still being used.

JDPX8121

Fig. 45 — Most Modern Cotton Harvesters Are Self-Propelled Pickers. Note the Machine's High Profile.

JDPX7670

Fig. 46 — Self-Propelled Cotton Stripper with Basket Partly Full of Cotton

Much like grain combines, cotton harvesters are complex machines with many moving parts to carry out several functions, As for any such machine, skill and caution are necessary for efficient and safe operation and service.

Many of the safety precautions discussed earlier for grain combines also apply to cotton harvesters. Please refer to them. Most of those precautions will not be repeated here. This section concentrates on potential hazards that are somewhat unique to cotton harvesters, in this order:

- Field operation
- Chokedowns
- Service
- Fire hazards

Field Operation and Dumping

Cotton harvesters are equipped with large overhead baskets to collect cotton as it is harvested. Not only are they quite high when picking cotton, but when the basket is raised for dumping cotton, it may be about 25 feet (7.6 m) high. The height of the machines calls for careful operation to avoid upsets and contact with electric wires.

The machinery operator's manual and safety signs will address specific concerns for your machine. Here are some of the precautions:

1. Lower the basket and check overhead clearance before driving the machine. When the basket is raised, the harvester becomes more top-heavy. That can cause an upset as the machine is turned or is moved across a field. Also check overhead clearance before driving near electric wires.

2. Reduce speed when moving over rough terrain. Cotton harvesters are top-heavy and can turn over on rough ground.

3. Keep your harvester safely away from ditches, creeks, and other steep, sloping ground. Keep end rows smooth and firm. Steep slopes and plowed turn-rows make turning difficult, and may cause an upset.

4. Reduce engine speed before braking or turning. Quick stops with the high-profile cotton pickers can result in the machine nosing over. Fast turns can result in upsets.

5. Remain seated when raising or lowering the basket on a tractor-mounted stripper. A sudden drop of the basket could result in a serious head injury to anyone standing in the wrong place on the operator's platform.

6. Keep everyone away from trailers in the field. Be sure there is no one in the trailer before dumping. A load of cotton falling into a trailer could seriously injure or suffocate someone trapped inside.

7. Be sure you're clear of electrical wires before raising or dumping a basket (Fig. 47). A raised basket may reach a height of 25 feet (7.6 m), and if it contacts a power line, you could be electrocuted.

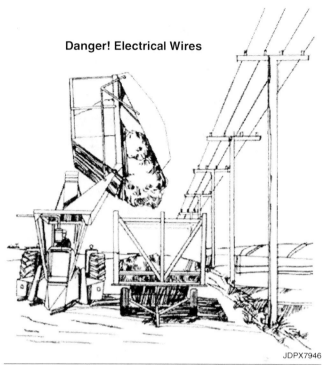

Danger! Electrical Wires

JDPX7946

Fig. 47 — Before Dumping Cotton, Be Sure No One Is in the Trailer, and Be Sure Basket Cannot Touch Electrical Wires When It Is Raised or Lowered

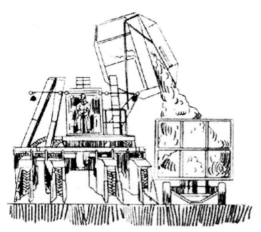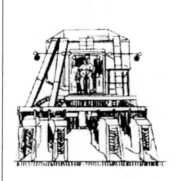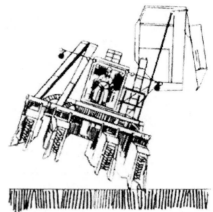

After Dumping Cotton...

Lower Basket Before Driving Away...

To Prevent An Upset

JDPX7947

Fig. 48 — Do Not Drive Harvester With Basket Raised and Don't Dump Toward a Downward Slope

8. Don't dump the basket toward a downward slope or grade. Dump from a level surface. A raised basket shifts weight toward the slope and could cause an upset (Fig. 48).

9. Be sure no one is near moving conveyors during unloading.

10. After dumping, lower the basket before driving away. Moving the harvester with the basket raised can cause an upset.

11. Wear personal protective equipment as appropriate. Harvesting cotton can be noisy and dirty. Also note that cab filters do not remove harmful chemicals from air that enters the cab. When working with agricultural chemicals, or in fields where they have been used, follow the chemical manufacturer's instructions for personal protection.

12. Do not allow extra passengers, especially children. Persons not required for the operation should not be around the machines.

Operator Presence Safety System Interlock

Disengage the power and shut off the engine before attempting to inspect, adjust, or unplug machinery. That is one of the basic farm equipment safety rules. Unfortunately, some people have disregarded that rule and were seriously injured or killed.

Some cotton harvesters have operator presence systems to make sure the picking unit is shut off before anyone works on or near moving parts. On a typical system, if the operator leaves the seat for more than a few seconds, picking units stop. The engine continues to run. The operator can re-engage power to the machine and picking units after returning to the seat, away from rotating parts.

A system such as just described may have a bypass control that enables an operator to engage power to rotate the picking units very slowly for inspection while standing on the ground (Fig. 49). To control drum rotation, the operator depresses a hand-held remote control switch connected to the seat safety system by an electrical cord or "tether." If your machine has such a safety system, use it as instructed and don't attempt to override it. (See "Safety Switch Interlock Systems" page 8).

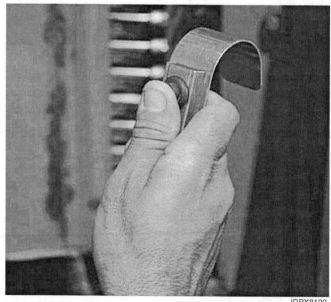

JDPX8122

Fig. 49 — With Operator Presence Safety System, Operator Uses Remote Control Switch (in Hand) to Slowly Turn Picker Drums for Inspection

Chokedowns

Many cotton harvester accidents could be prevented by avoiding chokedowns. You can do a more efficient job and reduce the risk of injury by keeping the main harvesting components — the picker drums, stripping rolls, and the reel on finger-type strippers — free of weeds, rocks, and stumps. When the machine is operating properly, you're safely in the operator station. To prevent chokedowns:

1. Practice good weed control and keep fields clear of rocks and stumps. A cotton harvester operating in a clean field is less likely to choke down.

2. Adjust all belts and chains on the harvesting unit according to instructions in the operator's manual. A belt that slips or a chain that jumps can allow the harvesting unit to choke down.

When a picker drum chokes, disengage all power and shut off the engine. Wait until all parts have stopped moving, and then remove the obstruction. Rotate the doffer manually or as otherwise instructed by the operator's manual until the obstruction can be removed. Never try to start the picker drum turning by engaging the power and pushing the spindles with your foot. If the obstruction suddenly breaks free and the picker unit turns under power, it will endanger any part of the body near the spindles (Fig. 50).

No! No! Wrong!

JDPX7948

Fig. 50 — Never Push or Kick Picker Spindles With Your Foot to Turn Drums With Power Engaged

After removing the obstruction, return to the operator's station, start the engine, and engage the power. If the clutch continues to slip, check for other obstructions or for bent picker bars or doffers.

NOTE: *Some operator presence safety systems with remote control enable you to slowly turn picking drums under power to be sure they are clear, and to check spindle damage. Follow operator's manual instructions for unplugging.*

If a cotton stripper chokes down, follow the basic safety practices: Disengage the power and shut off the engine before approaching moving parts. If you attempt to unplug stripping rolls, augers, or cleaners while they are moving, you risk serious injury. Don't take a chance!

Service

Cotton harvesters must be serviced regularly. These machines are unique in some ways, and proper maintenance is required for effective and safe operation. Follow these guidelines:

A primary safety precaution for servicing farm equipment is often repeated in this book: Disengage power, shut off the engine, and remove the key before working on a machine unless specifically instructed to do otherwise. Follow instructions in operator's manual. That applies especially to servicing and adjusting cotton pickers and strippers. Here are some other guidelines:

1. Wear close-fitting clothing and stay alert while lubricating picker bar cams, cam follower bearings, and spindle drive gears (Fig. 51). This job must be done while the picking unit is running on some machines. Check the operator's manual for the correct procedure.

Be Careful!

JDPX7949

Fig. 51 — Use Extreme Care When Lubricating Picker Bar Cams and Sun Gears if Operator's Manual for Older Machines Recommends Running Machine While Lubricating

Picking units are aggressive and require caution when servicing them. Some machines have on-board lubricating systems so you can lubricate row units from the operator station, or from the ground when equipped with a remote control (Fig. 52).

Fig. 52 — Checking Lubrication on Picking Unit With Remote Safety Control. Stand Clear

2. Disengage power, shut off the engine, take the key, and wait for all parts to stop moving before removing dust from V-belts on cotton harvesters. These belts often become covered with a fine dust glaze when harvesters are operated under dry conditions. They should be cleaned frequently.

3. Use the ladder and handrails when working on top of the basket (Fig. 53). A fall from the top of the basket could injure you seriously. If the basket lid opened, secure it to keep it from closing accidentally.

Fig. 53 — Use Handrails and Ladders When Working on Top of Basket

4. Before working around or under picker or stripper units, lower them to the ground or raise them and engage the safety stops. Shut off the engine and remove the key before approaching the row units. If safety stops are not provided support row units with solid blocks. Don't rely on hydraulics to support them.

5. Engage basket cylinder lock valves, safety lockouts, or stops before working under a raised basket. Avoid crushing injuries. Clear all persons away before lowering the basket.

6. After servicing, replace all shields and guards before starting the machine. Shields and guards are designed to protect you from moving parts, but they can do their job only if they are kept in place.

7. Make sure that the main gearshift lever is in park position and that no one is standing on or near the machine when you start it.

Fire Hazards

Before cotton is harvested, the plant is killed, either chemically or by frost. This means that highly flammable leaf trash and dead plant parts get mixed in with the lint. Cotton lint is also flammable. When such materials accumulate, fires can start easily.

Take these precautions to prevent fires during cotton harvesting:

1. Keep the engine clean (Fig. 54). Lint, leaf trash, and other dry materials on a hot engine could catch fire. Clean the area between hot engine parts and the hood periodically. Check exhaust pipes and muffler often for leaks. Hot exhaust gases or sparks could start a fire. Also handle fuel with special care to avoid fires.

Fig. 54 — To Avoid Fires, Keep Engine Free of Lint, Dust, and Trash. Shut Off Engine First!

2. Use only water or the manufacturer's moistening agent. Flammable petroleum-based moistening agents increase the fire hazard and may lower cotton grade. Note that moistening agents or spindle cleaners can be harmful if swallowed or if they contact eyes. Handle and store them as chemicals. Check the label.

3. Always dump the basket downwind into the trailer. Dumping it into the wind could result in cotton blowing back onto the engine and starting a fire.

4. Keep the doffer area free of lint and trash to avoid fire caused by friction between the spindles and trash (Fig. 55).

JDPX8125

Fig. 55 — Keep Doffer Area Clean. Friction Between Spindles and Trash Can Cause Fire

5. Prevent arcing at electrical terminals. Cotton lint or trash around the battery terminals and other electrical contacts can catch fire. Keep all terminals and contacts clean and tight.

6. Mount two large, multi-purpose (ABC) dry chemical fire extinguishers on machine for easy access. For type, size, and mounting locations, follow manufacturer's instructions.

7. If cotton ignites in the basket, dump the cotton. Don't dump it into the wind, and do not try to extinguish fire in the basket.

Transporting Cotton Harvesters

For safety in driving or transporting cotton harvesters on roadways, review the section in this chapter, "Moving Combines on Public Roads". Also review "Slow-Moving Vehicles" in Chapter 3, "Recognizing Common Machine Hazards"; and "Accidents on Public Roads" in Chapter 7, "Tractors and Implements". In addition, follow these recommendations:

1. Lower the basket for safe transport, and lock it so it can't be raised during transport.

2. Secure headers in raised position.

3. Check clearance before driving under electric wires, bridges, or through gates.

4. Anticipate the need to stop and turn so you can gradually slow down to avoid upset.

5. Follow operator's manual recommendations for your machine.

Test Yourself

Questions

1. (T/F) The best time to start getting a combine ready for harvest is right before harvesting begins.

2. Preparing a field for safe harvesting includes:

   a. Removing obstacles such as posts, stones, or stumps

   b. Preparing the edges of the fields for smooth turning and maneuvering of harvesting equipment

   c. Using adequate weed control practices

   d. All of the above

3. (T/F) Propulsion drive belts on combines run only when the transmission is in gear.

4. (T/F) When you are driving a self-propelled-combine on a two-lane road and making a right-hand turn into a field, the back of the combine will move to the left into oncoming traffic.

5. A safety switch that prevents starting the engine if the operator is not in the seat:

   a. Helps avoid unsafe start-up

   b. Helps keep people from getting hurt or killed

   c. Is an important safety feature

   d. All of the above

6. (T/F) Servicing a combine at the end of a day's operation can prevent rust on moving parts.

7. (T/F) When driving a combine on the road at transport speeds, you should raise the header as high as possible to avoid obstructions.

8. What may need to be changed when you attach a different header or header attachment?

   a. The pressure in the front tires

   b. The weight on the back of the combine

   c. All of the above

   d. None of the above

9. (T/F) When operating on hillsides, a combine will be more stable if the grain tank is full.

10. Requirements for safe and effective operation of corn pickers include:

    a. Keeping all safety shields and safety signs in place and in good condition

    b. Never attempting to inspect, clean, service, or adjust the picker while it is running

    c. Replacing snapping rolls as required for good feeding action

    d. All of the above

11. (T/F) Many cotton harvester accidents could be prevented by keeping all belts and chains properly adjusted.

12. A practice that helps reduce the fire hazard when working with a cotton harvester is:

    a. Keeping the engine free of lint and trash

    b. Using a petroleum-based moistening agent

    c. Dumping the basket upwind into the trailer

    d. All of the above

References

1. Fundamentals of Machine Operation, FMO — Combine Harvesting. 1991. Deere & Company, Service Publications Dept., FOS/FMO, Moline, IL 61265

2. Cotton Picker Safety. 1981. VHS video, 9 min. Order No. DSVHA81760EN. English and Spanish sound tracks on one tape. Deere & Company, Distribution Service Center, 1400 13th St., East Moline, IL 61244.

3. 9930 and 9960 Cotton Pickers Operation, Adjustments, and Safety. 1991. VHS video, 24 min. English, order No. DHPVA91010EN; Spanish, order No. DHPVA91010ES. Deere & Company, Distribution Service Center, 1400 13th St., East Moline, IL 61244.

Material Handling Equipment

12

JDPX8023

Introduction

More material are being moved on today's farms than ever before. This increased volume demands increased mechanization of material handling.

Many types of materials are moved, and many devices are used to move them. This chapter discusses safe operating procedures for some of the more common types of material handling equipment:

- Front-end loaders
- Skid-steer loaders
- Manure spreaders
- Grain bins
- Portable elevators and augers
- Farm wagons
- Blowers
- Silo unloaders
- Grinder-mixers
- Bale processors

Operator's Manuals and Safety Signs

For safe operation of material handling equipment, follow the safety instructions in the operator's manuals and safety signs for each machine you operate. Also, review earlier chapters in this book, including these:

- Chapter 3, "Recognizing Common Machine Hazards"
- Chapter 7, "Tractors and Implements"

Front-End Loaders

Loaders are versatile — they handle feed, manure, bedding, and other materials. Hydraulic fingertip controls permit the operator to load, lift, and control-dump large quantities of materials with ease and efficiency.

Loaders are potentially dangerous when not operated properly, because they change the center of gravity and stability of the tractor loader combination. This presents a potential tipping hazard to the operator.

Fig. 1 illustrates the way the center of gravity moves as the load is raised. Once the center of gravity is moved outside the base of stability, tipping will occur. Even raising an empty bucket high into the air will increase the chance of overturn.

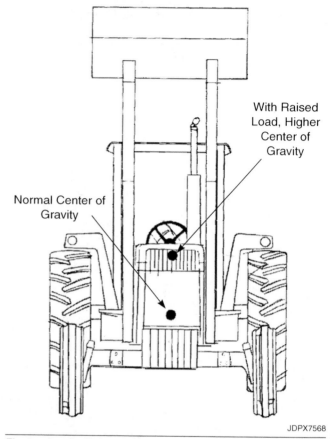

With Raised Load, Higher Center of Gravity

Normal Center of Gravity

JDPX7568

Fig. 1 — Loader Height Changes the Center of Gravity and Stability of the Tractor-Loader Combination

Since front-end loaders affect tractor stability, the tractor should be equipped with a ROPS (Rollover Protective Structure) and the seat belt should be used. Then, if there is an overturn, the operator has the protection of ROPS.

The following precautions can help you avoid not only overturns but also serious injury or death from other material handling accidents:

1. Keep the bucket low while carrying loads and while operating on inclines. Raising the bucket moves the center of gravity upward and forward. Bumps, holes, rocks, loose fill, or stumps can easily upset a tractor if the bucket is carried high. It also places undue stress on the loader frame. Drive uphill with loaded buckets rather than downhill, and stay off steep slopes to prevent bouncing and loss of control (Fig. 2). Also avoid turning or traveling on side slopes.

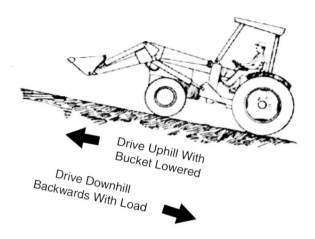

Fig. 2 — On Steep Inclines, Keep the Bucket Uphill From the Tractor to Prevent Upsets

2. Keep travel speed slow. Speed and height are the two most critical factors when working with loaders. Keep speed down when loading and transporting. Excess momentum causes the tractor to overturn in situations where there would ordinarily be little danger. In addition to the danger of overturning, you may damage the equipment, especially on rough ground. Be especially careful when making turns with the loader raised.

3. Use the recommended amount of ballast to give your tractor extra stability (Fig. 3). This will counterbalance the extra weight on the front of the tractor and help prevent tractor upsets. Also, set tractor wheels at the widest recommended width, and do not lift loads that exceed equipment limits. Check the operator's manuals for specific recommendations.

Fig. 3 — Use Ballast for Added Tractor Stability

4. Load the bucket evenly from side to side and keep within the normal capacity of the tractor and loader. This will help prevent upsets. The amount of material that can safely be loaded in the bucket will vary tremendously with its density. While it might be safe to fill the bucket with sawdust, the same volume of wet sand could cause a tractor upset. The capacity of your loader is specified in the operator's manual.

5. Never tow a tractor by attaching a tow chain or cable to the loader. Towing by the loader can overturn the tractor.

6. Remove the loader when it's not in use. Don't use a tractor with a loader when performing field operations — it's harder to handle, visibility is reduced, and fuel is wasted.

7. If you must use a tractor with loader to compact silage, use only a tractor with ROPS, and fasten the seat belt. Silage is slippery and unstable and can easily shift, causing an upset.

8. Avoid contact with overhead and underground electrical wires. Check overhead clearance. Also check underground utility locations before digging. If the machinery contacts electrical wires, stay in the seat. Back machinery away from wires before dismounting.

Operator controls are designed to be used by the operator while sitting in the operator's seat. Operating the controls from the ground or from behind the tractor causes erratic engagement of controls, and could result in an operator getting crushed or entangled in the mechanisms.

Remain at the controls until the hydraulic cycle is complete, and do not allow riders on the tractor or loader. Riders could fall off the loader while the tractor is moving. There usually is not enough time to get the tractor stopped before they are run over — to say nothing of the injuries received in the fall.

Do not use front-end loaders as hydraulic lifts or work platforms for people. Loader buckets can tilt, hydraulics can fail, and accidental control movement can dump people out.

Lower the bucket to the ground, shut off the engine, and remove the key before leaving the tractor. Then the loader cannot be accidentally lowered by children or others (Fig. 4). Also never allow anyone near or under a raised loader. Lower it to the ground for servicing. If it is ever necessary to work on a loader in raised position, secure it so it absolutely cannot fall.

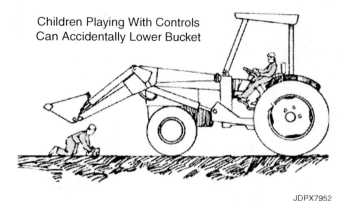

Children Playing With Controls
Can Accidentally Lower Bucket

JDPX7952

Fig. 4 — Operator Should Have Lowered the Bucket Before Leaving Tractor

When connecting hydraulic hoses, carefully check loader operation to make sure hoses are coupled to the proper tractor outlet. The loader must lift when you move the control to "lift" position. If it doesn't respond properly, you could be critically injured by unexpected action. Check operation according to operator's manual procedures.

Keep hydraulic connections tight. Repair leaky hoses or connections as soon as you discover them. A high-pressure stream of escaping hydraulic fluid can penetrate skin and cause serious infection or reaction, in addition to physical damage to internal flesh. Use a piece of cardboard or wood to detect pinhole leaks, not your hand. For more details, refer to hydraulic hazards in Chapter 3, "Recognizing Common Machine Hazards", and Chapter 7, "Tractors and Implements".

Use Proper Loader Attachments

Don't make the mistake of using a tractor-mounted loader with the wrong material handling attachment. That can result in a load falling and causing serious injury or property damage. There are several types of attachments available such as forklifts, buckets, silage clamps, and various bale handling devices. Use the right one.

Never handle round bales unless the loader is equipped for them. A bale can roll down from the raised loader, crushing the operator, and damaging the tractor (Fig. 5). Use only a manufacturer's approved attachment such as a bale fork, a grapple/clamp, or a hugger to securely hold bales in place. (See "Moving Large Round Bales." in Chapter 10, "Hay and Forage Equipment".)

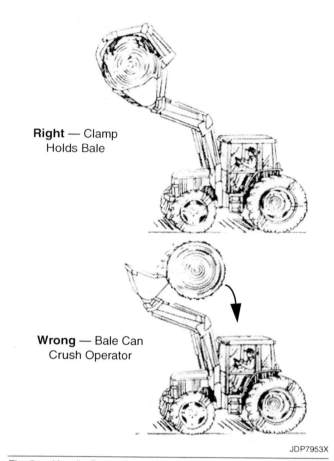

Right — Clamp Holds Bale

Wrong — Bale Can Crush Operator

JDP7953X

Fig. 5 — Handle Round Bales Only With Proper Attachments. A Falling Bale Can Crush Operator

Skid-Steer Loaders

Skid-steer loaders are compact and maneuverable. They can operate inside enclosures with low overhead clearance where tractor-mounted loaders cannot operate. They have a short, narrow wheelbase, which calls for caution to avoid upsets (Fig. 6).

JDPX7954

Fig. 6 — Avoid Overturns With Skid-Steer Loaders. Carry Loads Low and Slowly

Skid-steering is usually controlled by right and left hand levers which change speed and direction of the drive wheels on either side. Hence, the loaders can turn sharply or "spin" within their wheelbase. Foot pedals commonly control the lift arms and bucket (Fig. 7).

JDPX8128

Fig. 7 — Keep Area Clean. If Pedals or Levers Are Blocked or Bumped, You Could Be Injured

Modern skid-steer loaders are equipped with ROPS and a seat belt, as well as overhead protection against falling objects. Side screens keep the operator from reaching into the area where lift arms could crush or shear an arm or leg; and they protect against objects that might strike the operator.

Safety Information

Follow the safety instructions in the operator's manual and in safety signs on your loader Also review Chapter 3, "Recognizing Common Machine Hazards". You may also refer to "Skid-Steer Loader Safety Manual," from Equipment Manufacturer's Institute, 10 South Riverside Plaza, Chicago, IL 60606-3710.

Safe Operating Practices

Many safety precautions for skid-steer loaders are the same as for tractor-mounted loaders: Use care to avoid upsets; avoid contact with electrical wires; don't allow riders; watch out for others; and use proper attachments. Please study those precautions in the preceding section, "Front-End Loaders".

Since skid-steer loaders are designed for compactness and maneuverability, the operator is very close to the lift arms and bucket. Also, the operator gets on and off the machine near the controls. Hence, there are several special precautions for safe operation:

1. Keep the operator's cab free of dirt, chains, or other objects. They could interfere with safe operation of control levers and pedals, or they could cause you to fall or trip and accidentally move a control.

2. Never enter the cab or reach into it unless lift arms are lowered and attachments are flat on the ground. Raised lift arms and bucket can fall and crush you. Don't touch control levers or foot pedals as you enter. It may cause the machine to move and you could be injured (Fig. 8). The engine must be off.

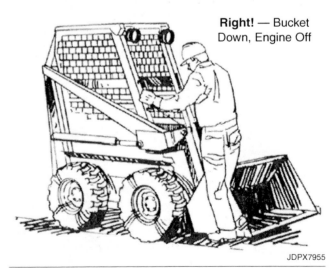

Right! — Bucket Down, Engine Off

JDPX7955

Fig. 8 — Enter or Exit Loader Only if Lift Arms and Attachment Are Lowered to the Ground

3. Before start-up, clear others from the area. People, especially children, can be blocked from your view and could be run over. Check all around the machine to be sure no one is on it or near it.

4. Always wear the seat belt (Fig. 9). The loader can move suddenly and violently on rough terrain or in quick stops. Without the belt, you can be pitched against the inside of the machine or out of it, where you can be run over or crushed by the bucket or lift arms.

Right! — Wear Seat Belt

JDPX7956

Fig. 9 — Always Wear The Seat Belt. Before You Unbuckle It, Lower Lift Arms And Stop Engine

5. Stay alert for holes, rocks, rough surfaces or other terrain hazards when moving the load. Since skid-steer loaders are so compact, such hazards can cause violent pitching or upsets.

6. Keep your entire body inside the operator's station (Fig. 10). Never reach or lean outside the cab. The lift arms or bucket could crush you as they move up or down or unexpectedly fall. Keep all protective screens or guards in place to prevent such serious injuries.

No! Wrong! — Bucket Can Fall

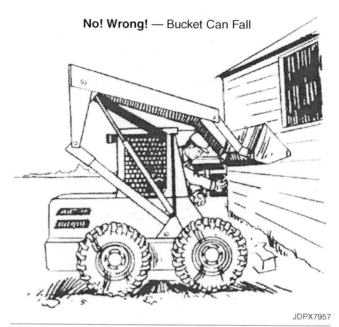

JDPX7957

Fig. 10 — Never Reach Or Lean Outside The Cab During Operation. You Could Be Crushed

7. Before unbuckling the seat belt and leaving the seat, lower the lift arms and attachment completely to the ground. Then, put controls in the park position, set the parking brake, stop the engine, and remove the key. Cycle the hydraulic controls to relieve pressure. If you bump controls and the lift arms are not secured, or if something fails, the lift arms and bucket could fall and crush you.

8. If you must service any part of the machine with lift arms raised, be sure lift arms and bucket are securely restrained or supported. Some machines have lift arm stops for that purpose. Follow the operator's manual.

9. Never stand or lean anywhere on the loader when the engine is running. As lift arms move up and down next to the vehicle, they create several pinch points and crush points, which can cause serious injuries.

10. Wear personal protective clothing and equipment as recommended in the operator's manual and as indicated by job conditions.

Safety Interlock Systems

Some skid-steer loaders have safety interlock (operator presence) systems. One type has a seat switch. It requires the operator to be in the seat with the belt fastened before the engine can be started or lift arms and bucket operated. Safety interlock systems are intended to prevent mistakes as discussed above. Check them frequently according to the operator's manual. If they don't work right, get them repaired at once. Don't bypass them!

Endgate Raised for Unloading

Heavy Objects Thrown Out Can Injure Anyone Nearby

JDPX7638

Fig. 11 — Beaters May Throw Heavy Objects Over 100 Feet (30 m). Keep Others Clear

Manure Spreaders

Manure spreaders can be classified into these types:

- Box-type with rear heaters

- V-tank with side discharge

- Flail

- Closed tank

Manure spreaders are generally drawn behind a tractor and powered by the PTO. Hence, the safe operating rules in Chapter 7, "Tractors and Implements", also apply to spreader operation. Review them. The following precautions apply primarily to manure spreaders:

- Only one person is needed to operate the machine. When two or more people are working together, one may turn on the machine and accidentally cause an injury to the other. Always check to see if anyone is in the machine or nearby, and clear them from the area before starting the engine and engaging the PTO. Be sure no one is near beaters, augers, or unloading impellers.

- Operate only when seated properly on the tractor. Never allow anyone to ride on the drawbar or hitch because of the great danger of falling or getting caught in the PTO, and never use the PTO shield as a step for mounting the tractor.

- Be concerned about anyone within 100 feet (30 m) of the sides and rear of the spreader. Ask people to move away before you engage the PTO so that they will not be hit by thrown objects (Fig. 11).

- Check often to see that PTO shields turn freely (Fig. 12). Never make this check when the PTO is in operation. You must make sure that the shield is not somehow bound to the shaft and becomes a hazard in which you could get caught. Keep hands, feet, and clothing away from all moving parts, and correct the situation immediately.

Disengage PTO, Shut Off Engine

Check to See if Shield Turns Freely

JDPX7958

Fig. 12 — Manure and Mud Can Prevent PTO Shields From Turning Freely. Check Them Often

Like most machines, manure spreaders get plugged from time to time, presenting a situation that could be dangerous.

Excessive plugging can be prevented if you:

1. Keep stones, boards, and other solid objects out of the spreader. These not only may plug the spreader, but could cause breakage or accidental injury to someone if they are hurled from the machine.

2. On flail-type spreaders, make sure flails and chains are loose before loading in freezing weather. If frozen, loosen them with a pry bar. Don't try to break them loose by repeatedly engaging the power supply. It is wise to follow this precaution even if the spreader looks OK, because you might find that chains and flails are frozen. Do not attempt to free them without taking safety precautions, such as disengaging the PTO and turning off the engine.

3. On beater-type spreaders, keep chains and beater mechanisms in good working order. Replace stretched chains. When dropping loads into the spreader, try not to hit the beater mechanism and chains with the bucket or drop heavy loads of frozen manure from excessive heights. Never get in a spreader to clean it while it's running. The gathering chains or hydraulic-powered push panel can force you into the beaters.

4. On V-tank spreaders, maintain the unloading auger and discharge expeller so they always operate freely. (V-tank spreaders handle liquid, slurry, or solid manure.)

5. In cold weather, clean inside of spreaders so manure won't freeze up moving parts.

Good maintenance helps prevent problems that could lead to accidents. If the spreader does become plugged:

1. Stop the tractor and park properly. Use the brakes or parking gear to keep the tractor in place. Never rely on gears to keep a heavy load from moving. On an incline, the weight of the tractor and spreader could create a force strong enough to turn the engine over and get the tractor moving.

2. Disengage the PTO, shut off the engine, remove the key, and set the parking brake before inspecting, unplugging, or doing any work on a spreader. If your loader has an endgate, close it. If it falls or lowers accidentally, it could cause crushing injuries.

3. Moving parts can cause serious injury to a person in the spreader when it starts up: conveyor chains or a hydraulic push-panel could force you into rotating beaters; a falling endgate could cause crushing injuries; in V-tank spreaders the auger in the bottom, and also the discharge expellers, could mangle someone severely. Allow no one in or near the spreader with the tractor engine running.

4. Determine and correct the cause of plugging. Perhaps a rock or chunk of frozen manure has lodged in the beater,

unloading auger, or discharge expeller. A drive chain could have broken, or maybe the spreader was overloaded. Clear all unloading mechanisms so they will move freely. If you remove shields or guards for inspection, unplugging, or service, replace them all before operation.

Maintain rotary flails. The peripheral speed and centrifugal force could cause rotating flails to come loose unless they're properly maintained. Replace flails when they become worn. Keep them bolted tightly, and check their condition regularly to detect danger. Never approach the machine while it's in operation.

Always use a locking safety hitch pin and safety chains to hitch the spreader to a tractor, For more hitching safety information, see Chapter 7, "Tractors and Implements".

Keep the jack stand in good working order. Before unhitching the spreader from the tractor, make sure the jack is locked securely in place. Never rely on an unlocked jack to hold the spreader. Even a slight movement of the tractor could make the spreader fall.

Closed-Tank Manure Spreaders

Liquid manure spreaders generally use a PTO-powered pump to fill large portable tanks and to spread manure. Here are some recommended procedures for their use:

1. Be sure the tractor is properly ballasted and capable of safely controlling and stopping the heavy load of liquid manure, The spreader should be securely hitched to the tractor with a locking safety hitch pin and safety chain.

2. Make sure the relief valve is operative to avoid excessive pressure buildup.

3. Before transport, securely fasten the covers to prevent leaking and spilling on public roads. Such leaking and spilling creates environmental pollution and is subject to regulation in some areas.

Additional dangers in handling liquid manure center on the manure storage system. Stored liquid manure produces toxic gases. They are most likely to become a problem when the manure is agitated.

Manure Gases

Gases are often given off when animal manure is stored. The most common gases given off from liquid manure are ammonia, carbon dioxide, methane, and hydrogen sulfide.

A person can suffocate in a manure building or pit because of gases given off by manure. The gases displace air until there is no oxygen to breathe.

Hydrogen sulfide is released rapidly and is most dangerous when liquid manure is first agitated. It has a foul odor, similar to rotten eggs. It causes headaches, dizziness, and nausea in concentrations as low as 0.5%. Exposure to a 1% concentration can result in unconsciousness or death. Although you can smell very low levels of hydrogen sulfide, continued exposure dulls your sense of smell and you may not know that you're in danger. Several deaths have been attributed to this gas.

Carbon dioxide is an odorless gas that is a normal part of the air we breathe. However, it exists in the air at a very low concentration (about 0.03%). When it's present in higher concentrations, it displaces the air so that less oxygen is available. Concentrations of 3% to 6% can cause heavy, labored breathing, drowsiness, and headaches. A 30% concentration can cause death by suffocation.

Methane gas is nontoxic. Concentrations as high as 50% cause only headaches. However, methane is highly flammable. It ignites readily, and methane-air mixtures can explode.

Here are just a few of the safety precautions:

1. Know the effects of each of the gases described above. Anytime you detect one or more of the symptoms, get to fresh air immediately. A delay could be fatal.

2. Do not rely totally on smell to detect the presence of hydrogen sulfide gas. It may be present in hazardous concentrations even though you can't smell it.

3. Provide maximum ventilation to keep gases away from people and animals whenever a tank is pumped or agitated. If a power failure has occurred and lasted for several hours, open all windows and doors and get all people and livestock out of the building before they suffocate (Fig. 13). It is best to have an emergency generator.

JDPX7959

Fig. 13 — Turn On the Ventilation System for Maximum Ventilation When Agitating Liquid Manure Tanks Under Buildings

4. Don't allow any smoking or other fire source in or around the liquid manure tank. Methane-air mixtures are explosive. Keep unauthorized people away.

5. Only a trained person, equipped with proper breathing apparatus, should enter a liquid manure transport tank or slurry storage silo. Have the local fire or sheriff's department come to the site. You can't "hold your breath" and rescue someone. Keep rescue equipment (rope, harness, respirator) near manure storage (Fig. 14).

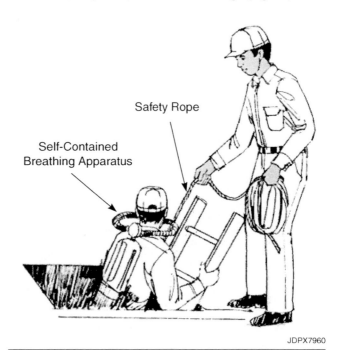

JDPX7960

Fig. 14 — Only a Trained Person With Breathing Apparatus Should Enter Liquid Manure Storage

Manure Drowning

People can also drown in liquid manure. Thick crusts on top of liquid manure in storage can appear solid, but a person will break through into the liquid. Sometimes a person just slips and falls into the liquid. That is one more reason to keep rescue equipment near liquid manure storage, and to have others at the site if you must enter the storage as discussed in preceding paragraphs.

Work around liquid manure storage can be hazardous. For more information on safety for its storage and removal, contact the state safety extension specialist at your land grant university, or your county Cooperative Extension office. Contact OSHA (Occupational Safety and Health Administration) for regulations affecting employees entering and working in confined air spaces.

Grain Bins

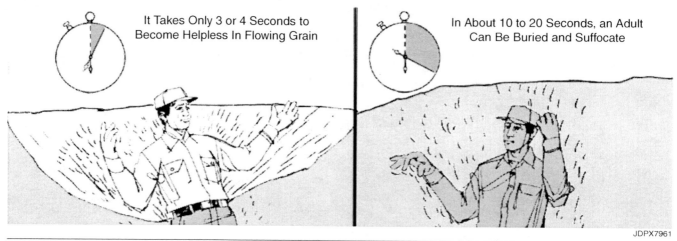

It Takes Only 3 or 4 Seconds to Become Helpless In Flowing Grain

In About 10 to 20 Seconds, an Adult Can Be Buried and Suffocate

JDPX7961

Fig. 15 — An Adult Could Be Completely Covered With Grain in 10 to 20 Seconds

The large grain storage bin has become a common sight on the American farm. Though grain bins appear to involve little risk. there are a number of potential hazards you should recognize.

Suffocation in Grain

One of these dangers is accidental suffocation in grain bins. This often happens during the unloading of the bin when the grain may bridge and you have to break the bridge to get the grain flowing again.

The unsuspecting person who enters a bin with the unloader running may sink in the flowing grain before realizing what has happened. It only takes 3 or 4 seconds to become helpless, a few seconds more to be submerged in the grain, and suffocation soon follows (Fig. 15).

The average adult takes up only about 7 cubic feet (0.2 cubic meters) of space. It only lakes about 10 seconds with an average auger to till this space. That/s how long it takes to be covered by flowing grain.

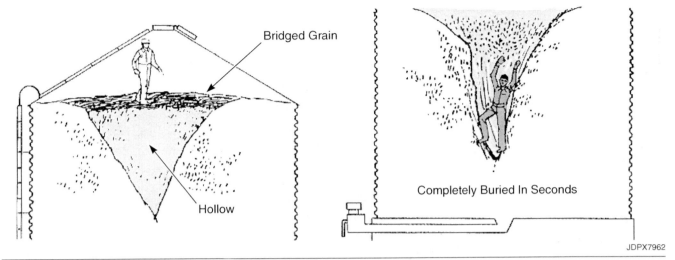

Fig. 16 — A Partially Unloaded Bin May Be a Trap, Because Bridged Grain Could Hide a Dangerous Cavity

Here are some safety measures to follow to prevent these accidents:

1. If grain bridges, shut off the unloader and use a pipe or some other long object to break the bridge and get the grain flowing again. Never enter a grain bin while the unloading operation is going on (Fig. 16). Bridged grain may look perfectly safe from the top, but it could hide a cavity that you could fall through and be submerged almost immediately. It only takes a few inches of grain covering you to cause suffocation.

2. If you must enter a bin, disconnect the power source and make sure no one can turn it on while you're inside. If the grain should start flowing for some reason, stay near the outer wall and keep walking until the grain flow stops or the bin is empty.

3. Install ladders and safely ropes in all bins. This will provide an exit if you need to get out, as well as a safe way of getting in. But remember that even if there is a ladder in the bin, you must be able to get to it. If you walk out to the center of the bin and get caught in flowing grain, you may not be able to reach a ladder on the side. Hang a safety rope from the center (Fig. 17). But, also take the precautions in paragraph No. 4.

4. If you must enter a bin, tie yourself with a rope and harness so you have a way of getting out. Have two extra people handy in case something happens — one to hold the rope and one to get extra help if necessary.

5. Another problem is being overcome by carbon dioxide. Carbon dioxide is given off by wet grain as it ferments. The CO_2 pushes oxygen out of the bin. Forced ventilation and breathing equipment can help.

6. If someone is buried without a rope, shut off the auger, turn on the blower to add fresh air, and call for emergency medical rescue. Open emergency grain doors, or cut or knock holes in the bin around its base. Assume the trapped person is alive and will need CPR.

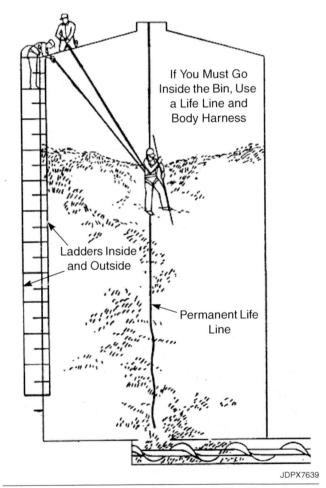

Fig. 17 — Wear a Harness and Rope and Have Two People to Help You When Entering a Bin

Dust and Fumes

Spoiled grain or grain in poor condition often gives off a lot of dust that can be dangerous to workers, especially those with certain allergies or asthma. Cleaning out a grain bin can produce enough dust to cause such a reaction. Use a filter respirator that will remove fine dust particles. Keep it handy, and wear it when working in a grain bin, even if it seems uncomfortable.

Safety Information and Regulations

For more information on safety and rescue involving grain bins, silos, and tanks, contact your state or county Cooperative Extension Service. For regulations that apply to employees, contact Occupational Safety and Health Administration (OSHA).

Portable Elevators and Augers

Elevators and augers are specialized devices used to lift and transport materials like grain, hay, and silage. The simplicity of elevators and augers often leads operators into thinking that they are safe. Unfortunately, these machines have become a major source of farm accidents and fatalities.

Operating Elevators and Augers

The primary danger in operating elevators and augers is getting caught in moving parts, such as auger flighting, belts, chains, or PTO shafts. Allow workers to operate this equipment only after they are made aware of the potential dangers and safety practices. A few minutes spent in examining the equipment and operator's manuals to learn the potential hazards is time well spent.

Keep children away from the elevator, whether it's stored or in use. Elevators are not intended as slides or seesaws, and children should never be allowed to climb on them.

Don't wear loose, floppy clothing when working around elevators or augers. Such clothing is easily caught and can pull the wearer into the machine before there is time to stop the machine or pull free.

Operate the machine only with guards or covers in place. This minimizes the chance of getting caught in moving parts or allowing objects to fall into the mechanisms. If your machine does not have guards over potentially dangerous moving parts, order them from the manufacturer. The investment could save your life.

Run your elevator or auger no faster than necessary to move the material efficiently from the inlet. For example, on a bale elevator, the bales should be able to be fed into the elevator end-to-end, that is, one bale touching the next. Using the elevator at excessive speeds with only one or two bales on it at a time puts unnecessary wear on the elevator and increases the chances of workers getting caught in the mechanism. The slower speed may also allow the elevator to be stopped before causing damage if a bale gets caught. Keeping the bales spaced evenly also helps to balance the load on the undercarriage and keep the elevator from upending.

Transporting and Positioning Elevators and Augers

Use extreme care when transporting portable elevators and augers. Transport them only in a lowered position with the safety locking device in place. Elevators become top-heavy when they are raised (the same as trying to carry a ladder in an upright position), so the proper way to maneuver them is by keeping them attached to the tractor. Keep in mind that the wheels do not follow in the tracks of the tractor on turns. In addition to possibly hitting a post or building, there is also the danger of the tractor tires hitting the elevator. So avoid sharp turns when towing an elevator or auger. Allow plenty of turning space (Fig. 18).

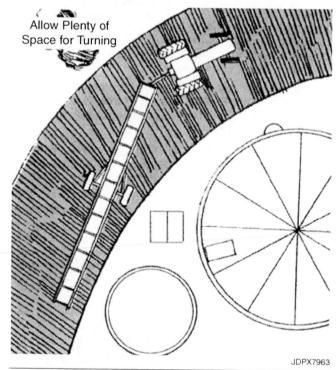

Allow Plenty of Space for Turning

JDPX7963

Fig. 18 — Allow Plenty of Room for Turning

Before moving or raising an elevator or auger, clear others from the area. Also, be sure it is empty to help avoid a top-heavy upset or excess strain on the lift cables.

Don't exceed safe travel speeds on public roads. Use a flag to mark the end of the machine, plus SMV emblem and reflectors. Also, use safety chains and pin the hitch securely. (See Chapter 7, "Tractors and Implements".) Follow local traffic regulations.

Check for power lines when moving or positioning a raised elevator or auger (Fig. 19). Keep the elevator clear of power lines. When rubber tires are the only point contacting the ground, you could become the perfect conductor to complete the circuit if the elevator contacts a power line.

Fig. 19 — Keep Elevator or Auger a Safe Distance From Power Lines — Don't Get Electrocuted

Check the condition of cables regularly. When cables fray or are partially cut, they may suddenly give way and allow the elevator to fall and cause injury to the operator or others. Determine and correct the cause of the faulty cable and replace it. Make sure the cable clamps are tight, and always keep at least three turns of cable on the windlass. This will help reduce tension on the cable clamps and reduce the chance of the cable pulling loose.

When raising or lowering an elevator, make sure the cable wraps properly and does not wind off the windlass. Check that the support arms are positioned properly. If one wheel is lower than the other when the elevator is lowered into position, the frame may tilt to one side so that when you raise it, the arms may not align properly. If proper care is not taken, the elevator may fall off the side of the frame and damage the elevator or injure the operator.

Most elevators and augers are equipped with a safety device to keep the elevator from being raised too high. Raising the elevator too high increases chances of upsetting. Do not alter this safety feature. If your elevator does not have this feature, install a safety stop and warn other workers of this potential hazard.

In addition to contact with power lines, collapse of the undercarriage is one of the most common causes of fatal accidents involving portable elevators.

Construction of the undercarriage of a typical elevator is shown in Fig. 20. The lower support is attached to the elevator and is free to rotate about its attaching point. The upper end of the upper support is moved along the bottom of the elevator to raise or lower the elevator to the desired height.

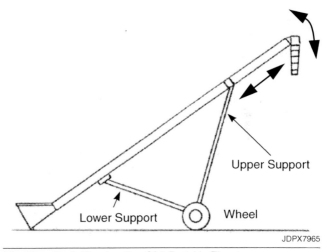

Fig. 20 — Basic Construction Of A Portable Elevator

Two types of collapses can occur if the elevator handler is not careful:

- Upend accident

- Cable or winch failure

The upend collapse-type accident usually occurs when the elevator is being moved by hand. For example, when the elevator is raised too high, as shown in Fig. 21A, the upper end becomes top-heavy. If the wheels hit a hole or obstruction, the handler may lose his or her grip on the elevator and the upper end will fall to the ground as shown in Fig. 21B. The supporting undercarriage collapses, as illustrated in Fig. 21B and C, when the wheels roll to the rear. Elevator safety stops, as shown in Fig. 21D, can help prevent such a collapse.

The cable or winch failure collapse-type accident occurs when the cable or winch fails to hold the elevator up, as shown in Fig. 21E. This can also happen if the elevator upends. The upper support slides out toward the end and allows the elevator to fall, as shown in Fig. 21F and G. Elevator safety stops, as shown in Fig. 21H can help prevent such a collapse.

Two Kinds of Elevator Collapse-Type Accidents:

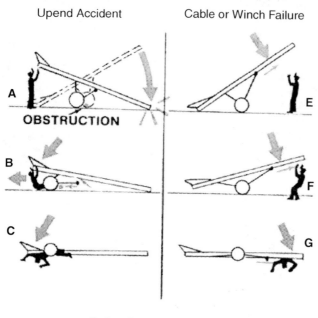

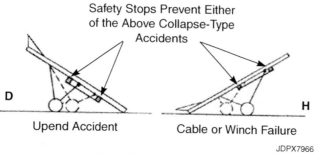

Fig. 21 — How Elevator Collapses Can Occur

To avoid such accidents, make sure the auger or elevator is empty before lifting or moving. Don't move it manually. Use a vehicle. When unhitching, lift slowly and keep the intake end low.

If an elevator begins to upend, do not hang on and try to stop it — get out of the way. Otherwise you may be crushed under a collapsed elevator. Safety tracks that don't allow the trough to be separated from the undercarriage can also prevent collapsing. Keeping the elevator attached to the tractor while maneuvering it will also prevent such an accident.

Remove any objects blocking the wheels when attempting to raise or lower an elevator. When the elevator is being raised or lowered, the wheels must be free to move if the elevator is attached to a tractor or truck. When the elevator is in position for operation, support both ends and block the wheels to prevent slippage or teetering that could cause the elevator to shift position or to nose over when the material reaches the top (Fig. 22).

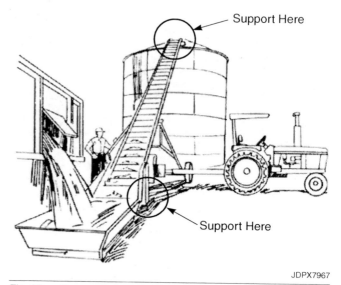

Fig. 22 — Support Both Ends of the Elevator When It Is in Use to Prevent Tipping or Loss of Control

Another danger involved in raising or lowering some elevators is the accidental release of the cable crank. Be very careful not to allow the crank to be released and spin freely. If this happens, arms and fingers can be broken or serious head injuries can result. It is nearly impossible to get out of the way in time to avoid being hit. Never try to stop a spinning crank (Fig. 23). Many elevators are equipped with safety clutches to keep this from happening. Make sure this clutch is maintained properly. Replace any worn or broken parts and check adjustments periodically as recommended in the operator's manual.

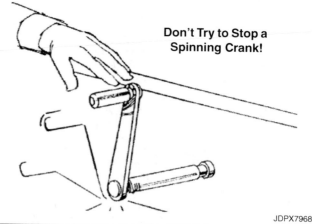

Don't Try to Stop a Spinning Crank!

Fig. 23 — Don't Try to Stop a Spinning Crank — You Could Be Seriously Injured

Do not ride or climb on the trough of a bale or grain elevator. The lack of handholds, coupled with metal surfaces that become slippery with wear, provides a hazard that could lead to a slip or fall. And your weight at the top of the elevator could cause it to tip forward. Add these to the danger of getting fingers, feet, or clothing caught in moving parts, and you have a set of circumstances just right for an accident.

Disengage the power source before removing chaff or caught bales from the ends of the elevator. Position the elevator so that the chaff or bales fall freely away from the end. This should eliminate the danger of getting entangled in the chains and sprockets at the end of the elevator.

Grain Intake Guards

Augers, in general, are reputed to be among the most potentially hazardous types of farm equipment. Injuries often involve the loss of a hand or foot. Extreme caution must be taken to prevent feet or hands from getting into an auger. The best way to do this is to use shields or guards.

Many portable grain augers have guards or shields to prevent feet and hands from accidentally getting into rotating auger flighting. (See "Shear Points" in Chapter 3, "Recognizing Common Machine Hazards".) One type of grain auger guard is a metal grating over the grain intake area (Fig. 24). It allows grain to flow, but it protects against accidental contact with the auger. There are also baffle-type guards.

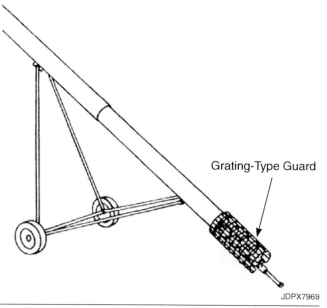

Grating-Type Guard

JDPX7969

Fig. 24 — Grain Intake Guards Let Grain Flow but Help Keep People Out. Keep Them in Place

If your machine has safety guards over the rotating auger, keep them in place and in good condition. If such guards are not in place, contact your equipment dealer or manufacturer. If guards are available, get them installed.

Even with intake or flighting guards in place, you should exercise caution to avoid injury.

Take precautions to prevent falling or slipping into the auger. Keep tools picked up and away from the auger, and keep the area clear of ice, mud, grease, or grain that could cause a fall, Never use your hand or foot to push material into a plugged auger. Always use a stick or rod. If a machine gets plugged, shut off the power before attempting to remove the obstruction.

Farm Wagons

There are several kinds of farm wagons — from flatbed wagons used for hand stacking and hauling bales to PTO-powered, self-unloading forage wagons.

Use a safety locking hitch pin. Many wagon accidents happen when a pin jumps out and releases a wagon. The wagon may turn over or climb a tractor tire and kill the operator. Also use safety chains, especially on public roads.

When loading wagons, keep the load distributed evenly, from side to side and front to rear. An evenly distributed load is easier to control and reduces strain on the frame.

Know the recommended load limits listed in your operator's manual and don't exceed them. These limits allow you to maintain control of the equipment. Excessive loads cause you to lose full control of the wagon.

Keep within the load limits posted on bridges. A large gravity box filled with shelled corn may weigh more than 19,000 pounds (8,626 kg) and could exceed the load limits for a bridge or the equipment. Keep the wheel bolts tight and use recommended tires and tire pressures.

Slow down when towing heavy loads to maintain steering and breaking control and to avoid fishtailing and upset. Be especially careful on slopes even when the wagon is equipped with brakes. Reduce speed on rough ground and at corners, and keep the tractor in gear when going downhill. A general rule is to use a gear no higher than the one you would use to pull the same load up the hill.

For more information on hitching and towing implements, and also preventing accidents on roads, see Chapter 7, "Tractors and Implements".

Feed and Forage Wagons

Shafts, belts, pulleys, beaters, chains, and the other working mechanisms on feed and forage wagons can be hazardous if you fail to follow safe operating practices. Keep people away from wagons while they're in operation. Even. when wagons are properly shielded, they are not foolproof. Some parts must be exposed to do work. Keep hands and feet away from all moving parts.

Emergency shutoff devices on wagons can't prevent contact with all moving parts. But some shutoffs can help you avoid more serious injury if you should get caught. Stay at the controls when the wagon is operating (Fig. 25). Never enter a wagon with the tractor engine or the wagon operating.

Fig. 25 — Stay at the Controls When the Wagon Is Operating. Stay Away From the Unloading Mechanisms

Before attempting to unclog any part of the machine, disengage all power, shut off the engine, and take the key. Don't just turn off the part you are trying to unclog. You or another person could accidentally turn it back on or get caught in another mechanism while trying to clean out the clog. Make sure the engine is off whenever you lubricate, adjust, or repair a machine.

You will reduce hazards if you maintain your wagon properly. Some things that are potentially hazardous are safety-trip mechanisms that don't work properly and guards that aren't in place. Keep safety mechanisms in perfect operating condition. Always keep guards and shields in place.

Grain Wagons and Carts

Grain wagons and carts require essentially the same precautions as other wagons: safety in hitching and unhitching, loading, towing, and traveling on public roads. However, there are some unique potential hazards in unloading grain wagons and carts, and also trucks.

Gravity-Unloading Wagons

When the unloading door in the bottom of a gravity-unloading wagon is opened, the grain pours out rapidly. A few hundred bushels (liters) of grain can be dumped in a few minutes. That can create the hazard discussed earlier for grain bins: grain suffocation or drowning.

When grain flows from a large gravity wagon, it can pull a person down like quicksand (Fig. 26). Adults may become helpless in a few seconds and completely covered in 10 or 20 seconds. Then they suffocate. Children will be overcome sooner. Most victims of grain wagon drowning are 16 years of age and younger.

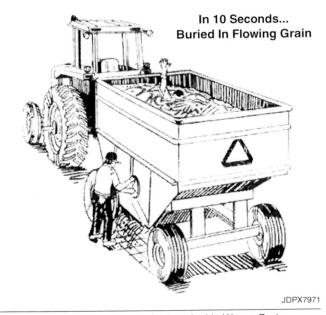

In 10 Seconds...
Buried In Flowing Grain

Fig. 26 — Never Get on Top of Grain or Inside Wagon During Unloading. You Could Drown

You can avoid grain wagon drowning accidents:

1. Allow no riders. Develop that habit. If people don't ride on a load of grain, they're less likely to be there during unloading.

2. Never allow anyone on top of the wagon, or in it, during unloading when equipment is running. Before you start unloading, check to be sure no one is in a dangerous area.

3. Persons who are not required for the unloading operation should not be allowed in the area, especially children. Don't allow children to work alone.

If someone is trapped in a grain wagon, cart, or truck, quickly remove grain from around the victim to allow breathing. Send for help such as an emergency rescue squad. Keep removing grain and block it away so the victim can breathe.

Auger-Unloading Grain Wagons and Carts

The primary precautions are to avoid grain drowning as described above, and to avoid entanglement in the unloading augers:

1. Don't allow anyone inside the cart or wagon while unloading or when the engine is running.

2. Disengage the power, shut off the engine, and remove the key before entering the tank.

Hoisted Grain Wagons and Trucks

When a grain bed or box is raised like a dump truck, the grain pours out. No one should be in or near a hoisted or dumping wagon or truck box as it is being raised or lowered. He or she could become a grain drowning victim or could be crushed. Keep everyone clear of the raised box to avoid being crushed if it falls or is lowered. Safely block a raised box if someone must get under it for any reason.

For more information on grain drowning rescue contact the state extension safety specialist at your land grant university or your county Cooperative Extension office. Get it before you need it.

Blowers

The most important safety precautions for working around forage blowers are:

- Stay clear of all moving parts.

- Keep all shields in place.

- Never climb into the hopper or use hands or feet to force forage into a blower.

Failure to follow precautions could result in your entanglement in moving parts.

Before lubricating, adjusting, or unplugging the blower, always:

- Disengage power.

- Shut off the tractor engine and take the key.

- Wait for the blower fan to stop.

If the blower is PTO-operated, fasten the blower hitch securely to the tractor drawbar. Vibration could cause the blower to move, causing the PTO shaft to separate and rotate dangerously; or it could bring the blower pipe down on you.

Blower Pipe

Another important aspect of blower safety is the assembly of the blower pipe, and safe procedures to solve plugging problems. Blower pipe should be assembled on the ground and then raised with a rope and pulley. Attach the rope to the mid-section of the pipe and make a half hitch near the top of the pipe.

Carefully raise the pipe into position, tie off the rope, and secure the pipe at top and bottom. Raise the deflector from inside the silo to eliminate the need for transferring the unit from the outside to the inside at the top. This reduces the chances of a fall. If using a telescoping pipe, always attach the smallest diameter section to the blower to reduce the chances of plugging.

Silo Unloaders

If you have a top-unloading silo unloader, you have to raise the unloader to the top of the silo before filling the silo. This should be a two-person operation. One person on the ground operates the winch that lifts the unloader, and the other watches the rising unloader and guides the person at the winch, or signals if something goes wrong (Fig. 27).

The person observing the unloader should be positioned at the top of the silo and watch the cable pulleys and electrical cables to make sure they don't tangle or get caught. Do not enter the silo while the unloader is being raised or suspended from the top, and never crawl out onto a suspended loader for any reason. It is generally unstable, and you could easily fall.

The person raising the unloader should not leave the controls for any reason while the machine is in operation.

Do not remove the safety catch or brake device from the winch. Without it, a slip could cause a spinning lift crank that could break an arm if an attempt is made to stop it as the unloader goes out of control and falls to the bottom of the silo.

If the silo is not covered, try to raise the unloader on dry and relatively calm days. This helps prevent accidental slips and falls that are more likely to happen on wet or windy days.

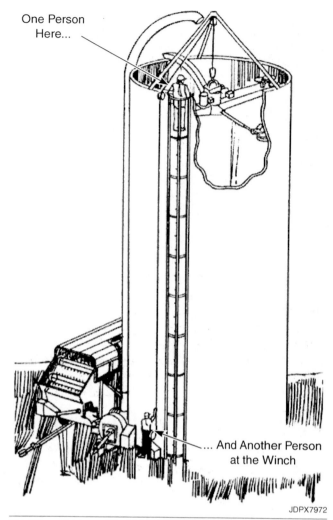

One Person Here...

... And Another Person at the Winch

JDPX7972

Fig. 27 — Raising a Silo Unloader Is a Two-Person Job

Have electrical installations performed by a qualified electrician; power company officials should be consulted to guarantee an adequate power supply.

Safety controls should be kept in repair and properly used. They include:

- Overload protection — manual reset type

- Emergency shutoff switch located on equipment or in feed room

- Coordinated switches that will stop the rest of the units automatically when one part of the system stops

- Portable, remote electrical control near the unloader to test-operate the machine with main control "locked out" (on newer machines)

To prevent accidental starting by someone else:

1. Before working in the silo, remove fuses and put them in your pocket, or lock out the circuit breaker.

2. Lock the main switch and attach a lockout tag or message that the machine is being serviced.

3. Make sure all circuit protection devices are of the manual reset type.

Perform seasonal check of the wiring for cracks, fraying, or bare wires, and replace defective wires.

The unloader may occasionally become plugged or buried, and you must go into the silo to unplug the machine. Before entering the silo, shut off the power supply. Never try to unplug an unloader while it's running — you could be injured. Some farmers find it convenient to have a set of walkie-talkie radios, one of which should be left with someone nearby for use in case assistance is needed (Fig. 28). However, because of the noise while the unloader is running, you may not be able to hear the other person.

Have Helper Nearby or Use Walkie-Talkies in Case Assistance Is Needed

JDPX7973

Fig. 28 — Be Able to Get Assistance if Necessary

If you must start the unloader while in the silo, take precautions to avoid getting caught in the silo or chute. This can happen if you start the unloader while in the silo and the feeder below fails to remove the silage from the bottom of the chute. Have a second person make sure that everything on the outside is working properly.

Some unloader setups have a portable, remote electrical control at the unloader location so you can momentarily apply power to test adjustments or repairs. Before using the control, be sure no one is near the gathering mechanism. The test control device must be in "locked out" position to prevent activation of the main system control.

Silo Gases

Grain silage can generate dangerous gases after silo filling time and for up to 2 weeks afterward. Nitrogen dioxide (NO_2) is the most dangerous of the gases. When inhaled, it can cause severe chemical pneumonia, or "silo filler's disease," and even death. Nitrogen dioxide has a pungent, sweetish odor. At high concentrations, a reddish-brown color may be visible. Follow these practices:

1. Don't enter a silo during the filling process until the blower has run for 30 minutes.

2. Don't allow anyone to enter the silo for any reason for 7 to 10 days after filling is completed, Even silos without roofs and feed rooms or sheds at the bottom of the silo chute can accumulate dangerous gases since the gases are heavier than air. Respirators won't help. Only a self-contained breathing apparatus is adequate.

3. When the silo is opened, run the blower again for a least 30 minutes before entering.

4. Don't enter a silo unless someone is on the outside to monitor your activity.

If you notice any signs of nausea or headache, get to fresh air, then see a doctor. Even if the exposure doesn't seem serious, problems can develop some time after exposure and cause lung congestion.

Grinder-Mixers:

Some livestock farmers use grinder-mixers to prepare their own feeds. These machines present many of the same potential hazards as other PTO-driven equipment. However, the grinding mechanism calls for special caution.

Beyond using standard safe PTO-operating procedures, there are a few other practices that can help reduce potential hazards of grinder-mixers.

Hammermills are provided for grinding various types of grain and hay into small particles of a desired size. These hammers revolve at extremely high speeds. Keep the hammermill cover in place whenever the hammermill is moving to avoid getting caught in the mechanism or being struck by thrown material (Fig. 29). Generally, the hammermill keeps on running for a short time after the power supply is shut off, so make sure it has stopped completely before opening the cover.

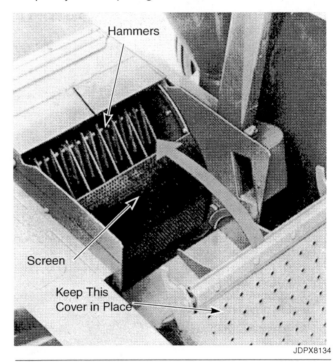

JDPX8134

Fig. 29 — Never Open Cover Until Hammermill Stops Completely. Keep It Securely Fastened

Observe additional precautions when feeding material into the grinder. Keep all shields in place, including the shield that keeps material from being thrown back out of the grinder. Failure to do this could result in injuries to the face or eyes. If no guard is provided over the auger in the concentrate-loading hopper, use extreme care to keep your hands away from the auger. Also, keep feed bags from getting caught and plugging the unit. Do not remove these auger guards unless repairs are necessary, and then replace the guards before operating the unit.

If a bag, flake of hay, or any other material gets caught in the machine, don't try to pull it out while the grinder is running. The machine can pull in your arm faster than you can let go. Disengage the power and shut off the engine before removing the obstruction.

To grind rectangular bales, use a bale feed attachment that feeds the hay into the grinder slowly to reduce the danger of pulling a hand or arm into the grinder (Fig. 30). Don't try to grind thick slices of hay that could plug the grinder. And never push plugged material into the grinder with your hand. The plugged material could give way suddenly, allowing you to slip into the grinding mechanism. Use a push stick to push the material to the grinder.

JDPX7974

Fig. 30 — Use a Bale Feeding Attachment to Feed Hay Into a Grinder Safely

Bale Processors (Tub Grinders)

Operating a bale processor requires the usual safety precautions for towed, PTO-operated machines. However, there are some unique concerns. The following paragraphs provide a few of the precautions.

A tilting tub on some machines picks up large round bales of hay or straw and raises them to a vertical position. Steel rotors in the bottom of the tub shred the bale as the tub and bale turn. Material is discharged through a chute for feeding or bedding livestock.

Before tilting the tub, be sure the hitch is secured to the tractor drawbar and that the area behind the machine is clear. When the tub is tilted back, it can force the hitch upward, and the tub could fall all the way backward if the hitch is not secured.

Before you work around the machine with the tub raised, engage the tub-tilt safety lock device so the tub cannot fall and crush someone. Disengage the power and shut off the engine (Fig. 31).

JDPX7975

Fig. 31 — Before Working on Machine With Tub Tilted, Engage the Tub-Tilt Safety Lock

To avoid entanglement in the drive that turns the bale tub, or in any other moving part, stay clear of the machine when it is operating.

Rocks or other objects from a bale can be thrown out the discharge chute with enough force to injure someone. Keep people away from the discharge area, especially when the chute is operating in the raised position.

Test Yourself

Questions

1. The position of a tractor-loader's center of gravity:

   a. Is always in the same place

   b. Changes depending on the tractor's position

   c. Changes as the position of the loader changes

   d. Has nothing to do with safe operation

2. Significant dangers when using a loader are:

   a. Overloading

   b. Traveling with the leader too high

   c. Handling round bales without proper attachments

   d. All of the above

3. (T/F) When using a skid-steer loader, it is not necessary to wear the seat belt, because you are protected by the operator's cab.

4. (T/F) If the chains in a manure spreader are frozen down, the best procedure for loosening them is to engage the PTO slightly a number of times to break them loose.

5. The four most common gases produced by stored liquid animal manure are:

   a. Ammonia, carbon dioxide, methane, and hydrogen sulfide

   b. Ammonia, carbon dioxide, carbon monoxide, and methane

   c. Ammonia, carbon monoxide, methane, and hydrogen sulfide

   d. Carbon dioxide, carbon monoxide, methane, and hydrogen sulfide

6. Which precaution can you ignore if you must enter a grain bin?

   a. Stay near the edge of the bin

   b. Secure yourself with a rope and harness

   c. Lay a board on the grain to walk on

   d. Have someone watch you from outside the bin

7. Dangers involving portable elevators and augers include:

   a. Getting caught in moving parts

   b. Elevator collapse

   c. Contacting overhead power lines

   d. All of the above

8. If the intake portion of a portable auger elevator is equipped with a grating-type guard, what common hazard will it protect against?

   a. Shear point

   b. Wrap point

   c. Pinch point

   d. All of the above

9. (T/F) Most grain wagon suffocations involve youths age 16 and younger.

10. Safety devices that should be used with a silo unloader

   a. Emergency shutoff switch

   b. Manual reset overload protection

   c. Both of the above

   d. None of the above

11. (T/F) Silo gases are easily detected when entering silos if they are in sufficient quantity to be dangerous.

References

1. Animal Handling Safety Considerations. 1992. (Includes manure pit gases and silo gases.) Cooperative Extension Service, University of Missouri, Columbia, MO 65211.

2. Don't Drown in a Grain Wagon. Pm-1334a. 1989. Cooperative Extension Service, Iowa State University. Ames, IA 50011.

3. Safe Storage And Handling Of Grain. 1983. Cooperative Extension Service, University Of Missouri, Columbia, MO 65211.

4. Manure Storage Safety: ASABE EP470. 1992. American Society Of Agricultural and Biological Engineers. 2950 Niles Rd., St. Joseph, MI 49085-9659.

5. Working In Confined Spaces. NIOSH Publication No. 80-106. National Institute For Occupational Safety and Health, Mail Stop C-13, 4676 Columbia Pkwy, Cincinnati, OH 45226.

Farm Maintenance
Equipment

13

JDPX8058

Introduction

A number of labor-saving devices are available for general utility and farm maintenance operations. Examples include:

- Rotary mowers (cutters)
- Posthole diggers
- Post drivers
- Chain saws
- Lawn and garden equipment
- Utility haulers

Many of these items are used only occasionally, with many months passing between uses. It's important to consider the safety precautions necessary for this equipment, because operators tend to forget procedures that aren't followed repeatedly.

The most important sources of safety precautions are the operator's manual and safety signs for each machine. Read them and follow them. Also, review Chapter 7, "Tractors and Implements" and Chapter 3, "Recognizing Common Machine Hazards".

Rotary Mowers (Cutters)

Rotary mowers or cutters are used for cutting and shredding stalks, clipping pastures, clearing underbrush, and mowing roadbanks, lawns, and waterways (Fig. 1). Some can cut brush up to 3 inches (76 mm) in diameter. Know the job you're going to do and use the right type of rotary mower for it. This helps you do the job safely and prolongs mower life.

JDPX8037

Fig. 1 — When Operating Rotary Cutters, Keep Others Away. Flying Objects Can Be Dangerous

Safety Practices for Rotary Mowers

There are many common guidelines for mowing, whether you use a lawn and garden tractor or a row-crop tractor. When handled improperly or carelessly, either type can become involved in a serious accident.

Use a tractor equipped with ROPS and seat belt, and fasten the seat belt. In addition to protection in an overturn, the seat belt can keep you from being bounced off the tractor and run over by the cutter.

When cutting tall grass, weeds, or brush, watch for objects like tree stumps, tin cans, rocks, and other obstacles that could be hit or thrown by the mower blades.

Keep people out of the area in which you're working. If anyone comes near, shut off the mower (cutter) and the tractor. Debris can be thrown hundreds of feet (meters), and can cause serious injury. (Review "Thrown Object Hazards" in Chapter 3, "Recognizing Common Machine Hazards".)

Don't allow riders (Fig. 2). Riders have been pitched off and crushed by tractors and mangled by rotary cutters. A rider can also distract the operator, causing an accident. Avoid those accidents. Allow no riders.

NO! WRONG! NO RIDERS!

JDP7976X

Fig. 2 — Don't Allow Riders. They Can Be Pitched Off Tractor and Run Over by Cutter

Make sure the PTO is not engaged before starting the tractor. Engage the clutch and increase engine speed slowly. Operate the cutter at recommended speeds. Never overspeed a 540 rpm machine with a converter coupling on a 1000 rpm PTO tractor. It could damage the mower, and failed parts could fly out, injuring people.

Anytime you hit an obstruction or hear a strange noise, disengage the PTO, shut off the tractor engine, and take the key before investigating the cause. Be sure to set the brakes before dismounting. Many rotary mowers have blades that continue to rotate for some time after the PTO is disengaged. To avoid injury, be sure the blades have stopped before approaching the mower.

Be very careful when turning sharp corners with drawn machines (Fig. 3). The rear tractor wheels could catch the mower frame and throw the mower toward you. Use front-wheel weights when operating three-point-hitch mounted mowers on rough terrain and during transport. If weights aren't used, the front wheels may bounce and cause you to lose steering control.

JDPX7977

Fig. 3 — Do Not Turn Sharply With Pull Mowers. The Rear Tractor Wheels Catch the Mower Frame and Throw the Mower

You can increase your safety and mowing efficiency by making sure that the mower is level and that the rear tires of the tractor are set wider than small mower frames. Wide-set tires will not press down the material about to be cut as much and will provide greater tractor stability (Fig. 4).

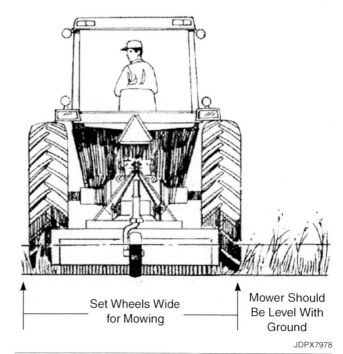

Set Wheels Wide for Mowing — Mower Should Be Level With Ground

JDPX7978

Fig. 4 — Increase Tractor Stability by Setting the Tires to Maximum Width

Don't try to cut brush with a rotary mower designed only for forage, because you may be exposed to hazards caused by machine failure. When you buy a mower, be sure to find out what materials it can safely cut. You'll need heavy-duty blades for any type of saplings. Small models with heavy blades may not be able to do the job, because the wood may clog the mower as it goes through the chute. Your job may require a mower especially designed for clearing brush.

Before leaving the cutter unattended, lower it to the ground, disengage the PTO, engage the parking brake or put the tractor in the park position. Then shut off the engine and remove the key. Don't risk injury to someone else.

Safety Maintenance

Before operating your mower, become familiar with its operation, adjustments, and maintenance procedures. The best way to learn is to study the operator's manual carefully. Be sure you know all the recommended safety precautions.

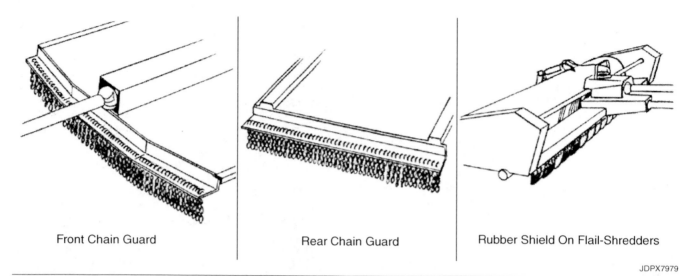

| Front Chain Guard | Rear Chain Guard | Rubber Shield On Flail-Shredders |

JDPX7979

Fig. 5 — Keep Shields And Safety Chain Guards In Place

Begin your preoperational check by making sure the PTO is disengaged and the engine is shut off. Look for loose nuts and bolts. Inspect the blades of mowers for nicks and dullness. You should be able to feel a sharp edge along the cutting portion of the blade. If the blades feel smooth and rounded, take time to sharpen them. Blade sharpness is a key to efficient mowing. Stalk cutting dulls blades quickly because of the abrasive action of soil. When the blades become too dull for additional sharpening, replace them because they are ineffective. They can also be dangerous since mowing will be more difficult, and hazardous conditions may develop as a result of such problems.

If you must work underneath the machine to check, sharpen, or replace the blades or blade holders, or for other service, support it with sturdy stands or blocks. Don't rely on the hydraulic system to hold it up. Always make sure the tractor can't be started by anyone when that service is being performed. Remove the key.

Rotary mowers are often equipped with runners and safety chain guards (Fig. 5). To avoid excessive wear when cutting to short heights, keep the mower just high enough off the ground so it doesn't ride on the runner shoes. The chain guards reduce the possibility of objects being thrown from under the mower and seriously injuring or killing someone in the area. Make sure chain guards are maintained and kept in place. If they must be removed or raised for certain crop conditions, make a special effort to keep people out of the area when mowing. Be sure to replace or readjust the chain guards as soon as you finish the crop.

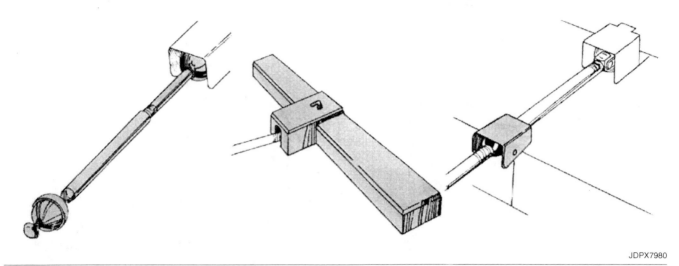

JDPX7980

Fig. 6 — Keep Powershaft Shields in Good Condition

Power transmission shafts must be protected by shields or guards (Fig. 6). Keep shields and guards in place on the machine — don't leave them in the shop. Always replace shields or guards after maintenance or repair jobs are completed.

Transport Safety

Before transporting mower, disengage the PTO so blades won't rotate in raised position. Transport the machine fully raised and locked in position to prevent accidental lowering.

Posthole Diggers

Posthole diggers are tractor-powered augers that drill holes 6 to 12 inches (15 to 30 cm) or more in diameter. Special auger sizes are also available from most manufacturers. Let's look at the best ways to dig straight, clean, and uniform holes and to do it safely. Here are some basic practices for using posthole diggers:

1. Shift the transmission into the park or neutral position.

2. Set tractor brakes before digging. If the tractor rolls, the auger shaft will be bent.

3. Run the digger as slowly as possible to keep it under control.

4. Dig the hole in small steps (Fig. 7). Dig down several inches, then bring the auger up to let the soil clear. Repeat this procedure until the desired depth is reached. This allows better control of the auger and can prevent difficulties that could lead to an accident.

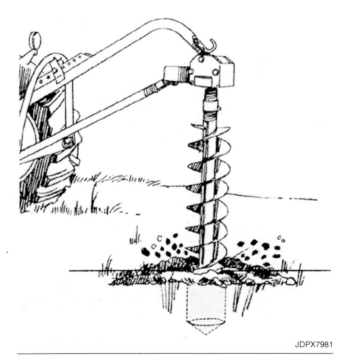

JDPX7981

Fig. 7 — Dig Hole in Small Steps, Bringing the Auger Up to Let Soil Clear. Stay Away When Auger Is Turning

5. Never wear loose clothing, long hair, or drawstrings when operating posthole diggers. They can get entangled in rotating parts.

6. Keep everyone clear of the digger and power shafts when the auger is turning. Keep power shaft shields in place.

7. Stop the auger and lower it to the ground when you are not digging.

Lodged Augers

Removing a lodged auger can be dangerous work. Be careful. If the auger gets stuck in wet clay, stones, or roots, disengage the PTO immediately and turn off the engine. Turn the auger backwards with a large wrench (Fig. 8). Then attempt to raise the auger with the hydraulic system. Do not attempt to raise the auger while turning it with a wrench. You could be injured if the PTO was accidentally engaged or the hydraulic system suddenly raised the auger!

JDPX7982

Fig. 8 — When Stuck, Stop Engine, Disengage PTO and Turn Auger Backwards With a Wrench

Shear Pin

An important safety device on your digger is the shear pin or shear bolt (Fig. 9). It's designed to break when there's too much force on the digger, much as a fuse blows when there's too much electric current on a circuit. It can prevent a sudden failure that could injure you. Be sure to use genuine shear pins and or shear bolts—not hardened bolts. Use only pins or bolts of proper specified length. Pins or bolts that are too long can catch clothing and cause serious injury or death.

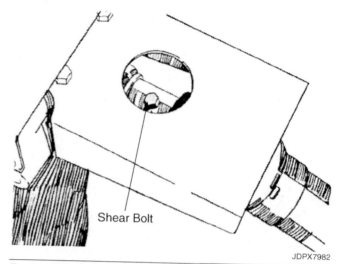

Shear Bolt

JDPX7982

Fig. 9 — The Shear Bolt or Shear Pin Is an Important Safety Device — Use Only the Correct One.

Post Drivers

The post driver can be hazardous if not operated properly (Fig. 10). Follow the manufacturer's instructions to avoid injury. Here are some general safety precautions:

1. Always shut off the engine and lower the hammer before attempting to adjust or lubricate the driver.

2. Never place your hand between the top of the post and the hammer. You could be severely injured if the hammer fell or was tripped accidentally.

3. Never exceed recommended hammer strokes per minute (usually no more than 20). Faster operation may damage the driver and injure the operator.

4. When driving posts, use a safety fork or guide to steady the post and prevent injury if the post should break or be deflected.

5. Keep your hands away from posts that are about to be driven.

6. Adjust or lubricate the driver only when the hammer is in the lowest position to avoid being injured by the hammer if it falls or is tripped.

7. Put all shields in place before operating the driver.

8. Store your driver safely with the hammer in the lowered position, or use safety locks to prevent accidental lowering.

9. Wear safety glasses during operation.

Fig. 10 — Post Drivers Can Be Hazardous if Not Operated Properly

Chain Saws

Chain saws are common on farms, ranches, and urban areas for trimming trees, clearing storm damage, or cutting firewood (Fig. 11). They involve several hazards. To avoid injury, chain saw users should study the operator's manual, especially before the first use and before trying new techniques.

Fig. 11 — Avoid Cutting Trees on Windy Days. Sudden Winds and Changes in Direction Can Be Dangerous

"Chain Saw Safety Manual" from Deere & Company provides detailed instructions. "Chain Saw Safety", published by U.S. Consumer Product Safety Commission, is another source.

Consider the following:

- The weather
- Fire
- The tree
- The saw
- The operator

The Weather

Wind can create very serious hazards when cutting down trees. Even on sunny days, the wind can come up suddenly or change direction unexpectedly, causing a tree to fall in the wrong direction — perhaps onto power lines. buildings, or people. Avoid cutting trees on windy days, or use these days for limbing or trimming only.

Rain, snow, and ice may lead to slips and falls. Whenever the weather presents a hazard, wear protective clothing and work slowly and carefully. If possible, put the job off until conditions improve.

Fire

Periods of hot, dry weather make leaves and grass a fire hazard, especially when it's windy out. Spilled fuel adds to the danger. Refuel on bare ground (Fig. 12). Gradually release pressure in fuel tank before removing the cap completely for refueling. Move at least 10 feet (3 m) away from the fuel source before starting the saw. A faulty muffler can provide a spark that could set off a fire. Don't let dry combustible material contact a hot muffler.

Fig. 12 — Refuel on Bare Ground to Reduce the Fire Hazard

The Tree

Certain trees can be dangerous to inexperienced operators. Lumberjacks have coined some expressions that describe problem trees:

- Widow maker — This is a tree with broken or dead limbs. A limb doesn't have to be very big or very high on a tree to be capable of causing a serious injury if it suddenly fell on you.

- Spring pole — This is a sapling that's bent and held down under tension by another tree. If the spring pole is cut or the other tree is removed from it, the sapling can snap up with tremendous force and seriously injure anyone nearby.

- Schoolmarm — This is a tree with a prominent fork in the trunk, or two trees grown together at the base, making it difficult to predict which way it will fall. Until you've had plenty of experience or instruction, don't attempt to cut trees like these. And don't try to cut any tree with a diameter greater than the length of your chain saw blade. This requires special techniques, and you could be injured if the saw kicks back at you.

The Saw

Use a chain saw with the safety features that help you avoid injury, including hand guards; exhaust muffler (spark arresting muffler for dry conditions or if required by law); trigger and throttle lockout; bumper spikes; also a chain brake, chain catch, and other devices discussed later under "Kickback."

With extended use, chain saw noise and vibration can cause hearing loss, fatigue, and swelling of the hands. To reduce these potentially harmful effects:

1. Wear ear protection.

2. Take periodic rest breaks.

3. Select a saw that has low noise and vibration characteristics.

A chain saw must be properly maintained to be safe. This includes sharp teeth, correct chain tension, proper lubrication, and a properly tuned engine. If you notice that the chain tends to walk sideways while cutting or the discharge shows a fine powder instead of wood chips, your saw needs sharpening. Keep the engine adjusted so it remains running but the chain stops moving when the throttle is released. Your operator's manual is the best source of maintenance information — use it.

All chains stretch with use. Most of this stretch will occur during the first half hour of operation. Your operator's manual will show you the best way to break in your saw for efficient cutting action and longer chain life. Keep the chain properly tensioned to keep it from jumping off the guide and injuring you.

The Operator

The most important ingredient in chain saw safety is the operator. Operate the saw safely and use personal protective equipment (Fig. 13). This protective equipment should be provided for the head, ears, eyes, feet, legs, and hands.

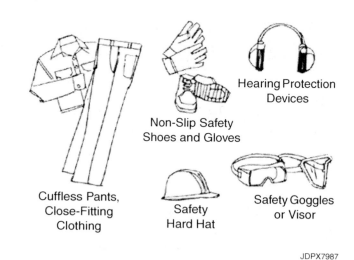

Hearing Protection Devices

Non-Slip Safety Shoes and Gloves

Cuffless Pants, Close-Fitting Clothing

Safety Hard Hat

Safety Goggles or Visor

JDPX7987

Fig. 13 — Use Protective Equipment When Operating a Chain Saw. Equipment Should Meet ANSI Z87.1 Standard

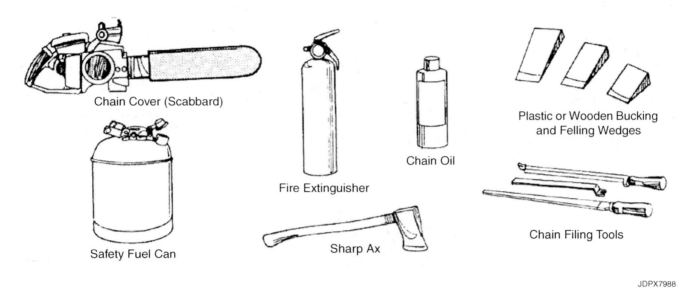

Chain Cover (Scabbard)

Fire Extinguisher

Chain Oil

Plastic or Wooden Bucking and Felling Wedges

Safety Fuel Can

Sharp Ax

Chain Filing Tools

JDPX7988

Fig. 14 — Use Proper Equipment to Help You Do the Job Safely

When using a chain saw, take along the following equipment, as shown in Fig. 14.

- A carrying case for the saw or a scabbard for the guide bar

- A supply of fuel in a UL-approved can (this is the safest storage container)

- Oil for the chain oiler

- Some plastic or wooden wedges (not hard metal ones)

- A sharp single-blade ax

- Tools for chain maintenance

- A fire extinguisher or shovel

The correct position for starting a chain saw is shown in Fig. 15. Never allow another person to help you start a chain saw. If either of you slips or lets go, the other could be cut.

Pull Starter Rope Straight Up to Start Engine

Chain Must Be Free of Obstacles

JDPX7563

Fig. 15 — The Chain Saw Will Move as Soon as the Engine Starts. Make Sure You Hold the Front Handle With the Left Hand and Place Your Right Foot Through the Rear Handle

Kickback

Causes of Saw Blade Kickback

Blade Nose Strikes
Another Object

Improper Starting of
Bore

Top Or Blade Nose Touches
Bottom Or Side Of Kerf During
Reinsertion

Special Chain, Bar, and Tip Guard

JDPX7564, JDPX7989,

Fig. 16 — Avoid Situations That Can Cause Kickback. Use Safety Features

Use extreme caution to be sure the chain does not contact limbs or logs other than the one you want to cut, strike nails or stones, or touch the ground when it's operating. The saw will jump back if the chain at the nose of the bar touches anything. This is called kickback.

Kickback has many causes. A chain that's misfiled or loose is more likely to kick back. A saw may kick back if you start a cut with the saw chain moving too slowly. (Start each cut at full throttle.) See other kickback situations in Fig. 16. When a running saw kicks back, it can be fatal.

These safety features will help prevent or reduce kickback and injury: low-kickback saw chain and guide bars; a guard on the tip of blade to protect against pinching and kickback; a chain brake to stop chain if saw kicks back; front and rear hand guards. There are also chain catches mounted on the power head of the saw to shorten any backward swing of the chain if it breaks.

General Safety Precautions

Note that a chain brake does not prevent kickback. It only stops chain rotation after kickback to help prevent injury. Even with a chain brake, depend on your own good sense and safe cutting methods to prevent kickback. Make sure the chain brake stops the chain in the time specified in the operator's manual.

Stand to the side of the saw to avoid being cut by kickback. Never stand directly in back of the chain saw (in line with the blade) while you are cutting.

Be sure the bumper is against the tree while sawing, or the chain riding across the tree may jerk the saw out of your hands.

Don't carry a chain saw when it is running. If you fall, the saw could spin around and cut you severely. Carry the saw with the guide bar to the rear and the muffler away from you. Use the scabbard for transport.

Always hold the saw firmly with both hands on the handles when it is running. Shut it off before setting it down.

Before attempting the following operations, get the feel of your saw. Make a few trial cuts on small logs supported off the ground so the chain clears the ground. Let the saw do the cutting. You don't need to apply extra pressure.

Felling

Only after you have mastered steady and even cutting should you attempt to fell a tree.

Check the situation carefully before felling a tree. Take note of the larger branches and wind direction to determine how the tree will fall. Be sure you have a clear area around the tree in which to work, and an open pathway for an escape route (Fig. 17).

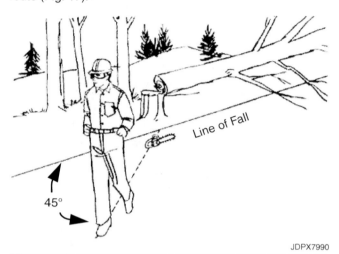

Fig. 17 — Plan a Safe Retreat Before Felling a Tree

When you must control the direction of the fall, notch the tree first. Then use a pulley and rope to control fall (Fig. 18). This is a situation where you should work with another person.

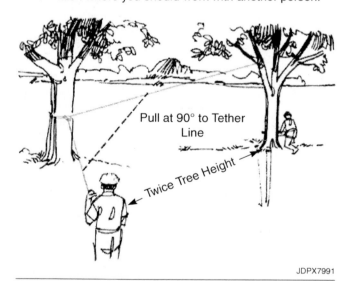

Fig. 18 — Use a Pulley and Rope to Control the Direction of a Fall

Remove dirt and stones from the trunk of the tree where the cut will be made.

Examine trees for loose or dead limbs before felling. If such limbs appear to be a hazard, remove them before felling the tree.

When felling a tree:

1. Cut through trees less than 8 inches (200 mm) thick with one cut.

2. On larger trees, make the notch cut on the side of the tree on which it is expected to fall (Fig. 19). It should have a depth of approximately one-third the diameter of the tree. Make the lower notch cut first. This keeps the chain from binding and being pinched by the wedge of wood while the notch cut is made.

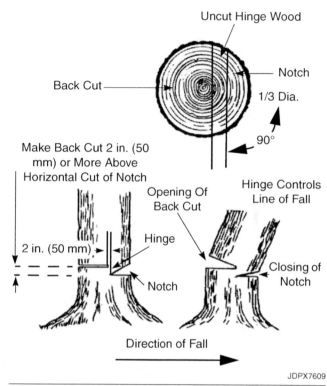

Fig. 19 — Felling a Tree

3. Make the felling or back cut at least 2 inches (50 mm) higher than the horizontal notching cut. The felling cut should be kept parallel with the horizontal notching cut. Cut it so that wood fibers are left to act as a hinge, keeping the tree from twisting and falling in the wrong direction.

4. Keep the guide bar in the middle of the cut so the cutters returning in the top groove don't recut the wood. Don't twist the guide bar in the groove. Guide the saw into the tree — don't force it. The rate of feed will depend on the size and type of timber.

5. Do not cut through the hinge fibers. The tree could fall in any direction — maybe in the direction in which your are retreating.

6. Remove the saw from the cut, carefully avoiding kickback, and shut it off before the tree falls. The tree will begin to fall as the felling cut approaches the hinge fibers. Check your retreat path (Fig. 20), put the saw down, and move quickly away to safety.

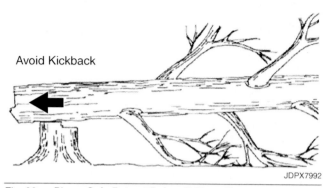

Avoid Kickback

Fig. 20 — Plan a Safe Retreat Before Felling a Tree

Limbing

Most chain saw accidents happen during limbing operations. Leave the larger lower limbs to support the log off the ground to aid bucking cuts (Fig. 21). Prune the smaller limbs in one cut by starting at the bottom end of the tree. Undercuts should be used on limbs supported by branches to keep from binding the chain.

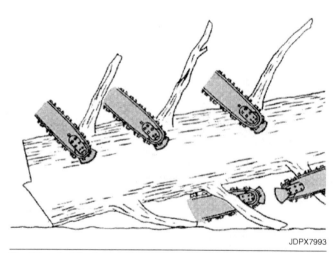

Fig. 21 — Let the Lower Limbs Support the Log for Limbing

Keep the tree trunk between you and the chain saw. Don't cut limbs from the side of the tree where you are standing. Also, watch for springback of the limb being cut or a second branch held by the limb being cut. Relieve springback tension before beginning the cut.

Bucking

Make sure you have good footing and can get out of the way if the log should start to roll. On sloping ground, stand above the log rather than below it. If possible, raise the log clear of the ground by using limbs, logs. or chocks. To avoid pinching the guide bar and saw chain in the cut and splintering the log at the finish of the cut, use the following procedures (Fig. 22).

1. When the log is supported along its entire length, cut it from the top (overbuck) (Fig. 22A).

2. When the log is supported on one end, cut one-third of the diameter from the underside (underbuck). Then make the final cut by overbucking the upper two-thirds to meet the underbucking cut (Fig 22B).

3. When the log is supported on both ends, cut one-third of the diameter from the top (overbuck). Then make the final cut by underbucking the lower two-thirds to meet the overbucking cut (Fig. 22C).

A   Cut From Top (Overbuck)
Avoid Cutting Earth

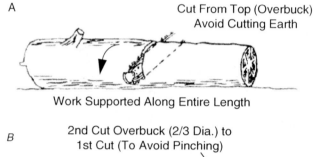

Work Supported Along Entire Length

2nd Cut Overbuck (2/3 Dia.) to
B   1st Cut (To Avoid Pinching)

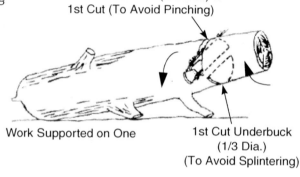

Work Supported on One   1st Cut Underbuck
(1/3 Dia.)
(To Avoid Splintering)

C   1st Cut Overbuck (1/3 Dia.)
(To Avoid Splintering)

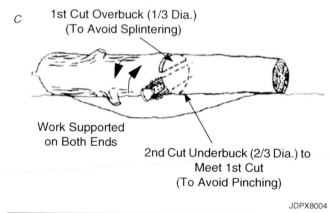

Work Supported
on Both Ends

2nd Cut Underbuck (2/3 Dia.) to
Meet 1st Cut
(To Avoid Pinching)

Fig. 22 — Use These Bucking Procedures for Safety

Trimming

When removing a limb on a standing tree, hoist the saw with a rope. Don't carry the saw while climbing. You need to use both hands to climb safely.

It is best to use a safety rope around the tree, fastened securely to your waist to avoid falls.

Topping

Topping is a technique for cutting off the top part of a tree while it's still standing (Fig. 23). It's a difficult procedure, and should be attempted only by highly skilled loggers or arborists.

Fig. 23 — Topping a Tree Is a Job for Professionals — Don't Try It!

Lawn and Garden Equipment

Lawn and garden tractors were first used primarily for lawn mowing. However, several other attachments have made them popular for various maintenance chores. Now, there are lawn tractors and riding mowers dedicated primarily to lawn mowing.

These vehicles are smaller than regular farm equipment, but their safe operation calls for similar precautions. The tractors are often used with towed or mounted implements. Each machine has an engine, uses fuel, has moving parts, carries an operator, and can be upset.

In the following section, we will discuss some of the primary safety issues involving lawn and garden tractors, lawn tractors, ride-on and walk-behind mowers, and tractor attachments.

For specific safety instructions, read and follow the operator's manual and safety signs on all your lawn and garden equipment.

Upsets (Overturns)

Upsets or overturn accidents often result in crushing injuries when the machine rolls onto the operator. People are often mutilated by whirling blades or other parts of mounted or towed implements. You can prevent rollovers by following these rules:

1. Don't drive where machines could tip or slip. Watch for holes, rocks, or similar hidden hazards. Avoid steep slopes. If you must shift your weight on the seat to avoid tipping, you're asking for trouble.

2. Travel up and down slopes or hillsides (Fig. 24). Traveling across a slope increases the chance of tipping. If there isn't enough traction to back up a slope, it's too steep to drive down. You could lose traction and roll over. If you must mow steep slopes, use a walk-behind mower, and mow across the slopes (Fig. 25).

Fig. 24 — Mow Up and Down on Slopes With Lawn and Garden Tractors and Riding Mowers—Not Across

Fig. 25 — Use Walk-Behind Mowers on Slopes Too Steep for Lawn and Garden Tractors or Ride-On Mowers

3. Slow down and be careful in sharp turns to prevent tipping or loss of steering and braking control. Be especially cautious when you turn on any hillside or slope. Plan ahead to avoid sudden starts, stops, and turns.

4. Don't exceed the load limits for your machine, and use ballast for stability. Use the proper hitch point, wheel widths, and counterweights or wheel weights. Check the operator's manual for each situation.

No Riders!

Lawn and garden tractors, lawn tractors, and riding lawn mowers are small and appealing. Children like to be around them, and adults often try to please children by giving them rides (Fig. 26). That is a serious mistake. Many children have had hands, feet, and limbs mangled when they slipped or bounced from parents' laps under tractor wheels or mower blades. Others have been killed. When mowing, keep children in the house or otherwise completely away from the mowing area. Don't let children play on or around lawn and garden equipment, even when it's not in use. Don't let them think of those machines as toys. Teach them to respect all machines and tools. Never allow riders!

No! Wrong!
No Riders!

JDPX7996

Fig. 26 — Never Allow Passengers. Children Can Fall Off and Be Mangled by Mower Blades

Safety Interlock System

Safety interlock systems on lawn and garden equipment help prevent many serious accidents. Some systems have a seat switch that requires the operator to be safely in the seat to start the engine. If the operator leaves the seat, the machine or mower blades shut down. Some interlocks prevent engine start-up if transmission, PTO, or mower is engaged, or if the park brake is not engaged.

Interlocks on walk-behind mowers also stop or prevent blade operation if you release or do not engage the safety control device.

Safety interlocks help you avoid mistakes. Check your interlock system according to the operator's manual. If the system doesn't work right, get it repaired. Don't disconnect it. And don't try to fool it, for instance, by putting a weight on the tractor seat.

Rotary Lawn Mowers

Besides upset or rollover accidents discussed on previous pages, many personal injury accidents with various types of ride-on and walk-behind lawn mowers result from:

- Contact with rotating blades
- Thrown objects striking people nearby

You can prevent such accidents.

Blade Contact Injuries

Injuries from rotary mower blades may occur when people try to check or unclog a mower as the blades are whirling under the deck, when an operator slips under the mower, or when a child is run over. Take these precautions:

1. Use mowers with proper safety shields and guards. Look for evidence that a mower meets established safety standards, such as the triangular OPEI label (Fig. 27).

JDPX8005

Fig. 27 — Use Outdoor Power Equipment That Meets Safety Standards

2. Keep all the safety shields and guards in place and in good condition. If you remove one for any reason, replace it before mowing.

3. Wear substantial closed-toe shoes when mowing. They can help prevent slipping and can reduce injury from the machines.

4. Don't allow people to put their hands or feet anywhere near the mower deck or discharge chute when the machine is running. Before you dismount, stop the engine and be sure any attachments have stopped moving.

5. Stop the mower if children are near the machine. They can be run over and mangled by the blades. Make sure they're in a safe area, such as in the house. Watch for them!

6. Shut the machine down when you leave it, so others won't start it. Follow the shutdown procedures in the operator's manual. Lower attachments to the ground and remove the key.

7. Check the operator's manuals for other specific precautions for your machines.

Thrown Object Injuries

Mowers can hurl objects great distances, causing eye injuries and other tragedies. (See Chapter 3, "Recognizing Common Machine Hazards".)

To prevent such injuries:

1. Clear the area to be mowed. Before mowing, check for rocks, sticks, or other objects that could be thrown by the blades. Even with guards in place, objects are sometimes thrown or deflected from the mower.

2. Keep mower deck guards, discharge chute, or bag in place. They help prevent the mower from throwing objects great distances.

3. Stop the mower if anyone comes near.

Walk-Behind Mowers

Push-type or walk-behind rotary mowers require about the same precautions as ride-on mowers to avoid blade contact or thrown object injuries. However, there are other concerns:

Mow across slopes, instead of up and down, so you don't slip under the mower blades. Wear closed-toe shoes for traction and for some protection against injury. Anti-slip safety shoes are ideal. Bare feet or sandals are definitely unsafe. Before working on the mower, disconnect the spark plug wire so the engine can't start up and turn the blade.

Mower Fires

Mower fires damage property, and they can injure people. Handling fuel and the accumulation of fuel, oil, and trash on hot surfaces requires a safety discipline to prevent fires around powered lawn mowing equipment.

Avoid spilling fuel on hot engine surfaces, and don't handle fuel near sparks or flames. That means no smoking! Also, keep the machine clean. When fuel, oil, and trash accumulate around hot surfaces, they can catch fire. Clean machines look better, too.

Front-End Loaders

Several manufacturers offer hydraulic front-end loaders as attachments for their larger lawn and garden tractors. The low overall height and width allows farm operators to scrape and load in spaces too small for larger farm tractors with front-end loaders and even some skid-steer loaders.

Use the following safety practices when using front-end loaders on lawn and garden tractors:

1. Avoid overloading. Check the operator's manual for the load capacity of your particular unit (Fig. 28). Overloading can cause loss of steering control. You must add ballast for stability. There are several types such as wheel weights, three-point hitch ballast, or other. Check the operator's manual for type and amount.

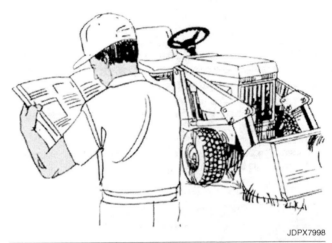

JDPX7998

Fig. 28 — Follow Recommended Bucket Load Capacity and Ballast in the Operator's Manual

2. Back down ramps and steep inclines and keep the bucket low. Otherwise, a hole or sudden bump could cause the center of gravity to shift, causing a forward rollover or loss of control.

3. Watch where you're going. Don't watch the bucket. Watching for spills should not be necessary if you have loaded the bucket properly and are using a safe speed for the ground you're moving over.

4. Operate the hydraulic loader controls only from the operator's seat. Don't attempt it while standing beside the tractor. You could upset the equipment, and you could be injured by the loader cylinders and lever arms.

5. Use the loader only for materials it was designed to handle. Do not allow people to ride in the bucket. Their weight could overload the unit or they could fall and be injured. Never use the loader for such things as removing fence posts. Do not hook a chain to the bucket to pull something.

6. Lower the bucket to the ground when you leave the machine. This will keep the bucket from accidentally falling onto unsuspecting children or others (Fig. 29).

JDPX7999

Fig. 29 — Lower the Bucket Before Leaving the Machine

7. For hydraulic pressure and leaks, follow the precautions outlined in Chapter 3, "Recognizing Common Machine Hazards", and Chapter 7, "Tractors and Implements".

For more complete information on front-end loader safety, review "Front-End Loaders" in Chapter 12, "Material Handling Equipment".

Blades

In farm maintenance, a blade can be used to push gravel or light fill dirt, snow, or manure across concrete. Blades can be mounted at the front, rear, or center of a lawn and garden tractor. They can be controlled manually or with a hydraulic assist. Most lawn and garden tractor blades are designed for light work, and operators should consider the following practices when using them:

1. Watch for obstructions. Avoid pipes, curbs, roots, and other obstructions that could be hit during operation. Hitting heavy objects can cause loss of control and throw the operator from the tractor seat.

2. Be prepared to stop. This will help reduce blade damage and prevent throwing you if you hit a hidden obstruction.

3. Keep operating speeds moderate. This is critical, especially when your blade is locked into a solid position for digging. A stiff blade that hits an object results in a sudden, jolting stop. Most blades can move up and down to follow the terrain in a float position. Others have a spring trip that allows the blade edge to pivot in a small

arc, reducing the shock to the operator and preventing damage to the blade. Both of these safety features require an alert operator who is prepared to stop.

4. Lower the blade when you leave the tractor to prevent accidental lowering that might occur when children or others tamper with the controls.

5. Remember that normal operation of a front-mounted blade can affect your steering ability. On loose surfaces like manure or snow, a blade can reduce steering control, and you could have an accident. Good operators keep this in mind and reduce their speed to maintain control of travel direction.

6. Block the blade if it must be raised while you make adjustments. Hands and feet can be injured or crushed if the blade falls on them during adjustment or repair (Fig. 30).

JDPX8000

Fig. 30 — Block the Blade Before Making Repairs or Adjustments in the Raised Position

7. Maintain recommended tire pressure to give the best traction. Equal tire pressure on the rear drive wheels also helps give uniform traction. Different tire pressures on the drive wheels can cause the tractor to pull to one side, and precision and control are reduced. You may also need tire chains and wheel weights for extra traction. Check your operator's manual for specific recommendations.

Other Lawn and Garden Equipment

There are many different types and sizes of lawn and garden, or grounds care, equipment and attachments used on farms and ranches. They include tillers, carts, snow blowers, front-mounted mowers, trimmers, and others. This book doesn't address them all, but it offers these general safety precautions that apply to almost any machine:

1. Select equipment with features for safe operation and service. Check for indications that they meet safety standards such as ANSI, SAE, ASABE, or OSHA (Fig. 31).

JDPX8005

Fig. 31 — Use Outdoor Power Equipment That Meets Safety Standards

2. Keep all safety features and safety signs in place and in good condition.

3. Read and follow the safety instructions in the operator's manual and safety signs.

4. Operate machines a safe distance from others, and don't allow riders.

Utility Haulers

Fig. 32 — Utility Haulers Are Work Machines That Call for Careful, Adult Operators

In the 1980s, manufacturers introduced utility haulers to carry light loads on farms and ranches. They were somewhat like golf carts with flotation tires and cargo boxes.

Utility haulers have four, five, or six wheels, and weigh up to 1,000 pounds (454 kg). These versatile, "go anywhere" machines can carry loads of several hundred pounds (kilograms), but they're much less costly to buy and operate than a pickup truck (Fig. 32).

In some respects, utility haulers resemble recreational ATVs (all-terrain vehicles), but they are designed as off-road work machines. Their uses include light hauling chores on farms and ranches, in nursery and landscaping operations, and also on golf courses. They are not high-speed, recreational vehicles.

Careful, Mature Operators

Utility haulers travel about 15 mph (24 km/h) or more. They are low-to-the-ground and maneuverable. They can travel almost anywhere, and can be enjoyable to operate.

However, utility haulers are not recreational vehicles. They are work machines that can be involved in serious accidents if operators are careless. Safe operation calls for adults who won't take unnecessary risks or engage in horseplay. Some manufacturers insist that operators should be at least 16 years of age.

Read the Operator's Manual

You should prepare to operate a utility hauler by studying the safety instruction in the operator's manual and safety signs on the machine. The following paragraphs address some primary safety concerns.

No Riders!

Some haulers have a seat for a person besides the operator. But, if there is no passenger seat, don't allow a rider (Fig. 33). No seat means no rider, not in the cargo box, not anywhere! Extra passengers may distract the driver; they can affect the center of gravity causing an upset; and they can fall off.

Fig. 33 — *If There Is No Passenger Seat, There Must Not Be Any Extra Riders, Anywhere!*

Avoiding Rollovers

Utility haulers can be overturned by careless operation. Because of their intended uses, most haulers aren't equipped with rollover protective structures (ROPS). If an 800-pound (363 kg), or heavier, machine rolls onto you, it can cause crushing injuries. People have been killed. Follow these precautions to avoid a rollover.

1. Get a "feel" for the hauler before you work with it. Practice driving on level or moderate terrain, making turns and stops at different speeds to learn how the machine handles in different situations.

2. Don't drive too fast for the conditions. Speed can cause upsets, even on level ground.

3. Slow down before turning, especially on a slope. Plan ahead to avoid sudden or sharp turns or stops, which can cause an overturn.

4. Reduce speed and use caution on slopes, rough ground, and with loads. Some machines will freewheel going downhill, so apply the brakes. Don't start or stop suddenly when going uphill or downhill. That can cause upsets to the rear, or to the front and side.

5. Watch out for holes, rocks, or other hidden hazards that could cause upsets. Also, stay a safe distance from drop-offs, or banks that could collapse, causing an overturn.

6. Never engage in "horseplay".

Carrying Loads

Loads affect a hauler's center of gravity and stability, as well as steering and braking ability. A load should not exceed the manufacturer's recommended weight and size. Check the operator's manual.

Distribute loads evenly inside the cargo box (Fig. 34). If a load is too high or is concentrated at the rear or side, it can cause the machine to tip over. Keep loads below and behind the front cargo barrier to avoid upsets, and so loads won't move forward onto the operator. Secure loose loads so they won't shift or fall off.

Fig. 34 — *Distribute Loads Evenly Inside Cargo Box to Prevent Upset. Secure Loose Loads*

Raising the Cargo Box

Before raising the cargo box to dump a load, be sure the rear wheels are not too close to the edge of a loading dock or an embankment that could collapse and cause an upset. Set the parking brake so the machine will not roll back.

If you raise the cargo box to work on the machine, be sure it is safely blocked up so it cannot fall on someone. Lower the box before leaving the machine unattended.

Towing Loads

If you tow a load, such as a utility cart, first load the hauler's cargo box to provide proper traction and braking ability. But, don't tow a load heavier than recommended.

Travel on Roads

Utility haulers are generally not equipped with approved lights, reflectors, and SMV emblem required for highway use, so they shouldn't be driven on public roads. See "Accidents On Public Roads" in Chapter 7, "Tractors and Implements". Also check state and local regulations.

ATV-Type Vehicles

Some recreational-type ATVs are also used for farm and ranch chores. Their safe operation at work calls for many of the same safety precautions discussed for utility haulers.

ATV manufacturers recommend the vehicles not be ridden by anyone under age 18 and that all riders take a training course. Some states regulate use of ATVs. That includes age of riders, licensing, registration, where and how they can be operated, and use of helmets. Check your state and local regulations.

Test Yourself

Questions

1. (T/F) Objects thrown by rotary-type mowers are generally moving too slowly to be dangerous.

2. (T/F) Having the rear tires of the tractor set wider than the mower frame increases the stability of the tractor, but presses down more of the material to be cut.

3. (T/F) Holes should be dug in small steps when using a posthole digger.

4. Recommended safety practices for operating a post driver include:

   a. Shutting off the engine and lowering the hammer before adjusting or lubricating the driver

   b. Using your hands to steady the post

   c. Always keeping the driver operating at over 20 hammer strokes per minute

   d. All of the above

5. A tree with broken or dead limbs is called a

   a. Spring pole

   b. Schoolmarm

   c. Widow maker

   d. Hazard type 1

6. (T/F) A rope and pulley can be used when you must control the direction a tree is to be felled.

7. One of the most common and serious accidents with lawn and garden equipment involves:

   a. Tire blowouts

   b. Upsets

   c. Brake failure

   d. Sharpening blades

8. (T/F) It is safest to push a walk-behind mower up and down rather than across steep slopes and inclines.

9. (T/F) Safety interlock systems are the cause of many lawn and garden tractor or rotary mower accidents.

10. In utility hauler operation, acceptable actions include:

   a. Allowing a passenger only in a passenger seat provided by the manufacturer

   b. Fastening loose loads in the cargo box behind and below the cargo barrier

   c. Slowing down before turning

   d. All of the above

References

1. Chain Saw Safety. Consumer Information Guide. 1984-421-506-814. U. S. Consumer Product Safety Commission, Washington, DC 20207

2. Chain Saw Safety Manual. OMM95226 Issue 19. 1989. Deere & Company.

3. Industrial/Agricultural Mowers Safety Manual. Equipment Manufacturers Inst., 10 S. Riverside Plaza, Chicago, IL 60606-3710.

4. A Mowing Safety Lesson (14 min.), and A Positive Safety Attitude (10 min.) about outdoor power equipment. 1990. Included in John Deere Consolidated Safety Video. Order # DKVHU89580EN (English), or DKVHU90565FS (Spanish).

5. Safety First With AMTs. Video (9 min.). Deere & Company.
   Order # DKVHC91580EN (English), or DKVHC91580ES (Spanish).

Appendix

JDPX8063

Glossary

A

ABSORPTION RATE — Rate at which a chemical enters the body.

ACCIDENT FREQUENCY RATE — Number of accidents per 1,000,000 hours of use of a specified machine. Gives a more complete accident picture than numbers of accidents only or accidents per worker.

ACCUMULATOR — A container that stores fluids under pressure as a source of hydraulic power It may also be used as a shock absorber.

ACUTE CHEMICAL EXPOSURE — Toxicity — Serious, single dose of or exposure to a chemical.

AIR MONITORING — The sampling for and measuring of pollutants (chemicals, dust, etc.) in the atmosphere.

AMMONIA GAS (NH_3) — Colorless gas given off by liquid manure. It can cause coughing; irritate the eyes, lungs, or throat; and may be fatal.

ANEMOMETER — Instrument used to determine wind speed. Helpful when using pesticide sprays.

ANHYDROUS AMMONIA — Pressurized ammonia without water, used as a nitrogen fertilizer. It can cause skin irritation, burns, and blindness. Prolonged inhalation can lead to suffocation.

ANTIDOTE — Remedy that counteracts the harmful effects of a poison.

ARTIFICIAL RESPIRATION — Method of trying to restore normal breathing to someone who has stopped breathing.

AVICIDE — Agent that destroys birds.

B

BACTERICIDE — Substance that destroys bacteria.

BODY PROTECTION — Special clothing designed to protect a worker. Includes aprons, rubber or plastic garments, and certain types of padding.

BUMP CAP — A hard shell cap, without an interior suspension system, designed to protect against low impact.

C

CARBON DIOXIDE (CO_2) — Colorless, odorless gas given off by liquid manure. Small concentrations (3–6%) can cause drowsiness and headache. Stronger concentrations (30%) can cause death by suffocation.

CARBON MONOXIDE (CO) — Colorless, odorless deadly gas present in engine exhaust.

CAUTION — Signal word that denotes a reminder of safety practices, or directs attention to unsafe practices that could result in personal injury if proper precautions are not taken. Also used on chemical labels to indicate low to moderate toxicity.

CENTER OF GRAVITY — Point about which all parts exactly balance one another. On conventional two wheel-drive tractors, it is located above the axle and behind the midpoint of the tractor's length.

CENTRIFUGAL FORCE — Force that resists any change in direction from travelling in a straight line. Centrifugal force is a major factor in tractor upsets.

CHOLINESTERASE TEST — Blood test to determine activity of cholinesterase, an enzyme that affects nerve impulse transmission. Regular tests are recommended for workers who handle organophosphates.

CHRONIC CHEMICAL EXPOSURE — Accumulation of a chemical in the body that causes a mild, slow poisoning.

CLASS A FIRE — Fire involving the burning of ordinary combustibles like paper or wood.

CLASS B FIRE — Fire in which grease, gasoline or flammable liquids are burned.

CLASS C FIRE — Electrical fire.

CLASS D FIRE — Flammable metals fire.

CONTAMINATION — Pollution of a person or object by a chemical or some other substance.

CRUSH POINTS — Are created when two objects move toward each other or one object moves toward a stationary object.

CUTTING POINTS — Are created by a single object moving rapidly or forcefully enough that it cuts a relatively soft object.

D

DANGER — Signal word that denotes an extreme intrinsic hazard exists which would result in high probability of death or irreparable injury if proper precautions are not taken. Also, a word used on the labels of highly toxic chemicals.

dB(A) — Sound level in decibels read on the A-scale of a sound level meter. The A-scale discriminates against very low frequencies (as does the human ear) and is therefore better for measuring general sound.

DECIBEL (dB) — A unit for measuring the relative intensity of sound.

DERMAL CHEMICAL EXPOSURE — A chemical entering the body by absorption through the skin.

DISCOMFORT INDEX — Formula that relates temperature to humidity, used to determine acceptable working conditions.

DOFFER — A device with lugs attached to it, used for the purpose of removing cotton from spindles in the operation of a cotton picker.

DOUBLE INSULATED — A method of encasing electrical tools so that the operator cannot touch metal parts that may carry electricity.

DRIFT — Spray particles carried on the wind, often creating a hazard.

E

EAR PROTECTION — Any device used to protect the ears from loud noises — generally earplugs or earmuffs.

ELECTRIC SHOCK — Passage of an electric current through a person's body.

EMETIC — A substance that induces vomiting.

EPA NUMBER — The number assigned to a chemical regulated by the U.S. Environmental Protection Agency.

ERGONOMICS — The study of human characteristics for the appropriate design of living and work environments.

EXPOSURE — Contact with a chemical, biological, or physical hazard.

EYE PROTECTION — Devices used to protect the eyes from dust, fumes, chemicals, etc. Some of these devices are safety glasses, goggles, and face shields.

F

FACE PROTECTION — A shield to protect the face from splashing, dust, chaff, sparks, etc.

FAIR LABOR STANDARDS ACT — Law that prohibits youth under age 16 from working at certain farm jobs. See LAWS in the following section.

FERTILIZER — Material (manure or chemical) used to make soil more fertile. See ANHYDROUS AMMONIA, MANURE.

FERTILIZER BURN — Skin irritation caused by the hygroscopic action of dry fertilizer.

FIELD OF VISION — Area a person is able to see from one position, normally about 170 degrees.

FISHTAIL — Back-and-forth weaving action of trailed equipment.

FLAMMABLE LIQUID — Any liquid having a flash point below 100°F (37.8°C).

FLASH POINT — The lowest temperature at which a liquid gives off enough vapor to form an ignitable mixture with air and produce a flame when a source of ignition is present.

FOOT PROTECTION — Special safety shoes designed to protect feet from sharp or falling objects.

FREEWHEELING PARTS — The parts continue to move, after the power is shut off, because of their own inertia or the inertia of other moving parts connected to them.

FUNGICIDE — Agent used to kill fungi.

G

GROUND FAULT CIRCUIT INTERRUPTER (GFCI) — A device that measures the amount of current flowing to and from an electrical source. When a difference is sensed, indicating a leakage of current that could cause an injury, the device quickly breaks the circuit.

GROUNDING — The procedure used to carry an electrical charge to ground through a conductive path.

GUARD — An enclosure that prevents entry into the point of operation of a machine or renders contact harmless.

H

HAND PROTECTION — Various types of gloves designed to protect the hands.

HAND SIGNALS — Standardized system of gestures and hand movements for use in machinery operation, published by the American Society of Agricultural and Biological Engineers.

HAZARD — Dangerous object or situation that has the potential to cause injury.

HAZARDOUS MATERIAL — Any substance or compound that has the capability of producing adverse effects on the health and safety of humans, including injury or death, and damage or pollution of land, air, or water.

HEAD PROTECTION — Headgear designed to protect the head against bumps and falling objects. Two common types are available: hard hats and bump caps.

HEARING CONSERVATION — The prevention or minimizing of noise-induced hearing loss through the use of hearing protection devices, engineering control, audiometric tests and training.

HERBICIDE — Substance used to kill weeds.

HERTZ (Hz) — Cycles per second.

HYDROGEN SULFIDE (H_2S) — Flammable, poisonous, bad-smelling gas given off by liquid manure, especially during agitation. Very small amounts can cause nausea, dizziness, and a dulled sense of smell, and 1% concentrations can cause unconsciousness or death.

HYGROSCOPIC — Readily taking up and retaining moisture. A characteristic of dry fertilizer that can cause skin irritation.

I

INHALATION CHEMICAL EXPOSURE — Chemical entering the body through the respiratory system.

INSECTICIDE — Agent used to kill insects.

INSPECTION — Monitoring function conducted to locate and report existing and potential hazards having the capacity to cause accidents in the workplace.

J

JACKKNIFE — A tractor and a hitched implement doubling up to form an angle of 90 degrees or less.

K

KICKBACK — Sudden, violent, backward and upward action of a chain saw.

L

LD_{50} VALUE — Number of milligrams of pesticide per kilogram of body weight that kills 50% of a group of test animals. The lower the LD_{50} value, the more toxic the chemical.

M

MAN-MACHINE COMMUNICATION — By use of controls, gauges, observation, and other means, man (a person) can understand how a machine is performing.

MAN-MACHINE SYSTEM — An arrangement of people and machines that interact to accomplish some task or work.

MANURE — Livestock excreta, often stored as a liquid. Liquid manure may give off poisonous gases, especially when agitated. See AMMONIA GAS, CARBON DIOXIDE, HYDROGEN SULFIDE, METHANE.

METHANE (CH_4) — Colorless, odorless gas given off by liquid manure. Methane is nontoxic, but highly flammable.

MITICIDE — Agent that destroys mites.

O

ORAL CHEMICAL EXPOSURE — Chemical entering the body through the mouth.

OVERBUCK — Cut a limb or log from above.

P

PERSONAL PROTECTIVE EQUIPMENT — Clothing and devices, designed to protect workers from injury or illnesses.

PESTICIDE — Agent used to destroy pests, such as insects, rodents, and weeds. Most pesticides are poisonous to humans, and some can be fatal. See AVICIDE, BACTERICIDE, FUNGICIDE, HERBICIDE, INSECTICIDE, MITICIDE, RODENTICIDE.

PESTICIDE SAFETY TEAMS — Group organized to help deal with serious chemical emergencies.

PINCH POINT — Place where two parts of a mechanism move toward one another, creating a potentially hazardous situation.

PINHOLE LEAK — Liquid under high pressure escaping through an extremely small opening.

PULL-IN POINTS — Are created when attempting to remove material while the machine is running — people may be pulled into the moving parts of the machine and seriously injured.

R

REACTION TIME — Interval of time between a stimulus or signal and a person's response to it,

RESPIRATORY PROTECTION — A device worn over the nose or mouth to protect the respiratory tract. Some devices also supply the wearer with oxygen. Respiratory protective devices include various types of respirators and self-contained breathing devices.

RODENTICIDE — Substance that kills rodents.

ROLLOVER PROTECTIVE STRUCTURE (ROPS) — Protective frame or cab designed to limit most tractor upsets to 90°, and to protect the operator in upsets beyond 90°.

S

SAFETY-ALERT SYMBOL — Symbol used on machines to draw attention to potential hazards and safety messages, and in operator's manuals to highlight safety instructions. It means ATTENTION! BECOME ALERT! YOUR SAFETY IS INVOLVED. It consists of an exclamation point within an equilateral triangle.

SAFETY SHOES — Footwear that meets ANSI safety standards for protection.

SCHOOLMARM — Tree with prominent fork in the trunk, or two trees grown together at the base.

SELF-CONTAINED BREATHING APPARATUS — A respiratory protective device that provides a supply of respirable air or oxygen, carried by the wearer.

SHEAR POINTS — Are created when the edges of two objects are moved toward or next to one another closely enough to cut a relatively soft material.

SHEAR LINE — Point at which the ground near a ditch or river begins to cave in when subjected to a heavy weight, such as a tractor or self-propelled machine.

SHIELD — Protective device or covering to protect people from moving machine parts or other hazardous areas.

SILO GAS — Dangerous gases (primarily carbon dioxide and nitrogen dioxide) that can build up to lethal levels inside silos and are almost impossible to detect.

SLOW-MOVING-VEHICLE EMBLEM (SMV) — A fluorescent and reflective yellow-orange, equilateral triangular emblem attached to the rear of tractors and other farm machines used on public roads. Its unique shape and colors identify machines traveling slower than 25 mph (40 km/h) in daylight and at night.

SPONTANEOUS COMBUSTION — Fire caused by chemical action within the substance that catches fire.

SPRING POLE — Sapling bent over and held down under tension by another tree.

STABILITY — Balance and resistance to upsets of a tractor or machine.

STORED ENERGY — Energy confined just waiting to be released.

SUFFOCATION — Death caused by lack of oxygen.

SYNERGISM (CHEMICAL) — Combined action of two compounds that produces an effect greater than the individual effects of the compounds added together. May lead to a serious hazard if two pesticides are mixed by an inexperienced person.

T

TORQUE — A twisting or turning force that is a factor in tractor upsets.

TOXICITY — Poisonous effect of a chemical or some other substance.

TRANSPORT POSITION — Narrowed machine position for highway travel.

U

UNDERBUCK — To cut a limb or log from underneath.

UNIVERSAL SYMBOLS — Picture symbols used on machine controls throughout the world.

V

VISIBLE COLOR SPECTRUM — A series of colors from red to violet that merge into one another.

W

WARNING — Signal word that denotes a hazard exists which can result in injury or death if proper precautions are not taken. Also used on labels of chemicals of medium toxicity.

WASTE — Any solid, liquid, or contained gaseous material that you no longer use. (See Hazardous Material.)

WIDOW MAKER — Tree with dead or broken limbs.

WRAP POINT — Exposed machine component that rotates or rotating shaft where a person's clothing or other material may begin wrapping around it, resulting in injury.

Abbreviations, Acronyms, and Symbols

ANSI — American National Standards Institute

ASABE — American Society of Agricultural and Biological Engineers

ATV — All-terrain vehicle

cm — Centimeter

CPSC — Consumer Product Safety Commission

cps — Cycle per second

CPR — Cardiopulmonary resuscitation

dB — Decibel

EPA — Environmental Protection Agency

ESA — Endangered Species Act

FIFRA — Federal Insecticide, Fungicide and Rodenticide Act

ft-lb — Foot-pound

GFCI — Ground-fault circuit interrupter (Also GFI ground-fault interrupter)

HCS — Hazard Communication Standard

ha — Hectare

hp — Horsepower

IDLH — Immediately dangerous to life and health

ISO — International Organization for Standardization

kg — Kilogram

km/h — Kilometers per hour

kPa — Kilopascal

kW — Kilowatt

L — Liter

LP — Liquefied petroleum

m — Meter

m^3 — Cubic meter

mg — milligram

mg/m^3 — Milligrams per cubic meter

mm — Millimeter

m/s — Meters per second

mph — Miles per hour

MSDS — Material safety data sheet

NFPA — National Fire Protection Association

NH_3 — Ammonia

NIFS — National Institute for Farm Safety

NIOSH — National Institute for Occupational Safety and Health

NRR — Noise reduction rating

NO_2 — Nitrogen dioxide

NSC — National Safety Council

OSHA — Occupational Safety and Health Administration

PPE — Personal protective equipment

ppm — Parts per million

psi — Pounds per square inch

PTO — Power take-off

RCRA — Resource Conservation and Recovery Act

ROPS — Rollover protective structure

rpm — Revolutions per minute

SAE — Society of Automotive Engineers

SMV — Slow-moving vehicle

µ/L — Microns per liter

UL — Underwriters Laboratory

USDA — United States Department of Agriculture

V — Volt

W — Watt

WPS — Worker Protection Standard

Laws

Introduction

By the time you have read this far in the book, you will be able to recognize common and specific farm machinery hazards and you will be trying to eliminate them. You also will know and use safe work practices, like those found in the previous chapters.

Laws are only supplemental to sound safety efforts because there is much that cannot be written into safety standards. But laws can be significant, so let us look at two important federal laws that concern farm owners, managers, their families, and their employees.

This section is intended to explain in broad terms the concept and effect of OSHA and Hazardous Occupations Order for Agriculture. It is not intended as a legal interpretation of the laws and should not be considered as such.

U.S. Public Law 91-596 — Known as OSHA

The Williams-Steiger Occupational Safety and Health Act of 1970 seeks "... to assure so far as possible every working man and woman in the Nation safe and healthful working conditions and to preserve our human resources..."

Under the Act,

Any farmer/rancher who employs one or more employees has the legal responsibility to assure a safe and healthful working place for all his/her employees. This is carried out through establishing safety standards that apply to agriculture, conducting inspections to determine if the employer is complying with the safety and health standards, and issuing citations and fines for violations.

Agricultural employers have the general duty to supply their employees a workplace free from all recognized hazards that cause or are likely to cause death or serious injury to employees. This duty is commonly referred to as the general duty clause.

Employer Responsibilities

Some of the specifically recognized responsibilities in the law include the following:

1. Comply with all agricultural safety and health standards. Currently, this includes the following agricultural standards:

- Storage and handling of anhydrous ammonia

- Temporary labor camps

- Pulpwood logging

- Slow-moving vehicle (SMV) emblem

- Rollover protective structures (ROPS)

- Guarding of farm field equipment, farmstead equipment, and cotton gins

- Field sanitation

The following are brief descriptions of the agricultural standards:

- Anhydrous Ammonia. Covers the container construction, location and installation, valves and fittings, and safety relief devices. Standards most applicable to farmers are those for nurse tanks on farm vehicles and on applicators.

- Temporary Labor Camps. Covers site selection, building construction, space, ventilation and heating. It also prescribes sanitation requirements for cooking and eating spaces, water supply, laundry and bathing facilities, toilets, refuse disposal, and insect and rodent control.

- Pulpwood Logging. Covers environmental conditions, clothing and personal protective devices, first aid, hand tools, explosives, stationary and mobile equipment, machinery guards, mufflers, and guylines.

- Slow-Moving Vehicles. Prescribes use of the triangular SMV emblem on the rear of all farm vehicles or towed equipment while traveling at speeds of (40km/h) 25 miles per hour or less on public roads.

- Rollover Protective Structures. Establishes the general requirements for rollover protective structures and seat belts on agricultural tractors. Specifies ROPS test procedures, performance requirements, and employee operating instructions.

- Machinery Guarding. Covers the application of guards and shields for protection from the hazards associated with moving machinery parts on farm field equipment, farmstead equipment, and cotton gins. Includes a list of safe operating practices for periodic employee instruction.

- Field Sanitation. Requires that fresh drinking water, toilet and hand washing facilities be easily accessible and in close proximity to all employees engaged in field hand-labor operations.

Copies of current standards can be obtained from the nearest regional OSHA office. (See list at end of this section.)

OSHA allows for each state to develop and operate its own state occupational safety and health program. These programs must be at least as stringent as the Federal OSHA Act. Contact your state's Department of Labor and/or Department of Health for specific details on the agricultural safety and health requirements for your state.

2. Inform workers of their rights and obligations.

The poster, Safety and Health Protection on the Job (OSHA2003), must be posted where all employees can see it. Possible locations may be near time cards or in the farm shop, machinery storage area, lunch room, or some place where all employees report to work. Employers may not discriminate against employees who properly exercise their rights under the Act.

3. Maintain Records.

Any agricultural employer with 11 or more employees must maintain the required OSHA records of work-related injuries and illnesses and post the annual summary (OSHA 200 Form) during the entire month of February each year. Additionally, they must report to OSHA within 48 hours any accident that results in one or more fatalities and/or hospitalization to five or more employees.

4. Inspections.

A Department of Labor Compliance Officer may inspect any farm/ranch that employs 11 or more employees at any reasonable hour. An inspector should be accompanied by an employee and employer representative on the walk-around inspection. The Compliance Officer may issue any of three types of citations:

- Serious violation involves a high probability of accident resulting in death or serious injury. An example would be working by an open PTO shaft or working near machinery with guards removed.

- Non-serious violation is a situation where there is injury potential, but where the injuries should not result in death or total disability.

- De minimis violation does not have direct or immediate relationship to safety or health. This could be the lack of a privacy door on a commode.

All citations must be posted at or near the location of the violation.

5. Training.

The ROPS and machinery guarding standards require that every employer must train all new employees at the time of employment and all employees at least annually in the safe operation and servicing of tractors, field implements, and farmstead equipment.

The Hazardous Occupations Order for Agriculture

The Hazardous Occupations Order, under the U.S. Department of Labor's Fair Labor Standards Act (FLSA), makes it unlawful to hire youth under the age of 16 to perform certain work activities that are deemed to be particularly hazardous.

Exemptions

The Hazardous Occupations Order for Agriculture exempts youth in the following situations:

- Youth under the age of 16 working on a farm/ranch owned or operated by the youth's parent or legal guardian.

- Youth ages 14 and 15 who have successfully completed an approved 4-H or FFA training program, passed a written exam, and demonstrated the ability to safely operate tractors and certain machinery can receive a training certificate that allows them to operate such tractors and certain machinery. (See Chapter 7, "Tractors and Implements".)

- Youth who are employed under a cooperative student-learner agreement between the employer and their school.

Agricultural Work Classified as Hazardous for Youth Under 16:

1. Tractors — Operating a tractor of over 20 PTO horsepower (14.9 kW), or connecting or disconnecting an implement or any of its parts to or from such a tractor. (See "Exemptions" in previous section for youth ages 14 and 15.)

2. General Machinery — Operating or assisting to operate (including starting, stopping: adjusting, feeding, or any other activity involving physical contact associated with the operation of) any of the following machines: corn picker, cotton picker. grain combine, hay mower, forage harvester, hay baler, potato digger, mobile pea viner, feed grinder, crop dryer, forage blower, auger conveyor, the unloading mechanism of a nongravity-type self-unloading wagon or trailer, power posthole digger, power post driver, or non walking-type rotary tiller. (See exemptions in previous section for youth ages 14 and 15.)

3. Specialized Machinery — Operating or assisting to operate (including starting, stopping, adjusting, feeding, or any other activity involving physical contact associated with the operation of) any of the following machines: trencher or earth moving equipment, forklift, potato combine. power-driven circular saw, band saw, or chain saw.

4. Livestock — Working on a farm in a yard (pen or stall) occupied by a bull, boar. or stud horse maintained for breeding purposes; or sow with suckling pigs, or cow with newborn calf (with umbilical cord present).

5. Woodlots — Felling, bucking, skidding, loading, or unloading timber with butt diameter of more than 6 inches (152 mm).

6. Ladders and Scaffolds — Working from a ladder or scaffold (painting, repairing, or building structures, pruning trees, picking fruit, etc.) at heights over 20 feet (6 m).

7. Transports — Driving a bus, truck, or automobile when transporting passengers, or riding on a tractor as a passenger or helper.

8. Toxic Atmospheres — Working inside a manure pit; a fruit, forage, or grain storage facility designed to retain an oxygen-deficient or toxic atmosphere; an upright silo within 2 weeks after silage has been added or when a top-unloading device is in operating position; or a horizontal silo while operating a tractor for packing purposes.

9. Chemicals — Handling or applying chemicals, including cleaning or decontaminating equipment; disposal or return of empty containers; or serving as a flagman for aircraft applying agricultural chemicals classified under the Federal Insecticide, Fungicide, and Rodenticide Act (7 U.S.C. 135 et seq.) as Category I of toxicity, identified by the word "poison" and the "skull and crossbones" on the label, or Category II of toxicity, identified by the word "warning" on the label.

10. Blasting — Handling or using a blasting agent, including but not limited to dynamite, black powder, sensitized ammonium nitrate, blasting caps, and primer cord.

11. Fertilizers — Transporting, transferring, or applying anhydrous ammonia.

For more specific information on the Hazardous Occupations Order for Agriculture or child labor regulations for your state, contact your local county Extension office, vocational agriculture instructor, or state Department of Labor.

OSHA Regional Offices

(Occupational Safety and Health Administration)

Region I

Connecticut, Massachusetts, Maine, New Hampshire, Rhode Island, Vermont
John F. Kennedy Federal Building, Room E340
Boston, Massachusetts 02203
(617) 565-9860

Region II

New Jersey, New York, Puerto Rico, Virgin Islands
201 Varick Street
Room 670
New York, New York 10014
(212) 337-2378

Region III

District of Columbia, Delaware, Maryland, Pennsylvania, Virginia, West Virginia
3535 Market Street
Gateway Building
Suite 2100
Philadelphia, Pennsylvania 19104
(215) 596-1201

Region IV

Alabama, Florida, Georgia, Kentucky, Mississippi, North Carolina, South Carolina, Tennessee
1375 Peachtree Street, NE
Suite 587
Atlanta, Georgia 30367
Phone (404) 562-2300
Fax (404) 562-2295

Region V

Illinois, Indiana, Michigan, Minnesota, Ohio, Wisconsin
230 South Dearborn Street
Room 3244
Chicago, Illinois 60604
(312) 353-2220

Region VI

Arkansas, Louisiana, New Mexico, Oklahoma, Texas
525 Griffin Street
Room 602
Dallas, Texas 75202
(972) 850-4145

Region VII

Iowa, Kansas, Missouri, Nebraska
City Center Square
1100 Main Street
Suite 800
Kansas City, Missouri 64105
(816) 426-5861

Region VIII

Colorado, Montana, North Dakota, South Dakota, Utah, Wyoming
1999 Broadway
Suite 1690
Denver, Colorado 80202-5716
(720) 264-6550

Region IX

Arizona, California, Guam, Hawaii, Nevada
71 Stevenson Street
San Francisco, California 94105
(415) 975-4310
(800) 475-4019 Technical Assistance
(800) 475-4020 Complaints
(800) 475-4022 Publications
(415) 975-4319 Fax

Region X

Alaska, Idaho, Oregon, Washington
1111 Third Avenue
Suite 715
Seattle, Washington 98101-3212
(206) 553-5930

EPA Regional Offices

(Environmental Protection Agency)

For additional information on laws and regulations for storage, handling, and disposal of agricultural hazardous materials, contact your county extension center, state Department of Agriculture, state Department of Natural Resources, state EPA office, or the Regional EPA office for your state:

EPA Region I

Connecticut, Maine, Massachusetts, New Hampshire, Rhode Island, Vermont
1 Congress Street
Boston, MA 02203
(617) 565-3420

EPA Region II

New Jersey, New York, Puerto Rico, Virgin Islands
290 Broadway
New York, NY 10007
(212) 637-3000

EPA Region III

Delaware, District of Columbia, Maryland, Pennsylvania, Virginia, West Virginia
841 Chestnut Street
Philadelphia, PA 191107
(215) 597-9800

EPA Region IV

Alabama, Florida, Georgia, Kentucky, Mississippi, North Carolina, South Carolina, Tennessee
345 Courtland Street, NE
Atlanta, GA 30365
(404) 347-4727

EPA Region V

Illinois, Indiana, Michigan, Minnesota, Ohio, Wisconsin
77 W. Jackson Boulevard
Chicago, IL 60604
(312) 353-2000

EPA Region VI

Arkansas, Louisiana, New Mexico, Oklahoma, Texas
1445 Rose Avenue
Dallas, TX 75202
(214) 665-6444

EPA Region VII

Iowa, Kansas, Missouri, Nebraska
726 Minnesota Avenue
Kansas City, KS 66101
(913) 551-7000

EPA Region VIII

Colorado, Montana, North Dakota, South Dakota, Utah, Wyoming
999 18th Street
Denver, CO 80202
(303) 293-1603

EPA Region IX

Arizona, California, Hawaii, Nevada, Guam, Marianas
75 Hawthorne Street
San Francisco, CA 94105
(415) 744-1702

EPA Region X

Alaska, Idaho, Oregon, Washington
1200 Sixth Avenue
Seattle, WA 98101
(206) 553-1200

Safety Education Materials

Texts

Fundamentals of Machine Operation:
Series of agricultural machinery texts which includes this text, Farm & Ranch Safety Management, Tractors, Tillage, Combine Harvesting, and others. Each book also addresses safety. John Deere Publishing, One John Deere Place, Moline, IL 61265.
Phone 1-800-522-7448.

Fundamentals of Service:
Series on servicing agricultural power machines. Includes Engines, Hydraulics, Tires and Tracks, Welding, and others. Books also address safety. John Deere Publishing, One John Deere Place, Moline, IL 61265.
Phone 1-800-522-7448.

Safety and Health for Production Agriculture: A college/university level textbook on production agriculture safety and health. 19943. American Society of Agricultural Engineers, 2950 Niles Rd., St. Joseph, MI 49085-9659.

Occupational Safety and Health Series: Four volumes and two study guides written to assist readers in establishing and maintaining safety and health programs for business and industry. National Safety Council, 1121 Spring Lake Drive, Itasca, IL 60143-3201.

Instructor's Guide

Farm & Ranch Safety Management Instructor's Guide assists teachers in planning agricultural safety lessons. The guide is divided into units corresponding to the chapters in this text. Each unit contains discussion questions, suggested teaching activities, transparency masters with notes, a camera ready laboratory exercise, a camera ready quiz, and an answer key for the quiz with explanations. Order No. FBM18504T ISBN 0-86691-187-1. John Deere Publishing, One John Deere Place, Moline, IL 61265.
Phone 1-800-522-7448.

Student Guide

Farm & Ranch Safety Management Student Guide contains quizzes corresponding to those in this text and a laboratory exercise suitable for homework or as an in-class project for each chapter. Photocopying large quantities of materials can be avoided if each class member is provided a Student Guide. Order No. FBM18604W, ISBN 0-86691-188-X. John Deere Publishing, One John Deere Place, Moline, IL 61265.
Phone 1-800-522-7448.

Manufacturers' Literature

Safety rules on specific agricultural machines are covered in the operator's manuals for those machines. Manuals are included with each new machine. If you need additional copies, contact your dealer or manufacturer's service department. Be sure to specify the make, model, year of manufacture, and serial number of the machine for which you need an operator's manual.

Some manufacturers also have special safety books, brochures, videos and other safety information. Contact your dealer or manufacturer's service department to determine how to obtain such materials.

John Deere Videos

Source: To order John Deere videos, contact John Deere Publishing, One John Deere Place, Moline, Illinois 61265.
Phone 1-800-522-7448.

John Deere Consolidated Safety Video — Contains eleven segments on farm safety including safety attitude, bypass starting, safety and labels, lawn mowers, combines, tractors, and others. It explains causes of injuries and encourages farm workers to take the time to be safe. 87 minutes. Order No. DKVHU89580EN (English) or No. DKVHU90565ES (Spanish), five segments, 54 minutes: Safety attitude, tractor safety (2 segments), bypass starting, and cotton picker safety.

Tractor Rollovers and ROPS — Discusses the importance of rollover protective structures (ROPS) on tractors. Rollover tests show the results of tractor rollovers with and without ROPS. Details are provided on installing ROPS on John Deere tractors manufactured since 1960. In addition, use of the foldable ROPS is demonstrated. 6:30 minutes. Order No. DKVHA92548EN.

Round Baler Safety Tips & Techniques for Large Round Balers — Explains how round baler accidents happen and interviews victims of entanglements. Discusses safety measures, including operating and adjusting, threading twine, handling bales with front-end loaders, hitching tractors, and attaching PTO shafts. Also discusses general operation and service, including belt repair, crop preparation, starting the bale, and procedures for making various sizes of bales. 31:11 minutes total (safety 9:11; operating tips 22:00). Order No. DSVHA90596EN.

Cotton Picker Safety — 1981 video describes safety precautions for cotton picker operation and maintenance by showing entanglements, clearing obstructions, general maintenance, safe operation and transporting. 9 minutes. Order No. DKVHA81760EN.

Cotton Picker Safety — 1990 video addresses operation, adjustment, and safety for John Deere 9930 and 9960 cotton pickers. Order No. DHPVA91010EN (English) or No. DHPVA91010ES (Spanish).

Organizations Involved in Agricultural Safety

There are many other sources of technical and nontechnical assistance in agricultural safety and health. The following organizations and agencies are among those committed to improving agricultural safety and health. This should not be viewed as a complete list, but as a starting point for assistance.

Assistance Key:

AV = Audio-Visual
D = Injury/Illness Data
E = Educational Programs
J = Journal
P = Publications
PD = Professional
PR = Promotions
R = Regulatory
S = Standards
T = Text
TA = Technical Development Assistance
M = Miscellaneous

American Farm Bureau Federation (AFBF)

600 Maryland Ave. SW
Suite 1000W
Washington, DC 20024

AFBF promotes agricultural safety and health through its state organizations. For assistance in determining what specific services or technical assistance are available in your state, contact your county or state Farm Bureau. AV, E, P, PR, R

American Society of Agricultural and Biological Engineers (ASABE)

2950 Niles Road
St. Joseph, MI 49085-9659
phone 269-429-0300
Fax: 269-429-3852
WWW.ASABE.org

ASABE is a professional/technical organization of national and international scope. Primary safety activities include technical programs and the development and publication of safety standards for farm equipment, structures, and storage, soil and water management systems. E, S, T, TA

Association of Equipment Manufacturers (AEM)

6737 W. Washington Street, Suite 2400
Milwaukee, WI 53214-5647
Phone: 414-272-0943
Fax: 414-272-1170

AEM is an organization of member companies who manufacture agricultural and construction equipment. They are active in proposing and promoting voluntary agricultural safety standards, creating and promoting safety education programs, and proposing and sponsoring safety research. P, PR, TA, M

Farm Equipment Manufacturers Association (FEMA)

1000 Executive Parkway, Suite 100
St. Louis, MO 63141-6369
Phone: 314-878-2304
Fax: 314-878-1742

FEMA is a trade organization of member companies who are smaller manufacturers of agricultural and related equipment. They are becoming increasingly involved in developing and promoting accident prevention management programs among their members and also in proposing and promoting voluntary safety standards. M

Farm Safety Association, Inc. (FSA)

75 Farquhar Street
Suite 101
Guelph, Ontario
N1H 3N4, Canada
Tel: 519.823.5600
Toll Free: 1-800-361-8855
Fax: 519-823-8880

FSA supports industry and government in developing standards, procedures, and regulations to improve health and safety for the Agriculture Industrial Sector. FSA offers a broad range of print, video, and digital products and services including training and consultation. These include a newsletter, statistical information, and guideline reporting forms. AV, E, P

Farm Safety 4 Just Kids (FS4JK)

11304 Aurora Ave.
Urbandale, IA 50322
Phone: 515-331-6506
Toll Free: 1-800-423-5437

FS4JK is a non-profit organization whose mission is to promote a safe farm environment to prevent health hazards, injuries, and fatalities to children and youth. Based in Iowa, they provide resources and training to individuals and communities in the United States and Canada to conduct farm safety awareness and educational programs. AV, P, PR

CropLife America (CLA)

1156 15th St. N.W.
Washington, DC 20005
Phone: (202) 296-1585
Fax: (202) 463-0474

CLA is the nation's largest trade organization for agriculture and pest management. They represent more than 80 developers, manufacturers, formulators and distributors of virtually all the crop protection products used by American farmers and growers. P, PR

National Farm Medicine Center (NFMC)

1000 N. Oak Avenue
Marshfield, WI 54449-5790

The Center has four emphasis areas within the field of production agricultural safety and health: clinical services, research, education, and community service. NFMC is developing into a nationally recognized center addressing the issues of child safety and health on farms and ranches. AV, U, E, J, P, PD, TA

National Institute for Farm Safety Inc. (NIFS)

NIFS is the largest organization of agricultural safety and health professionals in the U.S. and Canada. The organization provides opportunities for networking and supports professional development as well as educational and research activities related to agricultural safety and health. AV, E. P. PD, TA, M

National Institute for Occupational Safety and Health (NIOSH)

1600 Clifton Rd.
Atlanta, GA 30333
Phone (404) 639-3311

NIOSH is the U.S. federal agency responsible for occupational safety and health research, training, and development of criteria documents, hazardous substance lists, and certification of personal protective equipment. NIOSH has been active in the field of agricultural safety and health. D, E, J, A, PD, TA,

National Safety Council (NSC)

1121 Spring Lake Drive
Itasca, IL 60143-3201
Phone: (630) 285-1121
Fax: (630) 285-1315

NSC is the largest non-profit, non-government public service organization that develops safety and health educational programs and support materials. The Agricultural Division is the central focus for NSC's farm injury and illness prevention activities. Resources and educational opportunities are also available through the Youth Division. NSC and USDA co-sponsor National Farm Safety and Health Week. AV, D, E, P, PD, PR, R, T, TA

Occupational Safety and Health Administration (OSHA)

See Occupational Safety and Health Act and the listing of regional OSHA offices in the section "LAWS," earlier in this chapter.

OSHA is a U.S. federal agency that administers the Occupational Safety And Health Act, which applies to employee work places, including agricultural operations. E, R, S, P

United States Department of Agriculture

1400 Independence Ave., S.W.
Washington, DC 20250

The USDA has established agricultural safety and health programs as part of its ongoing Cooperative Extension Program in every state. For assistance in determining specific services and technical assistance available, contact your county Extension office and/or Cooperative Extension Service, or the safety extension specialist at your state land grant university. AV, D, E, P, R, PD,TA, M

ISBN 0-86691-352-1

9 780866 913522